WHAT EVERY CHEMICAL TECHNOLOGIST WANTS TO KNOW ABOUT...

Volume II

DISPERSANTS, SOLVENTS, AND SOLUBILIZERS

Compiled by

Michael and Irene Ash

Chemical Publishing Co., Inc.
New York, N.Y.

Chemical Publishing Co., Inc.

ISBN 0-8206-0327-9

Printed in the United States of America

PREFACE

This reference book is the second volume in the set of books entitled WHAT EVERY CHEMICAL TECHNOLOGIST WANTS TO KNOW . . . SERIES. This compendium serves a unique function for those involved in the chemical industry—it provides the necessary information for making the decision as to which trademark chemical product is most suitable for a particular application.

The chemicals included in this second book of the series have their major function as dispersants, solvents, and solubilizers, however, complete cross-referencing is provided for the multiple functions of all the chemicals.

The first section which is the major portion of each volume contains the most common generic name of the chemicals as the main entry. All these generic entries are in alphabetical order. Synonyms for these chemicals are then listed. The CTFA name appears alongside the appropriate generic name. The structural and/or molecular formula of the chemical is listed whenever possible. The generic chemical is sold under various tradenames and these are listed here in alphabetical order for ease of reference along with their manufacturer in parentheses. The *Category* subheading lists all the possible functions that the chemical can serve. Because of differences in form, activity, etc., individual tradenames of the generic chemical are used in particular applications more frequently. These are delineated in the *Applications* section. The differences in properties, toxicity/handling, storage/handling, and standard packaging are specified in the subsequent sections wherever distinguishing characteristics are known.

The second section of the volume TRADENAME PRODUCTS AND GENERIC EQUIVALENTS helps the user who only knows a chemical by one tradename to locate its main entry in section 1. The user can look up this tradename in this section of the book and be referred to the appropriate, main-entry, generic chemical name.

The third section GENERIC CHEMICAL SYNONYMS AND CROSS REFERENCES provides a way of locating the main entries by knowing only one of the synonyms. If the generic chemical is not in the volume, it will refer you to the volume in which it is contained.

The fourth section TRADENAME PRODUCT MANUFACTURERS lists the full addresses of the companies that manufacture or distribute the tradename products found in the first section.

The following is a list of the six volumes that comprise this series:

Volume I	Emulsifiers and Wetting Agents
Volume II	Dispersants, Solvents and Solubilizers
Volume III	Plasticizers, Stabilizers and Thickeners
Volume IV	Conditioners, Emollients and Lubricants
Volume V	Resins
Volume VI	Polymers and Plastics

This series has been made possible through long hours of research and compilation and the dedication and tireless efforts of Roberta Dakan who helped make this distinctive series possible. Our appreciation is extended to all the chemical manufacturers and distributors who supplied the technical information.

M. and I. Ash

NOTE

OTHER BOOKS BY MICHAEL AND IRENE ASH

A Formulary of Paints and Other Coatings, Volumes I and II
A Formulary of Detergents and Other Cleaning Agents
A Formulary of Adhesives and Sealants
A Formulary of Cosmetic Preparations
The Thesaurus of Chemical Products, Volumes I and II
Encyclopedia of Industrial Chemical Additives, Volumes I–IV
Encyclopedia of Surfactants, Volumes I–IV
Encyclopedia of Plastics, Polymers and Resins, Volumes I–IV
What Every Chemical Technologist Wants to Know About. . .
Volume I—Emulsifiers and Wetting Agents

ABBREVIATIONS

@ ... at
anhyd. ... anhydrous
APHA ... American Public Health Association
approx. ... approximately
aq. ... aqueous
ASTM ... American Society for Testing and Materials
avg. ... average
B.P. ... boiling point
Btu ... British thermal unit
C ... degrees Centigrade
CAS ... Chemical Abstracts Service
cc ... cubic centimeter(s)
CC ... closed cup
cm ... centimeter(s)
cm^3 ... cubic centimeter(s)
COC ... Cleveland Open Cup
compd. ... compound, compounded
conc. ... concentrated, concentration
cP, cps ... centipoise
cs, cSt ... centistokes
CTFA ... Cosmetic, Toiletry and Fragrance Association
DEA ... diethanolamine
disp ... dispersible, dispersion
dist ... distilled
DOT ... Department of Transportation
DW ... distilled water
EO ... ethylene oxide
equiv. ... equivalent
F ... degrees Fahrenheit
F.P. ... freezing point
FDA ... Food and Drug Administration
ft^3 ... cubic foot, cubic feet
g ... gram(s)
gal ... gallon(s)
HLB ... hydrophile-lipophile balance
insol. ... insoluble
IPA ... isopropyl alcohol
kg ... kilogram(s)
l, L ... liter(s)
lb ... pound(s)
M.P. ... melting point
M.W. ... molecular weight
max ... maximum
MEA ... monoethanolamine
MEK ... methyl ethyl ketone
mfg. ... manufacture
MIBK ... methyl isobutyl ketone
min ... minute(s)
min. ... mineral, minimum
MIPA ... monoisopropanolamine

misc. .... miscible

ml .... milliliter(s)

mm .... millimeter(s)

NF .... National Formulary

no. .... number

o/w .... oil-in-water

OC .... open crucible

PEG .... polyethylene glycol

pH .... hydrogen-ion concentration

pkgs .... packages

PMCC .... Pensky Marten closed cup

POE .... polyoxyethylene, polyoxyethylated

POP .... polyoxypropylene

PPG .... polypropylene glycol

pt. .... point

R&B .... Ring & Ball

RD .... Recognized Disclosure

ref. .... refractive

rpm .... revolutions per minute

R.T. .... room temperature

s .... second(s)

sol. .... soluble, solubility

sol'n. .... solution

sp.gr. .... specific gravity

SS .... stainless steel

std. .... standard

SUS .... Saybolt Universal seconds

TCC .... Taggart closed cup

TEA .... triethanolamine

tech. .... technical

temp. .... temperature

theoret. .... theoretical

TLV .... threshold limit value

TOC .... Taggart open cup

UL .... Underwriter's Laboratory

USP .... United States Pharmacopoeia

uv, UV .... ultraviolet

veg .... vegetable

visc. .... viscosity, viscous

w/o .... water-in-oil

wt .... weight

≈ .... approximately equal to

< .... less than

> .... greater than

≤ .... less than or equal to

≥ .... greater than or equal to

TABLE OF CONTENTS

Acetone (CTFA)

SYNONYMS:
Dimethyl ketone
2-Propanone
EMPIRICAL FORMULA:
C_3H_6O
STRUCTURE:

$$CH_3-\overset{\displaystyle O \atop \displaystyle \|}{C}-CH_3$$

CAS No.:
67-64-1
TRADENAME EQUIVALENTS:
Generically sold by:
Allied-Signal, Ashland, Eastman, Harwick, Shell, Union Carbide, Union Oil
CATEGORY:
Solvent, diluent, intermediate
APPLICATIONS:
Industrial applications: adhesives/cements (generic); paint, varnish, lacquer mfg. (generic); rubber (generic)
Industrial cleaners: precision equipment cleaner/drier (generic)
PROPERTIES:
Form:
Liquid (generic)
Color:
Colorless (generic)
Odor:
Characteristic, pleasant (generic)

Acetone (cont'd.)

Solubility:

Miscible with alcohol (generic)
Miscible with chloroform (generic)
Miscible with ether (generic)
Miscible with most oils (generic)
Miscible with water (generic)

M.W.:

58.08 (generic)

Sp.gr.:

0.791–0.793 (generic)

Density:

6.59 lb/gal (generic)

Visc.:

0.3075 cps (generic)

B.P.:

56.13 C (generic)

M.P.:

–94.9 C (generic)

Flash Pt.:

–9.4 C (OC) (generic)

Ref. Index:

1.3591 (20 C) (generic)

TOXICITY/HANDLING:

Practically nontoxic (generic)

STORAGE/HANDLING:

Flammable (generic)

Acetylated sucrose distearate (CTFA)

SYNONYMS:

Sucrose distearate, acetates

TRADENAME EQUIVALENTS:

Crodesta A-10, A-20 [Croda]

CATEGORY:

Dispersant, wetting agent, emulsifier

APPLICATIONS:

Cosmetic industry preparations: (Crodesta A-10); toiletries (Crodesta A-10)
Pharmaceutical applications: (Crodesta A-10)

PROPERTIES:

Form:

Solid (Crodesta A-20)

Wax (Crodesta A-10)
Color:
Yellow/white (Crodesta A-10)
Composition:
100% active (Crodesta A-20)
100% active, 3% monoester (Crodesta A-10)
Solubility:
Sol. in veg. oil (Crodesta A-10)
Sol. in veg./min. oil combinations (Crodesta A-10)
M.P.:
43–49 C (Crodesta A-10)
HLB:
1.0 (Crodesta A-20)
< 3.0 (Crodesta A-10)
Acid No.:
5.0 max. (Crodesta A-10)
Iodine No.:
1.0 max. (Crodesta A-10)
Saponification No.:
230–290 (Crodesta A-10)
Hydroxyl No.:
20 max. (Crodesta A-10)

Aminomethyl propanol (CTFA)

SYNONYMS:
2-Amino-2-methyl-1-propanol
AMP
Isobutanolamine
1-Propanol, 2-amino-2-methyl-
EMPIRICAL FORMULA:
$C_4H_{11}NO$
STRUCTURE:

$$CH_3—C(NH_2)(CH_3)—CH_2OH$$

CAS No.:
124-68-5

Aminomethyl propanol (cont' d.)

TRADENAME EQUIVALENTS:
AMP, AMP-95 [IMC]
CATEGORY:
Dispersant, solubilizer, neutralizer, stabilizer, emulsifier, catalyst, corrosion inhibitor, antifoam
APPLICATIONS:
Cosmetic industry preparations: emulsions (AMP, AMP-95); hairsprays (AMP)
Farm products: insecticides/pesticides (AMP, AMP-95)
Household detergents: (AMP, AMP-95)
Industrial applications: construction (AMP, AMP-95); dyes and pigments (AMP, AMP-95); paint mfg. (AMP, AMP-95); polymers/polymerization (AMP, AMP-95); water treatment (AMP, AMP-95)
PROPERTIES:
Form:
Liquid (AMP-95)
Solid (AMP)
Color:
Colorless (AMP-95)
APHA 20 (AMP)
Odor:
Low (AMP-95)
Composition:
95% active in water (AMP-95)
100% active (AMP)
Solubility:
Sol. in water (AMP, AMP-95)
Ionic Nature:
Anionic (AMP)
M.W.:
89.14 (AMP, AMP-95)
Sp.gr.:
0.928 (40/40 C) (AMP)
0.942 (AMP-95)
Density:
7.78 lb/gal (AMP)
7.85 lb/gal (AMP-95)
Visc.:
102 cp (30 C) (AMP)
147 cp (AMP-95)
F.P.:
–2 C (AMP-95)
B.P.:
165 C (AMP)

M.P.:
30 C (AMP)
Flash Pt.:
172 F (AMP)
182 F (TCC) (AMP-95)
Surface Tension:
36–38 dynes/cm (AMP-95)
TOXICITY/HANDLING:
Skin irritant; causes eye burns; protective goggles and gloves should be worn (AMP, AMP-95)
STORAGE/HANDLING:
Corrosive to copper, brass, and aluminum; store in iron or steel; combustible (AMP, AMP-95)
STD. PKGS.:
55-gal drums or bulk (AMP, AMP-95)

Aminotrimethylene phosphonic acid

SYNONYMS:
Amino tris (methylene phosphonic acid)
Nitrilotris (methylene) trisphosphonic acid
Phosphonic acid, [nitrilotris (methylene)] tris-
EMPIRICAL FORMULA:
$C_3H_{12}NO_9P_3$
STRUCTURE:

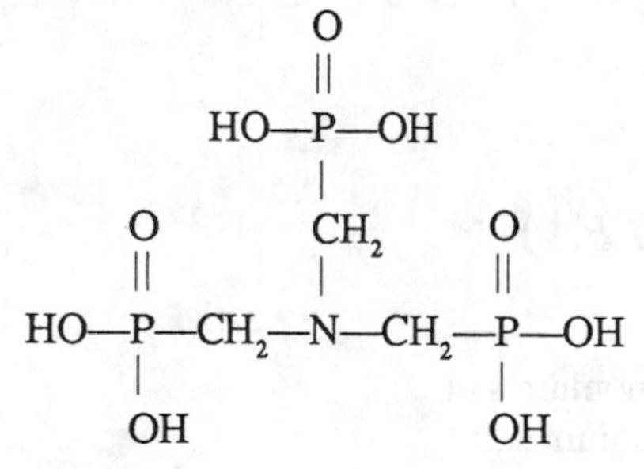

CAS No.:
6419-19-8
TRADENAME EQUIVALENTS:
Dequest 2000 [Monsanto]
Fostex AMP [Henkel]
Unihib 305 [Lonza]

Aminotrimethylene phosphonic acid (*cont'd.*)

CATEGORY:

Dispersant, scale inhibitor, corrosion inhibitor, sequestrant, conditioning agent

APPLICATIONS:

Industrial applications: heat exchange equipment (Unihib 305); industrial processing, aqueous systems (Dequest 2000; Unihib 305); metal processing (Dequest 2000); paper processing (Dequest 2000); textile/leather processing (Dequest 2000); water treatment (Fostex AMP; Unihib 305)

Industrial cleaners: (Dequest 2000)

PROPERTIES:

Form:

Liquid (Dequest 2000; Fostex AMP)

Clear low-viscosity liquid (Unihib 305)

Color:

Straw (Dequest 2000)

Pale yellow (Unihib 305)

Composition:

50% active (Fostex AMP)

50% active in water (Dequest 2000)

50% solids (Unihib 305)

Ionic Nature:

Anionic (Fostex AMP)

Stability:

Hydrolytically stable over a wide pH range and at elevated temps. (Unihib 305)

pH:

< 2 (1% sol'n.) (Unihib 305)

TOXICITY/HANDLING:

Mildly irritating to skin; moderate eye irritant; avoid prolonged/repeated skin contact; use with protective goggles (Unihib 305)

Ammonium cumenesulfonate (CTFA)

SYNONYMS:

Benzenesulfonic acid, (1-methylethyl)-, ammonium salt

(1-Methylethyl) benzenesulfonic acid, ammonium salt

EMPIRICAL FORMULA:

$C_9H_{12}O_3S \cdot H_3N$

STRUCTURE:

$(CH_3)_2CH$—[benzene ring]—SO_3NH_4

CAS No.:
37475-88-0
TRADENAME EQUIVALENTS:
Eltesol ACS60 [Albright & Wilson/Detergents]
Naxonate 6AC [Nease]
Reworyl ACS60 [Rewo Chemische Werke GmbH]
Ultra NCS Liquid [Witco Chem./Organics]
Witconate NCS [Witco Chem./Organics]
CATEGORY:
Hydrotrope, solubilizer, coupling agent, viscosity modifier, stabilizer, antiblocking agent, cloud point depressant
APPLICATIONS:
Household detergents: (Reworyl ACS60; Witconate NCS); dishwashing (Naxonate 6AC); heavy-duty cleaner (Eltesol ACS60; Naxonate 6AC); light-duty cleaners (Eltesol ACS60); liquid detergents (Eltesol ACS60; Naxonate 6AC; Ultra NCS Liquid); powdered detergents (Eltesol ACS60; Naxonate 6AC; Ultra NCS Liquid)
Industrial applications: adhesives (Naxonate 6AC); dyes and pigments (Naxonate 6AC); electroplating (Naxonate 6AC); lubricating/cutting oils (Naxonate 6AC); polymers/polymerization (Naxonate 6AC); printing inks (Naxonate 6AC); leather processing (Naxonate 6AC)
Industrial cleaners: sanitizers/germicides (Naxonate 6AC)
PROPERTIES:
Form:
Liquid (Naxonate 6AC; Reworyl ACS60; Witconate NCS; Ultra NCS Liquid); (@ 20 C) (Eltesol ACS60)
Color:
Pale pink (Eltesol ACS60)
Klett 80 max. (Naxonate 6AC)
Composition:
40% conc. (Ultra NCS Liquid)
60% active min. (Naxonate 6AC)
60% active min. in water (Eltesol ACS60)
60% conc. (Reworyl ACS60)
Solubility:
Sol. in water (Witconate NCS)
Ionic Nature:
Anionic (Naxonate 6AC; Reworyl ACS60; Witconate NCS; Ultra NCS Liquid)
Density:
1.10 g/cm³ (20 C) (Eltesol ACS60)
pH:
7.0–8.0 (Naxonate 6AC); (10% aq. sol'n.) (Eltesol ACS60)
STD. PKGS.:
55-gal lined drums or tankwagons (Naxonate 6AC)

Ammonium dodecylbenzenesulfonate (CTFA)

SYNONYMS:

Ammonium lauryl benzene sulfonate
Benzenesulfonic acid, dodecyl-, ammonium salt

EMPIRICAL FORMULA:

$C_{18}H_{30}O_3S \cdot H_3N$

STRUCTURE:

$CH_3(CH_2)_{10}CH_2—C_6H_4—SO_3NH_4$

CAS No.:

1331-61-9

TRADENAME EQUIVALENTS:

Conco AAS-50M [Continental]
Hetsulf 50A [Heterene]
Nansa AS40 [Albright & Wilson/Marchon]
Newcol 210 [Nippon Nyukazai]

CATEGORY:

Dispersant, detergent, emulsifier, wetting agent

APPLICATIONS:

Automobile cleaners: car shampoo (Nansa AS40)
Household detergents: (Nansa AS40); dishwashing (Nansa AS40); light-duty cleaners (Conco AAS-50M; Hetsulf 50A); liquid detergents (Conco AAS-50M; Nansa AS40)
Industrial cleaners: (Nansa AS40)

PROPERTIES:

Form:

Liquid (Conco AAS-50M; Newcol 210)
Slurry (Hetsulf 50A)
Opaque gel (Nansa AS40)

Color:

Golden yellow (Nansa AS40)

Composition:

40% active (Nansa AS40)
40% conc., contains isopropanol (Conco AAS-50M)
48% active (Hetsulf 50A)
50% conc. (Newcol 210)

Ionic Nature:

Anionic (Conco AAS-50M; Hetsulf 50A; Newcol 210)

pH:

6.0–6.5 (2% aq.) (Nansa AS40)
6.0–7.0 (5% aq. sol'n.) (Hetsulf 50A)

Biodegradable: (Hetsulf 50A)

Ammonium nonoxynol-4 sulfate (CTFA)

SYNONYMS:
Sulfated nonoxynol-4, ammonium salt
CAS No.:
31691-97-1 (generic); 63351-73-5
RD No.: 977060-75-5
TRADENAME EQUIVALENTS:
Alipal CO-436 [GAF]
Neutronyx S-60 [Onyx]
CATEGORY:
Dispersant, detergent, wetting agent, emulsifier, foaming agent
APPLICATIONS:
Automobile cleaners: car shampoo (Neutronyx S-60)
Cleansers: disinfectant hand cleaner (Neutronyx S-60)
Cosmetic industry preparations: (Alipal CO-436)
Farm products: insecticides/pesticides (Alipal CO-436)
Household detergents: all-purpose cleaner (Alipal CO-436); dishwashing (Alipal CO-436; Neutronyx S-60); glass cleaners (Neutronyx S-60)
Industrial applications: polymers/polymerization (Alipal CO-436); textile/leather processing (Alipal CO-436)

PROPERTIES:
Form:
Liquid (Neutronyx S-60)
Clear liquid (Alipal CO-436)
Color:
Varnish 5 max. (Alipal CO-436)
Odor:
Alcoholic (Alipal CO-436)
Composition:
58% active (Alipal CO-436)
60% conc. (Neutronyx S-60)
Solubility:
Sol. in water (Alipal CO-436)
Ionic Nature:
Anionic (Alipal CO-436; Neutronyx S-60)
Sp.gr.:
1.065 (Alipal CO-436)
Density:
8.9 lb/gal (Alipal CO-436)
Visc.:
100 cps (Alipal CO-436)
Stability:
Good (Alipal CO-436)

Ammonium nonoxynol-4 sulfate (cont'd.)

Surface Tension:
34 dynes/cm (1% sol'n.) (Alipal CO-436)

TOXICITY/HANDLING:
Eye irritant (Alipal CO-436)

Ammonium xylenesulfonate (CTFA)

SYNONYMS:
Benzenesulfonic acid, dimethyl-, ammonium salt
Xylene sulfonate, ammonium salt

EMPIRICAL FORMULA:
$C_8H_{10}O_3S \cdot H_3N$

STRUCTURE:

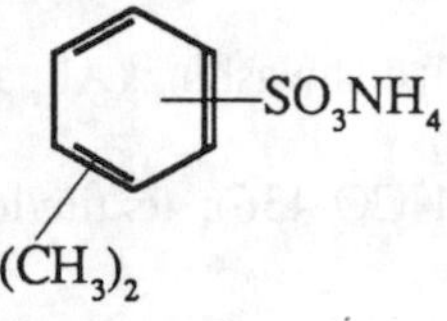

CAS No.:
26447-10-9

TRADENAME EQUIVALENTS:
Eltesol AX40 [Albright & Wilson/Marchon]
Hartotrope AXS40 [Hart Chem. Ltd.]
Naxonate 4AX [Nease]
Stepanate AM [Stepan]
Ultra NXS Liquid [Witco/Organics]
Witconate NXS [Witco/Organics]

CATEGORY:
Detergent, hydrotrope, solubilizer, coupling agent, cloud point depressant, viscosity modifier, solvent, freeze-thaw stabilizer, antiblocking agent, anticaking aid, processing aid, extraction reagent

APPLICATIONS:
Cosmetic industry preparations: shampoos (Witconate NXS)
Household detergents: (Witconate NXS); built detergents (Hartotrope AXS40); dishwashing (Naxonate 4AX; Stepanate AM); heavy-duty cleaner (Eltesol AX40; Naxonate 4AX; Stepanate AM); light-duty cleaners (Eltesol AX40; Hartotrope AXS40); liquid detergents (Naxonate 4AX; Ultra NXS Liquid); powdered detergents (Eltesol AX40; Naxonate 4AX; Ultra NXS Liquid; Witconate NXS)
Industrial applications: adhesives (Naxonate 4AX; Witconate NXS); aerosols (Witconate NXS); dyes and pigments (Naxonate 4AX); electroplating (Naxonate 4AX); industrial processing (Witconate NXS); lubricating/cutting oils (Naxonate 4AX;

Witconate NXS); paper mfg. (Witconate NXS); photography (Witconate NXS); pickling (Witconate NXS); printing inks (Naxonate 4AX; Witconate NXS); textile/ leather processing (Naxonate 4AX; Witconate NXS); wood pulping (Witconate NXS)

Industrial cleaners: metal processing surfactants (Witconate NXS); sanitizers/germicides (Naxonate 4AX); wax strippers (Stepanate AM)

PROPERTIES:

Form:

Liquid (Eltesol AX40; Naxonate 4AX; Stepanate AM; Ultra NXS Liquid; Witconate NXS)

Clear liquid (Hartotrope AXS40)

Color:

Colorless (Stepanate AM)

Pale yellow (Eltesol AX40)

Klett 30 (Witconate NXS)

Klett 50 max. (Naxonate 4AX)

Composition:

40% active (Hartotrope AXS40; Stepanate AM; Witconate NXS)

40% active min. (Naxonate 4AX)

41% active in water (Eltesol AX40)

45% conc. (Ultra NXS Liquid)

Solubility:

Sol. in water (Stepanate AM)

Ionic Nature:

Anionic (Eltesol AX40; Hartotrope AXS40; Naxonate 4AX; Stepanate AM; Ultra NXS Liquid)

Sp.gr.:

1.125 (Hartotrope AXS40)

Density:

1.1 g/cc (Witconate NXS)

9.05 lb/gal (Stepanate AM)

pH:

6.5–7.2 (1% aq.) (Stepanate AM)

6.5–8.5 (Naxonate 4AX)

7.0–8.5 (Eltesol AX40)

8.0 (Witconate NXS)

TOXICITY/HANDLING:

Avoid contact with skin and eyes (Hartotrope AXS40)

Avoid prolonged contact with skin; use normal safety precautions (Witconate NXS)

STORAGE/HANDLING:

Storage and transfer in stainless steel is recommended; store above 20 C to avoid crystallization; if crystallization occurs, heating will reliquefy (Witconate NXS)

Ammonium xylenesulfonate (cont'd.)

STD. PKGS.:

Drums, T/T (Hartotrope AXS40)
55-gal lined drums or tankwagons (Naxonate 4AX)

Butyl acetate (CTFA)

SYNONYMS:

Acetic acid, butyl ester
n-Butyl acetate
Butyl acetate, normal

EMPIRICAL FORMULA:

$C_6H_{12}O_2$

STRUCTURE:

$$CH_3\overset{\overset{\displaystyle O}{\|}}{C}—OC_4H_9$$

CAS No.:

123-86-4

TRADENAME EQUIVALENTS:

Generically sold by:
Ashland Chem., Eastman Chem., Union Carbide, Union Oil

CATEGORY:

Solvent

APPLICATIONS:

Industrial applications: cements (generic); paint mfg. (generic); rubber (generic)

PROPERTIES:

Form:

Clear liquid (generic)

Color:

Colorless (generic)

Odor:

Mild, nonresidual, fruity (generic)

Solubility:

Sol. in alcohols (generic)
Sol. in ether (generic)
Sol. in hydrocarbons (generic)
Slightly sol. in water (generic)

Sp.gr.:

0.8826 (20/20 C) (generic)

Density:

7.35 lb/gal (20 C) (generic)

F.P.:

–75 C (generic)

Butyl acetate (*cont'd.*)

B.P.:
126.3 C (generic)
Flash Pt.:
36.6 C (TOC) (generic)
Ref. Index:
1.2951 (20 C) (generic)
TOXICITY/HANDLING:
Skin irritant; toxic (generic)
STORAGE/HANDLING:
Flammable; moderate fire hazard (generic)

n-Butyl alcohol (CTFA)

SYNONYMS:
1-Butanol
n-Butanol
Butyl alcohol, normal
Butyric alcohol
Propyl carbinol
EMPIRICAL FORMULA:
$C_4H_{10}O$
STRUCTURE:
$CH_3CH_2CH_2CH_2OH$
CAS No.:
71-36-3
TRADENAME EQUIVALENTS:
Alfol 4 [Continental Oil]
Generically sold by:
Ashland Chem., Eastman Chem., Shell Chem., Union Carbide, Union Oil
CATEGORY:
Solvent, intermediate
APPLICATIONS:
Household detergents: (generic)
Industrial applications: (generic); cements (generic); dyes and pigments (generic); hydraulic fluids (generic); paint mfg. (generic)
PROPERTIES:
Form:
Liquid (Alfol 4; generic)
Color:
Colorless (Alfol 4; generic)

Odor:
Characteristic, penetrating, sweet (Alfol 4)
Typical, nonresidual (generic)
Composition:
99.3% active, 0.07% water (Alfol 4)
Solubility:
Misc. with alcohols (generic)
Misc. with ether (generic)
Limited sol. in water (Alfol 4); sol. 7.7% (20 C) (generic)
Sp.gr.:
0.809 (Alfol 4)
0.8109 (20/20 C) (generic)
Density:
6.76 lb/gal (Alfol 4); (20 C) (generic)
F.P.:
–89 C (generic)
B.P.:
117.7 C (Alfol 4; generic)
M.P.:
–128 C (Alfol 4)
Flash Pt.:
35 C (generic)
Ref. Index:
1.3993 (Alfol 4); (20 C) (generic)
TOXICITY/HANDLING:
Toxic on prolonged inhalation; eye irritant; absorbed by skin (generic)
STORAGE/HANDLING:
Flammable; moderate fire hazard (generic)

C_{12-15} *alcohols benzoate (CTFA)*

STRUCTURE:

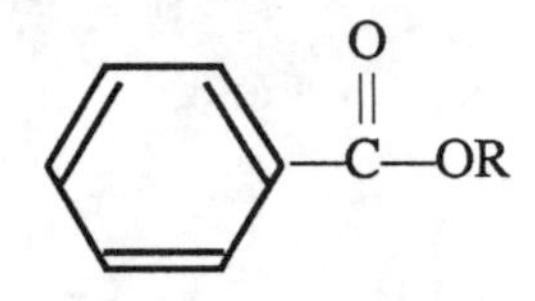

where R represents the C_{12-15} alkyl group

TRADENAME EQUIVALENTS:
Finsolv TN [Finetex]

CATEGORY:
Solubilizer, extender

APPLICATIONS:
Cosmetic industry preparations: (Finsolv TN); perfumery (Finsolv TN)
Pharmaceutical applications: sunscreens (Finsolv TN)

PROPERTIES:

Form:
Liquid (Finsolv TN)

Solubility:
Sol. in organic solvents (Finsolv TN)
Insol. in water (Finsolv TN)

TOXICITY/HANDLING:
Nonirritating to eyes; nonirritating and nonsensitizing to skin (Finsolv TN)

Calcium lignosulfonate (CTFA)

SYNONYMS:
Lignosulfonic acid, calcium salt

CAS No.:
8061-52-7

TRADENAME EQUIVALENTS:
Amberlig [Reed Lignin]
Glutrin [Reed Lignin]
Goulac [Reed Lignin]
Marasperse C-21 [Amer. Can]
Norlig A [Amer. Can]

TRADENAME EQUIVALENTS *(cont'd.)*:
- Lignosite 401, 1840, L [Georgia-Pacific]
- Lignosol B, BD, LC [Reed Ltd.]
- Lignosol SF, SFL [Reed Ltd.] (desugarized)
- PB-92 [Reed Lignin] (modified)

CATEGORY:

Dispersant, binder, emulsion stabilizer, deflocculant, soil and dust stabilizer, extender, reinforcing filler, wetting agent, emulsifier, tanning agent

APPLICATIONS:

Farm products: agricultural chemical formulations (Marasperse C-21); insecticides/herbicides/fungicides (Marasperse C-21; Lignosol B, BD)

Industrial applications: adhesives (Lignosite 1840; Lignosol B, LC); brick/refractory (Marasperse C-21; Lignosol B, BD); cement/concrete (Lignosol SF); ceramics/pottery (Glutrin; Goulac; Lignosol SFL; Marasperse C-21; Norlig A); coal/charcoal processing (Norlig A); construction (Lignosol B, BD); dyes and pigments (Marasperse C-21); foundry applications (Glutrin; Goulac; Lignosol B, BD); gypsum products (Lignosol BD, SF, SFL; Marasperse C-21); pelleting/granulating (Lignosite L); petroleum industry (Marasperse C-21; PB-92); plating baths (Lignosol BD); tanning (Lignosol B, BD, SF); wax emulsions (Lignosite 1840; Lignosol BD)

Industrial cleaners: (Marasperse C-21)

PROPERTIES:

Form:
- Liquid (Amberlig; Goulac; Lignosite 401, L; Lignosol B, LC, SFL; Norlig A)
- Powder (Glutrin; Lignosite 401; Lignosol BD, SF; Marasperse C-21; Norlig A; PB-92)
- Fine powder; also avail. as 50% sol'n. (Lignosite 1840)

Color:
- Light (Lignosite 1840)
- Yellow (Lignosol BD)
- Brown (Lignosol SF; Marasperse C-21)
- Black (Lignosol B)

Composition:
- 40% calcium lignosulfonate (Lignosite 401 (liquid), L)
- 48% conc. (Lignosol SFL)
- 50% active (Lignosol B, LC)
- 50% and 58% conc. (Amberlig; Norlig A (liquid))
- 70% calcium lignosulfonate (Lignosite 1840)
- 95% active (Lignosol BD, SF)
- 100% conc. (Glutrin; Lignosite 401 (powder); Marasperse C-21; Norlig A (powder); PB-92)

Solubility:

Completely sol. in water (Lignosol BD, SF; Marasperse C-21); dissolves in hot or cold water (Lignosite 1840, L)

Calcium lignosulfonate (*cont'd.*)

Ionic Nature:
Anionic (Amberlig; Glutrin; Goldlig; Goulac; Lignosite 401, L; Lignosol B, BD, LC, SF, SFL; Norlig A; PB-92)

Sp.gr.:
1.25 (Lignosol B); (30/4 C) (Lignosite L)

Density:
10.4 lb/gal (Lignosite L)
10.5 lb/gal (Lignosite 1840 (50% sol'n.)

Bulk Density:
23 lb/ft^3 (Lignosite 1840)
28–32 lb/ft^3 (Lignosol BD, SF)
35–40 lb/ft^3 (Marasperse C-21)

Visc.:
200 cps (80 F) (Lignosol B)

F.P.:
–10 C (Lignosol B)

B.P.:
105 C (Lignosol B)

Stability:
Stable to 250 C (Lignosol BD, SF)

pH:
Acid (Lignosol B)
4.5 (10% aq.) (Lignosite 1840); (27% sol'n.) (Lignosol BD)
5.5 (25% sol'n.) (Lignosite L)
6.0 (27% sol'n.) (Lignosol SF)
7.0–8.2 (3% sol'n.) (Marasperse C-21)

Surface Tension:
37 dynes/cm^2 (Lignosite L)

TOXICITY/HANDLING:
Avoid skin contact (Lignosite L)

STORAGE/HANDLING:
Can be stored in mild steel tanks if pH is maintained above 5.5; do not use or handle near strong oxidizing agents—toxic SO_2 may be released on decomposition (Lignosite L)

STD. PKGS.:
Drums or bulk (Lignosol B)
50-lb paper bags (Lignosol BD)

Caprylic/capric glycerides (CTFA)

SYNONYMS:

Glyceryl mono-di-caprylate/caprate

TRADENAME EQUIVALENTS:

Imwitor 742 [Dynamit-Nobel]

CATEGORY:

Coemulsifier, solubilizer, carrier

APPLICATIONS:

Pharmaceutical applications: lipophilic drugs (Imwitor 742)

PROPERTIES:

Form:

Soft crystalline mass (Imwitor 742)

Caprylic/capric triglyceride (CTFA)

SYNONYMS:

Mixed triester of glycerin and caprylic and capric acids
Octanoic/decanoic acid triglyceride

CAS No.:

RD No.: 977059-83-8

TRADENAME EQUIVALENTS:

Captex 300 [Capital City]
Hodag CC-33, CC-33-F [Hodag]
Lexol GT855, GT865 [Inolex]
Liponate GC [Lipo]
Miglyol 810, 812 [Dynamit-Nobel]
Myritol 318 [Henkel, Henkel KGaA]
Neobee M-5, O [PVO]
Softisan 378 [Dynamit-Nobel]
Standamul 318 [Henkel]

CATEGORY:

Solubilizer, solvent, emollient, thickener, viscosity control agent, spreading agent, penetrant, carrier, base, extender, stabilizer, diluent

APPLICATIONS:

Bath products: (Liponate GC); bath oils (Captex 300; Lexol GT855; Miglyol 810, 812; Standamul 318)

Cosmetic industry preparations: (Hodag CC-33, CC-33-F; Miglyol 810, 812; Myritol 318; Neobee M-5, O; Standamul 318); body oils (Standamul 318); creams and lotions (Captex 300; Lexol GT855; Liponate GC; Miglyol 810, 812; Neobee M-5, O; Softisan 378); makeup (Captex 300; Lexol GT855; Myritol 318); perfumery (Captex 300; Lexol GT865; Miglyol 810, 812); pigmented cosmetics (Captex 300; Lexol GT865; Liponate GC; Neobee M-5); skin preparations (Miglyol 810, 812);

sticks (Softisan 378)

Food applications: (Hodag CC-33, CC-33-F; Myritol 318; Standamul 318); beverages (Neobee M-5); dietetic products (Miglyol 810, 812); flavors (Captex 300; Lexol GT865; Miglyol 810, 812; Neobee M-5; Standamul 318); food additives (Neobee M-5, O)

Pharmaceutical applications: (Captex 300; Hodag CC-33, CC-33-F; Miglyol 810, 812; Myritol 318; Neobee M-5, O; Standamul 318); antibiotics (Lexol GT865); antiperspirant/deodorant (Standamul 318); injection products (Miglyol 810, 812); medicinals (Lexol GT865; Miglyol 810, 812; Neobee M-5, O); oral products (Miglyol 810, 812; Standamul 318); ointments (Miglyol 810, 812; Neobee M-5, O); suppositories (Miglyol 810, 812); topical products (Standamul 318); vitamins (Captex 300; Lexol GT865; Neobee O)

PROPERTIES:

Form:

Liquid (Hodag CC-33, CC-33-F; Lexol GT855; Liponate GC; Miglyol 810, 812)
Low-viscosity liquid (Captex 300; Neobee M-5, O)
Clear, low-viscosity liquid (Myritol 318; Standamul 318)

Color:

Colorless (Liponate GC)
Almost colorless (Miglyol 812; Standamul 318)
Gardner 3 max. (Miglyol 810)

Odor:

Odorless (Lexol GT855; Neobee M-5; Standamul 318)
Practically odorless (Neobee O)
Neutral (Miglyol 810, 812)

Taste:

Tasteless (Lexol GT855; Neobee M-5)
Neutral (Miglyol 810, 812)

Composition:

100% conc. (Hodag CC-33, CC-33-F; Myritol 318; Standamul 318)

Solubility:

Sol. in acetone (Lexol GT865; Miglyol 810, 812; Neobee M-5, O)
Sol. in alcohols (Captex 300; Lexol GT865; Miglyol 810, 812; Neobee M-5); sol. in anhydrous alcohol (Liponate GC)
Sol. in aliphatics (Captex 300)
Sol. in benzene (Miglyol 812)
Sol. in carbon tetrachloride (Miglyol 812)
Sol. in castor oil (Standamul 318); (@ 10%) (Myritol 318)
Sol. in chloroform (Miglyol 812); (20 C) (Miglyol 810)
Sol. in diethyl ether (20C) (Miglyol 810)
Sol. in 95% ethanol-SD40 (@ 10%) (Myritol 318); sol. in anhyd. ethanol (Standamul 318)
Sol. in ether (Miglyol 812)

Solubility *(cont'd.)*:
Sol. in isopropanol (20 C) (Miglyol 810)
Sol. in isopropyl myristate (Standamul 318); (@ 10%) (Myritol 318)
Sol. in MEK (Miglyol 812)
Sol. in min. oil (Lexol GT865; Liponate GC; Miglyol 810, 812; Neobee M-5, O; Standamul 318); (@ 10%) (Myritol 318)
Sol. in oils (Captex 300)
Sol. in oleyl alcohol (Standamul 318)
Sol. in organic solvents (Captex 300)
Sol. in light liquid paraffin (Miglyol 812)
Sol. in petroleum ether (Miglyol 812); (20 C) (Miglyol 810)
Disp. in silicone fluid (@ 10%) (Myritol 318)
Sol. in toluene (Miglyol 812); (20 C) (Miglyol 810)
Sol. in veg. oil (Lexol GT865; Liponate GC)
Insol. in water (Liponate GC); insol. (@ 10%) (Myritol 318)
Sol. in xylene (Miglyol 812)

Ionic Nature:
Nonionic (Hodag CC-33, CC-33-F; Miglyol 810, 812)

Sp.gr.:
0.93–0.96 (Neobee M-5)
0.938 (Neobee O)
0.94–0.95 (20 C) (Miglyol 810, 812)
0.950 (20 C) (Standamul 318)

Density:
0.95 g/ml (Myritol 318)

Visc.:
23 cps (Neobee M-5)
25 cps (Standamul 318)
27–30 cps (20 C) (Miglyol 810)
28–32 cps (20 C) (Miglyol 812)
30 cps (Neobee O)

Setting Pt.:
–5 C (Neobee M-5)
–4 C (Neobee O)

Solidification Pt.:
< –10 C (Myritol 318)

Gel Pt.:
–10 C max. (Standamul 318)

Cloud Pt.:
< –5 C (Myritol 318; Standamul 318)
0 C (Miglyol 810)

HLB:
12 (Myritol 318; Standamul 318)

Caprylic/capric triglyceride (cont'd.)

Acid No.:
0.1 max. (Liponate GC; Miglyol 810, 812; Myritol 318)
0.5 max. (Standamul 318)
Iodine No.:
0.5 (Neobee M-5)
0.5 max. (Miglyol 810; Myritol 318; Standamul 318)
1.0 max. (Miglyol 812)
8.0 (Neobee O)
Saponification No.:
300–315 (Neobee O)
325–355 (Liponate GC)
335–360 (Neobee M-5)
340–350 (Myritol 318; Standamul 318)
340–360 (Miglyol 810)
345–350 (Miglyol 812)
Hydroxyl No.:
5.0 max. (Standamul 318)
Stability:
Oxidation-stable (Miglyol 810, 812; Neobee M-5, O)
Excellent low-temp stability (Standamul 318)
Excellent low-temp. stability and resistance to oxidation (Myritol 318)
Ref. Index:
1.443 (Myritol 318)
1.4485–1.4505 (Miglyol 810)
Surface Tension:
32.0 dynes/cm (Neobee O)
32.3 dynes/cm (Neobee M-5)
Biodegradable: (Miglyol 810, 812)
STD. PKGS.:
374-lb net closed-head steel drums or bulk (Standamul 318)

Cocaminobutyric acid (CTFA)

SYNONYMS:
3-Aminobutanoic acid, N-coco alkyl derivs.
Butanoic acid, 3-amino-, N-coco alkyl derivs.
N-Coco-β-aminobutyric acid
STRUCTURE:

$$R{-}NH{-}\underset{\displaystyle CH_3}{\underset{|}{C}}HCH_2COOH$$

CAS No.:

68649-05-8

TRADENAME EQUIVALENTS:

Armeen Z, Z-9 [Armak]

CATEGORY:

Dispersant, antifogging agent, surfactant, corrosion inhibitor, stabilizer, emulsifier

APPLICATIONS:

Cosmetic industry preparations: (Armeen Z); shampoos (Armeen Z); shaving preparations (Armeen Z)

Industrial applications: dyes and pigments (Armeen Z-9); paint mfg. (Armeen Z); plastics (Armeen Z); printing inks (Armeen Z); rubber latex (Armeen Z); water treatment (Armeen Z)

PROPERTIES:

Form:

Liquid (Armeen Z-9)

Pumpable slurry (Armeen Z)

Color:

Yellow (Armeen Z-9)

Gardner 8 max. (Armeen Z)

Composition:

46–50% active; 10% isopropanol; 40% water (Armeen Z-9)

51–55% solids; 7% isopropanol (Armeen Z)

Solubility:

Disp. in benzene (Armeen Z)

Disp. in carbon tetrachloride (Armeen Z)

Sol. in ethyl acetate (Armeen Z)

Sol. in isopropanol (Armeen Z)

Insol. in kerosene (Armeen Z)

Insol. in petroleum ether (Armeen Z)

Sol. in water (Armeen Z-9); disp. (Armeen Z)

Ionic Nature:

Amphoteric (Armeen Z)

Sp.gr.:

0.90 (Armeen Z-9)

0.98 (Armeen Z)

Visc.:

247 cps (Armeen Z)

Pour Pt.:

65 F max. (Armeen Z)

Flash Pt.:

33 C (Abel-Pensky CC) (Armeen Z-9)

175 F (TCC) (Armeen Z)

Cocaminobutyric acid (cont'd.)

pH:

6.5–7.5 (10% aq.) (Armeen Z)

TOXICITY/HANDLING:

Skin irritant, severe eye irritant (Armeen Z-9)

STORAGE/HANDLING:

Avoid contact with strong oxidizing agents (Armeen Z-9)

STD. PKGS.:

200-l bung type steel drums (Armeen Z-9)

Coconut amine

SYNONYMS:

Amines, coco alkyl
Cocamine (CTFA)
Cocoamine
Coconut oil amine

STRUCTURE:

RNH_2
where R represents the coconut radical

CAS No.:

61788-46-3

TRADENAME EQUIVALENTS:

Adogen 160(D) [Sherex]
Amine KK [Kenobel]
Armeen C, CD [Armak]
Arosurf MG-160 [Sherex]
Kemamine P-650 [Humko Sheffield] (tech.)
Kemamine P-650D [Humko Sheffield] (distilled)
Lilamin 160 [Lilachim S.A.]
Lilamin 160D [Lilachim S.A.] (distilled)
Radiamine 6160 [Oleofina S.A.]
Radiamine 6161 [Oleofina S.A.] (distilled)

CATEGORY:

Corrosion inhibitor, stripping agent, emulsifier, flotation agent, dispersant, intermediate, acid scavenger, mold release, lubricant, bactericidal

APPLICATIONS:

Cosmetic industry preparations: (Radiamine 6160, 6161)

Industrial applications: chemical synthesis (Radiamine 6160, 6161); dyes and pigments (Radiamine 6160, 6161); ore flotation (Arosurf MG-160; Radiamine 6160, 6161); paint mfg. (Armeen C); petroleum industry (Adogen 160(D); Kemamine P-650, P-650D; Lilamin 160, 160D); plastics (Kemamine P-650, P-650D); printing inks (Lilamin 160, 160D); road building (Lilamin 160, 160D); rubber (Kemamine P-650, P-650D; Lilamin 160, 160D; Radiamine 6160, 6161)

PROPERTIES:

Form:

Liquid (Amine KK; Armeen C, CD; Arosurf MG-160; Kemamine P-650, P-650D; Lilamin 160; Radiamine 6160, 6161)

Color:

Gardner 1 max. (Kemamine P-650D)
Gardner 3 max. (Kemamine P-650; Lilamin 160)

Composition:

93% conc. (Kemamine P-650)
95% active (Lilamin 160)
97% conc. (Kemamine P-650D)
98% conc. (Amine KK)
100% conc. (Adogen 160(D); Radiamine 6160, 6161)

Solubility:

Sol. in acetone (Armeen C, CD)
Sol. in carbon tetrachloride (Armeen C, CD)
Sol. in chloroform (Armeen C, CD)
Sol. in ethanol (Armeen C, CD)
Sol. in isopropanol (Armeen C, CD)
Sol. in kerosene (Armeen C, CD)
Sol. in methanol (Armeen C, CD)
Sol. in common organic solvents (Kemamine P-650, P-650D)
Sol. in toluene (Armeen C, CD)

Ionic Nature:

Cationic (Amine KK; Kemamine P-650, P-650D; Radiamine 6160, 6161)

M.W.:

205 (Lilamin 160)

Sp.gr.:

0.781 (60 C) (Lilamin 160)
0.804 (25/4 C) (Armeen CD)
0.805 (25/4 C) (Armeen C)

Visc.:

2.75 cps (60 C) (Lilamin 160)
43.0 SSU (35 C) (Armeen CD)
44.2 SSU (35 C) (Armeen C)

M.P.:

16 C (Lilamin 160)
54–59 F (Armeen C)
57–63 F (Armeen CD)

Pour Pt.:

45 F (Armeen C)

Coconut amine (cont'd.)

55 F (Armeen CD)

Flash Pt.:

112 C (OC) (Lilamin 160)
230 F (Armeen CD)
240 F (Armeen C)

Fire Pt.:

250 F (Armeen CD)
270 F (Armeen C)

Iodine No.:

12.0 max. (Kemamine P-650, P-650D)
13.0 max. (Lilamin 160)

Coco/oleamidopropyl betaine (CTFA)

STRUCTURE:

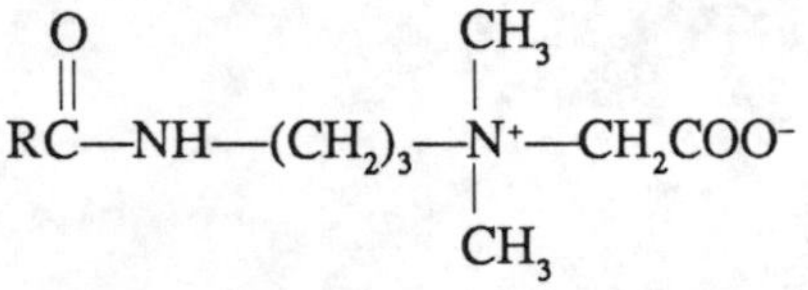

$$RC(=O){-}NH{-}(CH_2)_3{-}N^{+}(CH_3)_2{-}CH_2COO^{-}$$

where RCO^{-} represents a blend of coconut and oleic acid radicals

TRADENAME EQUIVALENTS:

Mirataine COB [Miranol]

CATEGORY:

Solubilizer

APPLICATIONS:

Cosmetic industry preparations: fragrances/natural oils (Mirataine COB)

PROPERTIES:

Solubility:

Sol. in water (Mirataine COB)

Cocoyl imidazoline (CTFA)

SYNONYMS:

Coco imidazoline
Coconut imidazoline
Cocoyl hydroxyethyl imidazoline
1-(2-Hydroxyethyl)-2-norcoco-2-imidazoline
1H-Imidazole-1-ethanol, 4,5-dihydro-2-norcocoyl-

STRUCTURE:

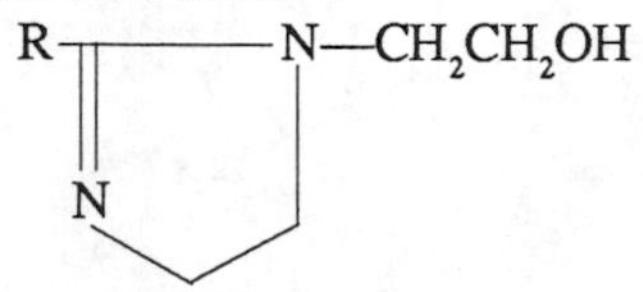

where R is derived from the coconut fatty radical

CAS No.:

61791-38-6

TRADENAME EQUIVALENTS:

Mackazoline C [McIntyre]
Miramine CC [Miranol]
Monazoline C [Mona]
Schercozoline C [Scher]

CATEGORY:

Dispersant, wetting agent, emulsifier, detergent, thickener, corrosion inhibitor, antistat, softener, bactericide, microbicide, lubricant

APPLICATIONS:

Automobile industry: upholstery/rugs (Schercozoline C)
Aviation industry: airplane domes (Schercozoline C)
Cosmetic industry preparations: (Schercozoline C); acid shampoos (Miramine CC)
Degreasers: (Miramine CC)
Farm products: agricultural oils/sprays (Monazoline C; Schercozoline C); fungicides/herbicides (Monazoline C; Schercozoline C)
Industrial applications: asphalt (Monazoline C); clays (Monazoline C); dyes and pigments (Monazoline C); lubricating/cutting oils (Miramine CC; Monazoline C; Schercozoline C); metalworking (Miramine CC; Schercozoline C); paint mfg. (Monazoline C; Schercozoline C); petroleum industry (Monazoline C); plastics (Monazoline C); surface treatment (Schercozoline C); textile/leather processing (Monazoline C; Schercozoline C)
Industrial cleaners: (Schercozoline C); acid cleaners (Schercozoline C); dairy/brewery cleaners (Monazoline C; Schercozoline C); polishes (Schercozoline C)

PROPERTIES:

Form:

Liquid (Mackazoline C; Miramine CC)
Liquid (may crystallize on aging) (Monazoline C)
Semisolid (Schercozoline C)

Color:

Amber (Miramine CC; Monazoline C)
Tan (Schercozoline C)

Odor:

Bland to faintly ammoniacal (Miramine CC)

Cocoyl imidazoline (cont'd.)

Composition:
90% imidazoline min. (Monazoline C; Schercozoline C)
99% active (Miramine CC)
100% conc. (Mackazoline C)

Solubility:
Sol. in acids (Monazoline C)
Sol. in chlorinated hydrocarbons (@ 10%) (Monazoline C)
Sol. in ethanol (@ 10%) (Monazoline C)
Sol. in most hydrocarbon solvents (Monazoline C)
Sol. in kerosene (@ 10%) (Monazoline C)
Sol. in min. oil (@ 10%) (Monazoline C)
Sol. in min. spirits (@ 10%) (Monazoline C)
Sol. in many organic solvents (Miramine CC)
Sol. in most polar solvents (Monazoline C)
Sol. in toluene (@ 10%) (Monazoline C)
Sol. in veg. oil (@ 10%) (Monazoline C)
Disp. in water (Miramine CC; Monazoline C)

Ionic Nature:
Cationic (Monazoline C; Schercozoline C)
Nonionic (Mackazoline C)

M.W.:
278 (Schercozoline C)
282 (Monazoline C)

Sp.gr.:
0.93 (Monazoline C)
0.940 (Miramine CC)

Density:
7.75 lb/gal (Monazoline C)

Acid No.:
1.0 max. (Monazoline C)

Alkali No.:
200–214 (Schercozoline C)
209 (Miramine CC)

Storage Stability:
May granulate on standing; may be cleared by heating to 60 C with gentle stirring to insure uniformity (Miramine CC)

pH:
10.5–12.0 (10% disp.) (Monazoline C)

Biodegradable: (Miramine CC; Monazoline C)

TOXICITY/HANDLING:
Severe primary skin irritant; may cause severe conjunctival irritation with corneal involvement (Miramine CC)

Decaglycerol decaoleate

SYNONYMS:
Decaglyceryl decaoleate
Polyglyceryl-10 decaoleate (CTFA)

CAS No.
RD No. 977057-94-5

TRADENAME EQUIVALENTS:
Aldo DGDO [Glyco]
Caprol 10G100 [Stokely-Van Camp]
Drewmulse 10-10-O [PVO Int'l.
Drewpol 10-10-O [PVO Int'l.]
Mazol PGO-1010 [Mazer]
Polyaldo DGDO [Glyco]
Santone 10-10-O [Durkee]

CATEGORY:
Emulsifier, solubilizer, dispersing agent, plasticizer, emulsion stabilizer, lubricant

APPLICATIONS:
Cosmetic industry preparations: (Drewmulse 10-10-O; Mazol PGO-1010); creams and lotions (Drewmulse 10-10-O); toiletries (Mazol PGO-1010)
Food applications: (Santone 10-10-O); food emulsifying (Drewpol 10-10-O; Mazol PGO-1010)
Industrial applications: lubricants/mold release compounds (Mazol PGO-1010)
Pharmaceutical applications: (Drewmulse 10-10-O; Mazol PGO-1010)

PROPERTIES:

Form:
Liquid (Aldo DGDO; Caprol 10G100; Drewmulse 10-10-O; Drewpol 10-10-O; Mazol PGO-1010; Polyaldo DGDO; Santone 10-10-O)

Color:
Amber (Aldo DGDO; Drewpol 10-10-O)
Gardner 8 (Drewmulse 10-10-O)
Gardner 8 max. (Mazol PGO-1010)

Composition:
100% conc. (Caprol 10G100; Polyaldo DGDO)

Solubility:
Sol. in min. oils (Drewmulse 10-10-O)
Sol. in peanut oil (Drewmulse 10-10-O)

Ionic Nature:
Nonionic

Decaglycerol decaoleate (cont'd.)

HLB:
< 1.0 (Santone 10-10-O)
2.5 (Caprol 10G100; Mazol PGO-1010)
3.0 (Aldo DGDO; Drewmulse 10-10-O; Drewpol 10-10-O; Polyaldo DGDO)
Acid No.:
10 max. (Mazol PGO-1010)
Iodine No.:
65–80 (Santone 10-10-O)
80 max. (Mazol PGO-1010)
Saponification No.:
160–190 (Drewpol 10-10-O)
165–180 (Santone 10-10-O)
165–185 (Aldo DGDO; Drewmulse 10-10-O; Mazol PGO-1010)
Hydroxyl No.:
25–45 (Santone 10-10-O)

Decaglycerol decastearate

SYNONYMS:
Decaglyceryl decastearate
Polyglyceryl-10 decastearate (CTFA)
CAS No.
RD No.: 977067-09-6
TRADENAME EQUIVALENTS:
Drewmulse 10-10-S [PVO Int'l.]
Drewpol 10-10-S [PVO Int'l.]
CATEGORY:
Emulsifier, solubilizer, dispersant, lubricant
APPLICATIONS:
Cosmetic industry preparations: (Drewmulse 10-10-S); emollient creams and lotions (Drewmulse 10-10-S)
Food applications: food emulsifying (Drewpol 10-10-S)
Pharmaceutical applications: (Drewmulse 10-10-S)
PROPERTIES:
Form:
Solid (Drewmulse 10-10-S; Drewpol 10-10-S)
Color:
Tan (Drewpol 10-10-S)
Gardner 8 (Drewmulse 10-10-S)
Solubility:
Sol. hot in min. oils (Drewmulse 10-10-S)

Sol. hot in peanut oils (Drewmulse 10-10-S)
Slightly sol. in propylene glycol (Drewmulse 10-10-S)

Ionic Nature:
Nonionic

HLB:
3.0 (Drewmulse 10-10-S; Drewpol 10-10-S)

Saponification No.:
150–190 (Drewpol 10-10-S)
160–180 (Drewmulse 10-10-S)

Decaglycerol octaoleate

SYNONYMS:
Decaglyceryl octaoleate

TRADENAME EQUIVALENTS:
Drewmulse 10-8-O [PVO Int'l.]
Drewpol 10-8-O [PVO Int'l.]
Mazol PGO-108 [Mazer]

CATEGORY:
Emulsifier, solubilizer, dispersant

APPLICATIONS:
Cosmetic industry preparations: (Drewmulse 10-8-O; Mazol PGO-108); emollient creams and lotions (Drewmulse 10-8-O); toiletries (Mazol PGO-108)
Food applications: food emulsifying (Drewpol 10-8-O; Mazol PGO-108)
Industrial applications: lubricants and mold release (Mazol PGO-108); plastics (Mazol PGO-108); synthetic fibers (Mazol PGO-108)
Pharmaceutical applications: (Drewmulse 10-8-O; Mazol PGO-108)

PROPERTIES:

Form:
Liquid (Drewmulse 10-8-O; Drewpol 10-8-O; Mazol PGO-108)

Color:
Amber (Drewpol 10-8-O)
Gardner 8 (Drewmulse 10-8-O)
Gardner 8 max. (Mazol PGO-108)

Solubility:
Sol. in min. oils (Drewmulse 10-8-O)
Sol. in peanut oil (Drewmulse 10-8-O)

Ionic Nature:
Nonionic

HLB:
3.4 (Mazol PGO-108)

Decaglycerol octaoleate (cont'd.)

4.0 (Drewmulse 10-8-O; Drewpol 10-8-O)

Acid No.:

10 max. (Mazol PGO-108)

Iodine No.:

75 max. (Mazol PGO-108)

Saponification No.:

150–180 (Drewpol 10-8-O)

155–175 (Drewmulse 10-8-O; Mazol PGO-108)

Diethylene glycol monobutyl ether

SYNONYMS:

Butoxydiglycol (CTFA)

2-(2-Butoxyethoxy) ethanol

Diethylene glycol butyl ether

Ethanol, 2-(2-butoxyethoxy)-

Monobutyl diethylene glycol ether

EMPIRICAL FORMULA:

$C_8H_{18}O_3$

STRUCTURE:

$C_4H_9OCH_2CH_2OCH_2CH_2OH$

CAS No.:

112-34-5

TRADENAME EQUIVALENTS:

Butyl Carbitol [Union Carbide]

Butyl Dioxitol [Shell]

Dowanol DB [Dow]

Ektasolve DB [Eastman]

Poly-Solv DB [Olin]

CATEGORY:

Solvent, coupling agent, coalescing agent, solubilizer, plasticizer, extraction solvent

APPLICATIONS:

Automotive industry applications: brake fluids (Butyl Dioxitol; Ektasolve DB; Poly-Solv DB)

Cosmetic industry preparations: (Poly-Solv DB); nail polish (Poly-Solv DB)

Farm products: agricultural sprays (Poly-Solv DB); insect repellents (Poly-Solv DB)

Household detergents: hard surface cleaner (Poly-Solv DB)

Industrial applications: adhesives (Poly-Solv DB); cutting oils (Butyl Dioxitol; Ektasolve DB); dyes and pigments (Butyl Carbitol; Butyl Dioxitol; Ektasolve DB; Poly-Solv DB); gums (Butyl Carbitol); paint/lacquer mfg. (Butyl Dioxitol;

Ektasolve DB; Poly-Solv DB); plastics (Poly-Solv DB); polishes and waxes (Poly-Solv DB); polymers/polymerization (Butyl Carbitol); printing inks (Butyl Dioxitol; Ektasolve DB; Poly-Solv DB); resins (Butyl Carbitol; Butyl Dioxitol; Ektasolve DB); textile/leather processing (Ektasolve DB; Poly-Solv DB)

Industrial cleaners: (Ektasolve DB); soaps (Butyl Carbitol; Ektasolve DB); solvent cleaners (Butyl Dioxitol)

PROPERTIES:

Form:

Liquid (Butyl Carbitol; Butyl Dioxitol; Poly-Solv DB)
Clear liquid (Ektasolve DB)

Color:

Colorless (Butyl Carbitol; Butyl Dioxitol)
APHA 10 max. (Poly-Solv DB)
Pt-Co 10 max. (Ektasolve DB)

Odor:

Faint butyl odor (Butyl Carbitol)
Faint, pleasant (Ektasolve DB)
Mild (Butyl Dioxitol)
Mild, characteristic (Poly-Solv DB)

Composition:

> 99% volatiles by vol. (Ektasolve DB)
100% active (Butyl Dioxitol; Poly-Solv DB)

Solubility:

Sol. in oils (Butyl Carbitol)
Sol. in many organic liquids (Dowanol DB); completely miscible with a wide variety of organic solvents (Butyl Dioxitol; Poly-Solv DB)
Sol. in water (Butyl Carbitol; Dowanol DB); completely sol. (Poly-Solv DB); completely sol. (20 C) (Ektasolve DB); completely miscible with water (Butyl Dioxitol)

M.W.:

162.22 (Poly-Solv DB)
162.23 (Ektasolve DB)

Sp.gr.:

0.949–0.952 (Butyl Dioxitol)
0.9536 (20/20 C) (Butyl Carbitol)
0.955 (20/20 C) (Ektasolve DB; Poly-Solv DB)

Density:

7.94 lb/gal (Ektasolve DB); (20 C) (Butyl Carbitol)
7.95 lb/gal (Poly-Solv DB)

Visc.:

0.0649 poise (20 C) (Butyl Carbitol)
6.5 cP (20 C) (Poly-Solv DB)

F.P.:

–76 C (Ektasolve DB)

–68.1 C (Butyl Carbitol)
–68 C (Poly-Solv DB)

B.P.:
230 C (760 mm) (Poly-Solv DB)
230 C min. (initial, 760 mm) (Ektasolve DB)
230.6 C (Butyl Carbitol)

Flash Pt.:
77.7 C (Butyl Carbitol)
111 C (COC) (Ektasolve DB)
116 C (COC) (Poly-Solv DB)

Fire Pt.:
117 C (Ektasolve DB)

Stability:
Stable (Butyl Dioxitol)

Ref. Index:
1.4316 (20 C) (Butyl Carbitol; Ektasolve DB; Poly-Solv DB)

Coefficient of Linear Expansion:
0.00088/°C (to 20 C) (Butyl Carbitol)
0.00091 (55 C) (Poly-Solv DB)

Specific Heat:
0.546 cal/g (20–25 C) (Butyl Carbitol)
2.283 J/g-°C (20 C) (Poly-Solv DB)

TOXICITY/HANDLING:
Contact with liquid may cause transient eye injury; prolonged skin contact may cause slight irritation and possible toxic effects from skin absorption; avoid prolonged breathing of vapors (Ektasolve DB)
Moderate eye irritant; mild primary skin irritant; avoid contact with eyes and skin; do not breathe vapors (Poly-Solv DB)

STORAGE/HANDLING:
Combustible (Butyl Carbitol)

STD. PKGS.:
55-gal (440 lb net) drums, tank cars, tank trucks (Poly-Solv DB)

Diethylene glycol monoethyl ether

SYNONYMS:
Diethylene glycol ethyl ether
Ethanol, 2-(2-ethoxyethoxy)-
Ethoxydiglycol (CTFA)
2-(2-Ethoxyethoxy) ethanol

EMPIRICAL FORMULA:

$C_6H_{14}O_3$

STRUCTURE:

$CH_3CH_2OCH_2CH_2OCH_2CH_2OH$

CAS No.:

111-90-0

TRADENAME EQUIVALENTS:

Carbitol Solvent [Union Carbide]
Dioxitol-High Gravity, -Low Gravity [Shell]
Dowanol DE [Dow]
Ektasolve DE [Eastman]
Poly-Solv DE (High Gravity), DE (Low Gravity) [Olin]

CATEGORY:

Solvent, coupling agent, intermediate

APPLICATIONS:

Automotive industry: brake fluids (Carbitol Solvent; Dioxitol-High Gravity; Ektasolve DE; Poly-Solv DE)

Cosmetic industry preparations: (Poly-Solv DE); nail polish (Poly-Solv DE)

Farm products: agricultural oils/sprays (Poly-Solv DE); insect repellents (Poly-Solv DE)

Household detergents: hard surface cleaner (Poly-Solv DE)

Industrial applications: adhesives (Poly-Solv DE); dyes and pigments (Carbitol Solvent; Dioxitol-High Gravity; Ektasolve DE; Poly-Solv DE); paint/stain mfg. (Carbitol Solvent; Dioxitol-High Gravity; Ektasolve DE; Poly-Solv DE); plastics (Poly-Solv DE); polishes and waxes (Ektasolve DE; Poly-Solv DE); printing inks (Ektasolve DE; Poly-Solv DE); resins (Carbitol Solvent; Dioxitol-High Gravity; Ektasolve DE); textile/leather processing (Carbitol Solvent; Dioxitol-High Gravity; Ektasolve DE; Poly-Solv DE)

Industrial cleaners: (Dioxitol-High Gravity); textile soaps (Carbitol Solvent)

PROPERTIES:

Form:

Liquid (Dioxitol-High Gravity, -Low Gravity; Poly-Solv DE (High Gravity), (Low Gravity))

Slightly viscous liquid (Carbitol Solvent)

Color:

Colorless (Carbitol Solvent; Dioxitol-High Gravity, -Low Gravity)

APHA 10 max. (Poly-Solv DE (High Gravity), (Low Gravity))

Pt-Co 10 max. (Ektasolve DE)

Odor:

Mild, characteristic (Poly-Solv DE)

Mild, pleasant (Carbitol Solvent)

Diethylene glycol monoethyl ether (cont'd.)

Composition:

72% diethylene glycol monoethyl ether, 28% ethylene glycol (Dioxitol-High Gravity)
75% diethylene glycol monoethyl ether, 25% ethylene glycol (Poly-Solv DE (High Gravity))
100% diethylene glycol monoethyl ether (Dioxitol-Low Gravity; Poly-Solv DE (Low Gravity))

Solubility:

Completely miscible with alcohols (Dioxitol-High Gravity)
Partially miscible with aliphatic hydrocarbons (Dioxitol-High Gravity)
Completely miscible with aromatic hydrocarbons (Dioxitol-High Gravity)
Completely miscible with esters (Dioxitol-High Gravity)
Completely miscible with other glycol esters (Dioxitol-High Gravity)
Completely miscible with glycols (Dioxitol-High Gravity)
Completely miscible with ketones (Dioxitol-High Gravity)
Sol. in many organic liquids (Dowanol DE); miscible with many organic solvents (Carbitol Solvent; Poly-Solv DE (High Gravity), (Low Gravity))
Sol. in water (Dowanol DE); completely sol. (Poly-Solv DE (High Gravity), (Low Gravity)); completely sol. (20 C) (Ektasolve DE); completely miscible (Dioxitol-High Gravity); miscible (Carbitol Solvent)

M.W.:

134.17 (Ektasolve DE; Poly-Solv DE (Low Gravity))

Sp.gr.:

0.986–0.990 (Dioxitol-Low Gravity)
0.989 (20/20 C) (Poly-Solv DE (Low Gravity))
0.990 (20/20 C) (Ektasolve DE)
1.021–1.027 (Dioxitol-High Gravity)
1.0253 (20/20 C) (Poly-Solv DE (High Gravity))
1.0272 (20/20 C) (Carbitol Solvent)

Density:

0.99 kg/l (20 C) (Ektasolve DE)
8.24 lb/gal (20 C) (Poly-Solv DE (Low Gravity))
8.53 lb/gal (20 C) (Poly-Solv DE (High Gravity))
8.55 lb/gal (20 C) (Carbitol Solvent)

Visc.:

4.59 cP (20 C) (Poly-Solv DE (Low Gravity))
6.9 cP (20 C) (Poly-Solv DE (High Gravity))

F.P.:

–90 C (Ektasolve DE)
–76 C (Poly-Solv DE (Low Gravity))
–75 C (Poly-Solv DE (High Gravity))

B.P.:

195 C (760 mm) (Poly-Solv DE (High Gravity))
195–202 C (Carbitol Solvent)

198 C (initial, 760 mm) (Ektasolve DE)
202 C (760 mm) ·(Poly-Solv DE (Low Gravity))

Flash Pt.:
85 C (TCC) (Poly-Solv DE (Low Gravity))
96 C (TOC) (Ektasolve DE); (COC) (Poly-Solv DE (High Gravity))
96.1 C (Carbitol Solvent)

Fire Pt.:
96 C (Ektasolve DE)

Ref. Index:
1.425 (Carbitol Solvent)
1.4273 (20 C) (Ektasolve DE; Poly-Solv DE (Low Gravity))
1.4297 (20 C) (Poly-Solv DE (High Gravity))

Coefficient of Linear Expansion:
0.00084 (55 C) (Poly-Solv DE (High Gravity), (Low Gravity))

Specific Heat:
2.308 J/g-°C (20 C) (Poly-Solv DE (High Gravity), (Low Gravity))

TOXICITY/HANDLING:
Mild primary skin and eye irritant; no toxic effect from saturated atmosphere; avoid contact with eyes and skin; do not breathe vapor (Poly-Solv DE (High Gravity), (Low Gravity))

STORAGE/HANDLING:
Combustible (Carbitol Solvent)

STD. PKGS.:
55-gal (450 lb net) drums, tank cars, tank trucks (Poly-Solv DE (Low Gravity))
55-gal (470 lb net) drums, tank cars, tank trucks (Poly-Solv DE (High Gravity))

Diethylene glycol monomethyl ether

SYNONYMS:
Diethylene glycol methyl ether
Ethanol, 2-(2-methoxyethoxy)-
Methoxydiglycol (CTFA)
2-(2-Methoxyethoxy) ethanol

EMPIRICAL FORMULA:
$C_5H_{12}O_3$

STRUCTURE:
$CH_3OCH_2CH_2OCH_2CH_2OH$

CAS No.:
111-77-3

Diethylene glycol monomethyl ether *(cont'd.)*

TRADENAME EQUIVALENTS:

Ektasolve DM [Eastman]
Methyl Carbitol [Union Carbide]
Poly-Solv DM [Olin]

CATEGORY:

Solvent, coalescing aid, diluent, intermediate

APPLICATIONS:

Automotive industry applications: brake fluids (Ektasolve DM; Methyl Carbitol; Poly-Solv DM)
Cosmetic industry preparations: (Poly-Solv DM); nail polish (Poly-Solv DM)
Farm products: agricultural sprays (Poly-Solv DM); insect repellents (Poly-Solv DM)
Household detergents: hard surface cleaner (Poly-Solv DM)
Industrial applications: (Methyl Carbitol); adhesives (Poly-Solv DM); dyes and pigments (Ektasolve DM; Poly-Solv DM); paint mfg. (Ektasolve DM; Poly-Solv DM); plastics (Ektasolve DM; Poly-Solv DM); polishes and waxes (Poly-Solv DM); printing inks (Ektasolve DM; Poly-Solv DM); resins (Ektasolve DM); textile/leather processing (Ektasolve DM; Poly-Solv DM)

PROPERTIES:

Form:

Liquid (Ektasolve DM; Methyl Carbitol; Poly-Solv DM)

Color:

Colorless (Methyl Carbitol)
Water-white (Ektasolve DM)
APHA 15 max. (Poly-Solv DM)

Odor:

Agreeable (Ektasolve DM)
Mild, characteristic (Poly-Solv DM)

Composition:

> 99% volatiles by volume (Ektasolve DM)

Solubility:

Miscible with many organic solvents (Poly-Solv DM)
Completely sol. in water (Poly-Solv DM); completely sol. (20 C) (Ektasolve DM); sol. (Methyl Carbitol)

M.W.:

120.15 (Ektasolve DM; Poly-Solv DM)

Sp.gr.:

1.021 (20/20 C) (Poly-Solv DM)
1.0211 (20/4 C) (Methyl Carbitol)
1.023 (20/20 C) (Ektasolve DM)

Density:

8.51 lb/gal (Ektasolve DM; Poly-Solv DM); (20 C) (Methyl Carbitol)

Visc.:

3.9 cP (20 C) (Poly-Solv DM)

F.P.:
–85 C (Ektasolve DM; Poly-Solv DM)

B.P.:
191 C min. (initial, 760 mm) (Ektasolve DM)
194 C (Methyl Carbitol); (760 mm) (Poly-Solv DM)

Flash Pt.:
93.3 C (Methyl Carbitol)
96 C (COC) (Ektasolve DM)

Fire Pt.:
96 C (Ektasolve DM)

Ref. Index:
1.4263 (20 C) (Ektasolve DM; Poly-Solv DM)
1.4264 (27 C) (Methyl Carbitol)

Coefficient of Linear Expansion:
0.00088 (55 C) (Poly-Solv DM)

Specific Heat:
2.149 J/g-°C (20 C) (Poly-Solv DM)

TOXICITY/HANDLING:
Causes transitory eye irritation; prolonged skin contact may result in some skin absorption; use with adequate ventilation to control vapor generated at elevated temps. (Ektasolve DM)
Mild primary skin and eye irritant; avoid contact with eyes and skin; do not breathe vapors (Poly-Solv DM)

STORAGE/HANDLING:
Combustible (Methyl Carbitol)
Combustible—keep away from heat and open flame (Ektasolve DM)

STD. PKGS.:
55-gal (470 lb net) drums, tank cars, tank trucks (Poly-Solv DM)

Diethylene glycol monoricinoleate

SYNONYMS:
PEG 100 monoricinoleate
PEG-2 ricinoleate (CTFA)
POE (2) monoricinoleate

Diethylene glycol monoricinoleate (*cont'd.*)

STRUCTURE:

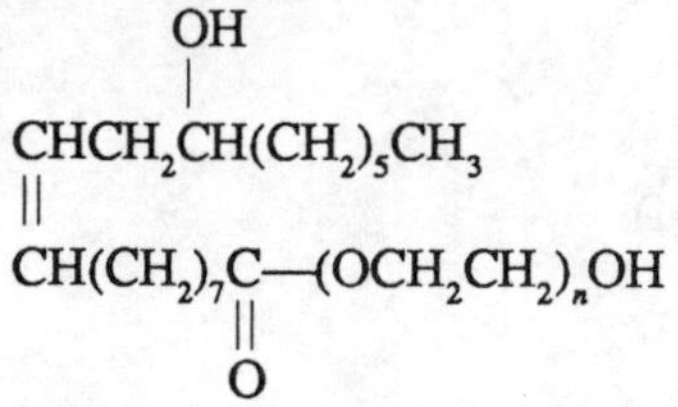

$$
\begin{array}{l}
\qquad\quad OH \\
\qquad\quad | \\
CHCH_2CH(CH_2)_5CH_3 \\
\| \\
CH(CH_2)_7C\text{—}(OCH_2CH_2)_nOH \\
\qquad\qquad \| \\
\qquad\qquad O
\end{array}
$$

where avg. $n = 2$

CAS No.:
5401-17-2; 9004-97-1 (generic)

TRADENAME EQUIVALENTS:
Pegosperse 100 MR [Glyco]

CATEGORY:
Dispersant, emulsifier, plasticizer

PROPERTIES:

Form:
Liquid (Pegosperse 100 MR)

Color:
Amber (Pegosperse 100 MR)

Composition:
100% conc. (Pegosperse 100 MR)

Solubility:
Sol. in acetone (Pegosperse 100 MR)
Sol. in ethanol (Pegosperse 100 MR)
Sol. in ethyl acetate (Pegosperse 100 MR)
Sol. in methanol (Pegosperse 100 MR)
Partly sol. in min. oil (Pegosperse 100 MR)
Miscible with naphtha (Pegosperse 100 MR)
Miscible with toluol (Pegosperse 100 MR)
Miscible with veg. oil (Pegosperse 100 MR)

Ionic Nature:
Nonionic

Sp.gr.:
0.98 (Pegosperse 100 MR)

Solidification Pt.:
< 0 C (Pegosperse 100 MR)

HLB:
4.8 ± 0.5 (Pegosperse 100 MR)

Acid No.:
< 10 (Pegosperse 100 MR)

Iodine No.:
65–71 (Pegosperse 100 MR)

Diethylene glycol monoricinoleate (cont'd.)

Saponification No.:
140–150 (Pegosperse 100 MR)
pH:
3–5 (5% aq. disp.) (Pegosperse 100 MR)

Dimethyl isosorbide (CTFA)

SYNONYMS:
1,4:3,6-Dianhydro-2,5-di-O-methyl-D-glucitol
DMI
D-Glucitol, 1,4:3,6-dianhydro-2,5-di-O-methyl
Isosorbide, dimethyl ether
EMPIRICAL FORMULA:
$C_8H_{14}O_4$
STRUCTURE:

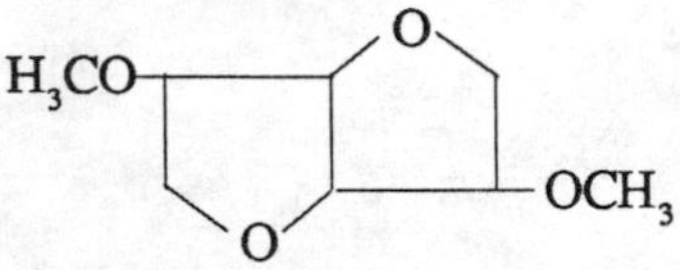

H_3CO — [bicyclic ring structure] — OCH_3

CAS No.:
5306-85-4
TRADENAME EQUIVALENTS:
Atlas G-100 [ICI Specialty Chem.]
CATEGORY:
Solvent
APPLICATIONS:
Cosmetic industry preparations: (Atlas G-100)
Pharmaceutical applications: (Atlas G-100)
PROPERTIES:
Form:
Liquid (Atlas G-100)
Composition:
100% conc. (Atlas G-100)
Solubility:
Misc. with all solvents (Atlas G-100)

Dipentene (CTFA)

SYNONYMS:

Cyclohexene, 1-methyl-4-(1-methylethenyl)-
DL-Limonene
1-Methyl-4-(1-methylethenyl) cyclohexene

EMPIRICAL FORMULA:

$C_{10}H_{15}$

STRUCTURE:

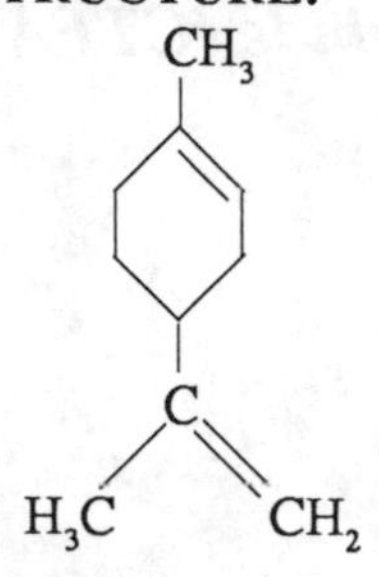

CAS No.:

138-86-3; 7705-14-8

TRADENAME EQUIVALENTS:

Unitene D [Union Camp; Harrisons & Crosfield]

CATEGORY:

Solvent, swelling agent, processing aid, tackifier

APPLICATIONS:

Industrial applications: resins (Unitene D); rubber (Unitene D)

PROPERTIES:

Sp.gr.:

0.86 (Unitene D)

B.P.:

173–188 C (Unitene D)

Dipropylene glycol monomethyl ether

SYNONYMS:

Dipropylene glycol methyl ether
Methoxy dipropylene glycol
1-(2-Methoxypropoxy)-2-propanol
Polyoxypropylene (2) methyl ether
Polypropylene glycol (2) methyl ether
PPG-2 methyl ether (CTFA)
2-Propanol, 1-(2-methoxypropoxy)-

EMPIRICAL FORMULA:

$C_9H_{16}O_3$

STRUCTURE:

$CH_3(OCHCH_2)_nOH$ (with CH_3 attached to the CH)

where avg. $n = 2$

CAS No.:

13429-07-7; 37286-64-9 (generic)

TRADENAME EQUIVALENTS:

Arcosolv DPM [Arco]

Dowanol DPM [Dow, Dow Chem. Europe SA]

Poly-Solv DPM [Olin]

CATEGORY:

Solvent, intermediate, solubilizer, plasticizer, dispersant, coupling agent

APPLICATIONS:

Cosmetic industry preparations: (Arcosolv DPM; Poly-Solv DPM); nail polish (Poly-Solv DPM)

Farm products: (Arcosolv DPM); agricultural oils/sprays (Poly-Solv DPM)

Household detergents: (Arcosolv DPM; Dowanol DPM); hard surface cleaner (Poly-Solv DPM)

Industrial applications: adhesives (Poly-Solv DPM); brake fluids (Poly-Solv DPM); dyes and pigments (Poly-Solv DPM); paint mfg. (Arcosolv DPM; Dowanol DPM; Poly-Solv DPM); plastics (Poly-Solv DPM); polishes and waxes (Poly-Solv DPM); printing inks (Arcosolv DPM; Dowanol DPM; Poly-Solv DPM); textile/leather processing (Poly-Solv DPM)

Industrial cleaners: (Arcosolv DPM)

PROPERTIES:

Form:

Liquid (Arcosolv DPM; Poly-Solv DPM)

Color:

APHA 15 max. (Arcosolv DPM; Poly-Solv DPM)

Odor:

Mild, pleasant, characteristic (Arcosolv DPM)

Mild, characteristic (Poly-Solv DPM)

Composition:

99+% active (Arcosolv DPM; Poly-Solv DPM)

Solubility:

Sol. in many organic liquids (Dowanol DPM); misc. with many organic solvents (Arcosolv DPM; Poly-Solv DPM)

Sol. in water (Arcosolv DPM; Dowanol DPM); misc. with water (Poly-Solv DPM)

M.W.:

148.2 (Arcosolv DPM; Poly-Solv DPM)

Dipropylene glycol monomethyl ether (cont'd.)

Sp.gr.:
0.950–0.953 (Arcosolv DPM)
0.954 (20/20 C) (Poly-Solv DPM)
Density:
7.91 lb/gal (Arcosolv DPM)
7.94 lb/gal (20 C) (Poly-Solv DPM)
Visc.:
3.6 cstk (Arcosolv DPM)
3.9 cP (20 C) (Poly-Solv DPM)
F.P.:
–83 C (Poly-Solv DPM)
–80 C (Arcosolv DPM)
B.P.:
187 C (7860 mm) (Poly-Solv DPM)
188 C (Dowanol DPM)
188.3 C (760 mm Hg) (Arcosolv DPM)
Flash Pt.:
75 C (TCC) (Arcosolv DPM)
169 C (TCC) (Poly-Solv DPM)
Ref. Index:
1.419 (Poly-Solv DPM)
1.422 (20 C) (Arcosolv DPM)
Surface Tension:
28.2 dynes/cm (Arcosolv DPM)
Conductivity:
3.63 K × 10^4 cal/cm^{2}s°C/cm (60 C) (Arcosolv DPM)
Specific Heat:
0.54 cal/g/°C (Arcosolv DPM)
TOXICITY/HANDLING:
Relatively low acute toxicity; may cause mild adverse effects on skin and eye contact; avoid ingestion, eye contact, prolonged skin contact, and inhalation of vapors (Arcosolv DPM)
Mild eye irritant; not a primary skin irritant; avoid contact with eyes and skin; do not breathe vapor (Poly-Solv DPM)
STORAGE/HANDLING:
Combustible liquid; store in carbon steel vessels; avoid contact with air when storing for long periods of time (Arcosolv DPM)
STD. PKGS.:
55-gal (440 lb net) drums, tank cars, tank trucks (Poly-Solv DPM)

Disodium cocoyl monoethanolamide sulfosuccinate

SYNONYMS:

Amides, coconut oil, N-[2-[(sulfosuccinyl) oxy] ethyl], sodium salts
Butanedioic acid, sulfo-, C-(2-cocamidoethyl) esters, disodium salts
Disodium cocamido MEA-sulfosuccinate (CTFA)
Sulfobutanedioic acid, C-(2-cocamidoethyl) esters, disodium salts
N-[2-[(Sulfosuccinyl) oxy] ethyl] coconut oil amides, sodium salts

STRUCTURE:

$$RC(=O)-NH-CH_2CH_2O-C(=O)CH(SO_3Na)CH_2C(=O)-ONa$$

where RCO- represents the coconut acid radical

CAS No.:

61791-66-0; 68784-08-7

TRADENAME EQUIVALENTS:

Mackanate CM [McIntyre]
Rewopol SBC 212, SBC 212/G [Rewo Chemische Werke GmbH]
Schercopol CMS-Na [Scher]

CATEGORY:

Solubilizer, base, detergent foaming agent, softener, surfactant

APPLICATIONS:

Bath products: bubble bath (Mackanate CM; Rewopol SBC 212; Schercopol CMS-Na); bath oils (Schercopol CMS-Na)
Cleansers: skin cleanser (Schercopol CMS-Na)
Cosmetic industry preparations: hair preparations (Schercopol CMS-Na); shampoos (Mackanate CM; Rewopol SBC 212, SBC 212/G; Schercopol CMS-Na); skin preparations (Schercopol CMS-Na)
Household detergents: carpet & upholstery shampoos (Mackanate CM); foam cleaners (Rewopol SBC 212, SBC 212/G); light-duty cleaners (Rewopol SBC 212, SBC 212/G)

PROPERTIES:

Form:

Liquid (Mackanate CM)
Low-visc. liquid (Rewopol SBC 212)
Clear liquid (Schercopol CMS-Na)
Granular (Rewopol SBC 212/G)

Color:

Yellow (Schercopol CMS-Na)

Odor:

Mild, characteristic (Schercopol CMS-Na)

Composition:

30% active (Schercopol CMS-Na)

Disodium cocoyl monoethanolamide sulfosuccinate (cont'd.)

40% conc. (Mackanate CM; Rewopol SBC 212)
95% conc. (Rewopol SBC 212/G)

Solubility:

Insol. in nonpolar solvents (Schercopol CMS-Na)
Sol. in most organic polar solvents (Schercopol CMS-Na)
Completely sol. in water (Schercopol CMS-Na)

Ionic Nature:

Anionic (Mackanate CM; Rewopol SBC 212, SBC 212/G; Schercopol CMS-Na)

M.W.:

471 (Schercopol CMS-Na)

Sp.gr.:

1.12 ± 0.01 (Schercopol CMS-Na)

Density:

9.3 lb/gal (Schercopol CMS-Na)

Visc.:

100 cps max. (Schercopol CMS-Na)

Cloud Pt.:

5.0 C (Schercopol CMS-Na)

pH:

5.0–7.0 (Schercopol CMS-Na)

STD. PKGS.:

55-gal poly-lined drums (Schercopol CMS-Na)

Disodium deceth-6 sulfosuccinate (CTFA)

STRUCTURE:

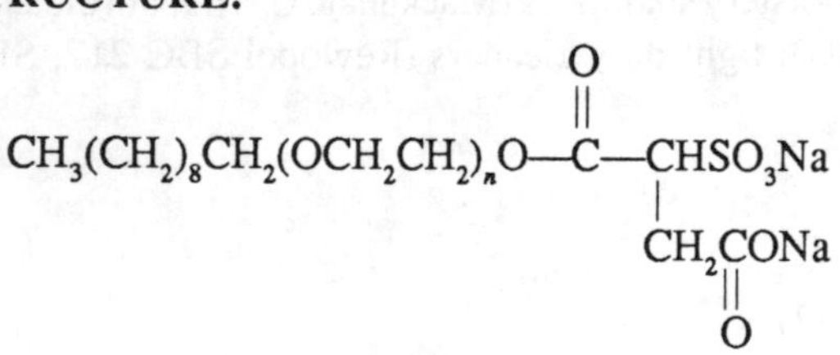

$$CH_3(CH_2)_8CH_2(OCH_2CH_2)_nO\text{—}\overset{\overset{O}{\|}}{C}\text{—}CHSO_3Na$$
$$\qquad\qquad\qquad\qquad\qquad |$$
$$\qquad\qquad\qquad\qquad\qquad CH_2\underset{\underset{O}{\|}}{C}ONa$$

where avg. $n = 6$

CAS No.

68311-03-5 (generic); 68630-97-7 (generic)

TRADENAME EQUIVALENTS:

Aerosol A-102 [Amer. Cyanamid]
Alconate D-6 [Alcolac]
Monawet TD-30 [Mona]

CATEGORY:
Emulsifier, solubilizer, foaming agent, dispersant
APPLICATIONS:
Industrial applications: plastics (Aerosol A-102; Alconate D-6); polymers/ polymerization (Aerosol A-102; Alconate D-6; Monawet TD-30); textile/leather processing (Aerosol A-102)
PROPERTIES:
Form:
Liquid (Alconate D-6; Monawet TD-30)
Clear liquid (Aerosol A-102)
Color:
Colorless to light yellow (Aerosol A-102)
Composition:
30% active (Aerosol A-102)
30% conc. (Monawet TD-30)
31% active (Alconate D-6)
Solubility:
Excellent sol. in water (Aerosol A-102)
Ionic Nature:
Anionic (Alconate D-6; Monawet TD-30)
Anionic-nonionic (Aerosol A-102)
Sp.gr.:
1.08 (Aerosol A-102)
Density:
9.01 lb/gal (Aerosol A-102)
Visc.:
≈ 40 cps (Aerosol A-102)
F.P.:
–4 C (Aerosol A-102)
Stability:
Excellent stability in acid, fair in base (Aerosol A-102)
Acid and salt tolerance (Monawet TD-30)

Disodium isodecyl sulfosuccinate (CTFA)

SYNONYMS:
Butanedioic acid, sulfo-, 4-isodecyl ester, disodium salt
Sulfobutanedioic acid, 4-isodecyl ester, disodium salt
EMPIRICAL FORMULA:
$C_{14}H_{26}O_7S \cdot 2Na$

Disodium isodecyl sulfosuccinate (cont'd.)

STRUCTURE:

$$C_{10}H_{21}O-\overset{\overset{\displaystyle O}{\|}}{C}-\underset{\underset{\displaystyle SO_3Na}{|}}{C}HCH_2\overset{\overset{\displaystyle O}{\|}}{C}-ONa$$

CAS No.
37294-49-8; 55184-70-8
TRADENAME EQUIVALENTS:
Aerosol A-268 [Amer. Cyanamid]
CATEGORY:
Emulsifier, solubilizer, surfactant
APPLICATIONS:
Industrial applications: plastics (Aerosol A-268)
PROPERTIES:
Form:
Clear liquid (Aerosol A-268)
Composition:
50% active (Aerosol A-268)
Solubility:
Excellent sol. in water (Aerosol A-268)
Ionic Nature:
Anionic
Sp.gr.:
1.19 (Aerosol A-268)
Density:
9.96 lb/gal (Aerosol A-268)
Visc.:
≈ 150 cps (Aerosol A-268)
F.P.:
–5 C (Aerosol A-268)
Stability:
Good (Aerosol A-268)
Surface Tension:
28 dynes/cm (Aerosol A-268)

Disodium myristamido MEA-sulfosuccinate (CTFA)

SYNONYMS:
Butanedioic acid, sulfo-, 2-[(1-oxotetradecyl) amino] ethyl ester, disodium salt
Disodium monomyristamido MEA-sulfosuccinate

SYNONYMS *(cont'd.):*
2-[(1-Oxotetradecyl) amino] ethyl ester of sulfobutanedioic acid, disodium salt
Sulfobutanedioic acid, 2 [1-oxotetradecyl) amino] ethyl ester, disodium salt

EMPIRICAL FORMULA:
$C_{20}H_{37}NO_8S \cdot 2Na$

STRUCTURE:

```
                    O
                    ||
CH3(CH2)12C—NH
                         |
                      CH2      O
                         |         ||
                      CH2O—CCHSO3Na
                                     |
                                   CH2C—ONa
                                           ||
                                           O
```

CAS No.:
37767-42-3

TRADENAME EQUIVALENTS:
Emcol 4100M [Witco/Organics]

CATEGORY:
Dispersant, wetting agent, foaming agent, detergent, emulsifer

APPLICATIONS:
Bath products: bubble bath (Emcol 4100M)
Cosmetic industry preparations: shampoos (Emcol 4100M)
Household detergents: carpet & upholstery shampoos (Emcol 4100M)
Industrial cleaners: textile scouring (Emcol 4100M)

PROPERTIES:

Form:
Creamy semisolid (Emcol 4100M)

Color:
White (Emcol 4100M)

Composition:
38% solids (Emcol 4100M)

Solubility:
Insol. in alcohol (Emcol 4100M)
Insol. in carbon tetrachloride (Emcol 4100M)
Insol. in perchloroethylene (Emcol 4100M)
Insol. in Stoddard solvent (Emcol 4100M)
Disp. in water (Emcol 4100M)

Ionic Nature:
Anionic (Emcol 4100M)

Disodium myristamido MEA-sulfosuccinate (*cont'd.*)

Sp.gr.:
1.01 (25/4 C) (Emcol 4100M)
Acid No.:
4.5 (Emcol 4100M)
pH:
6.5 (3% aq. disp.) (Emcol 4100M)
Surface Tension:
37.8 dynes/cm (0.05% solids) (Emcol 4100M)
TOXICITY/HANDLING:
Tests indicate minimal skin irritation; may cause moderate eye irritation (Emcol 4100M)
STORAGE/HANDLING:
Store in tightly closed containers in a cool, dry area (Emcol 4100M)

Disodium nonoxynol-10 sulfosuccinate (CTFA)

STRUCTURE:

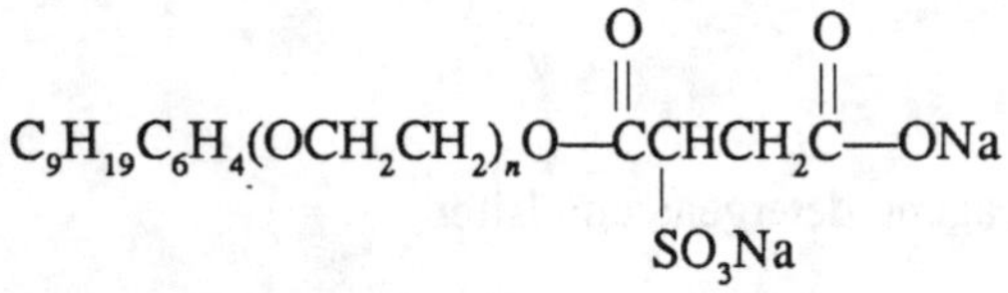

$$C_9H_{19}C_6H_4(OCH_2CH_2)_nO\text{—}\overset{O}{\overset{\|}{C}}\underset{\underset{SO_3Na}{|}}{C}HCH_2\overset{O}{\overset{\|}{C}}\text{—}ONa$$

where avg. $n = 10$
CAS No.:
67999-57-9 (generic)
RD No.: 977069-21-8
TRADENAME EQUIVALENTS:
Aerosol A-103 [Amer. Cyanamid, Cyanamid B.V.]
CATEGORY:
Solubilizer, emulsifier, dispersant
APPLICATIONS:
Industrial applications: germicides (Aerosol A-103); plastics (Aerosol A-103); polymers/polymerization (Aerosol A-103); surfactant-salt mixtures (Aerosol A-103)
Industrial cleaners: lime soap dispersing (Aerosol A-103)
PROPERTIES:
Form:
Liquid (Aerosol A-103)
Composition:
34% conc. (Aerosol A-103)
Solubility:
Generally insol. in organic nonpolar solvents (Aerosol A-103)

Partly sol. in organic polar solvents (Aerosol A-103)
Sol. in water (Aerosol A-103)

Ionic Nature:
Anionic

Disodium oleamido PEG-2 sulfosuccinate (CTFA)

SYNONYMS:

Butanedioic acid, sulfo-, 1(or 4)- [2- [2- [(1-oxo-9-octadecenyl) amino] ethoxy] ethyl] ester, disodium salt
Butanedioic acid, sulfo-, C-[2[[2- [(1-oxo-9-octadecenyl) amino] ethoxy] ethyl] ester, disodium salt
Disodium monooleamido PEG-2 sulfosuccinate
Sulfobutanedioic acid, 1(or 4)- [2- [2- [(1-oxo-9-octadecenyl) amino] ethoxy] ethyl] ester, disodium salt
Sulfobutanedioic acid, C- [2- [2- [(1-oxo-9-octadecenyl) amino] ethoxy] ethyl] ester, disodium salt

EMPIRICAL FORMULA:

$C_{26}H_{47}NO_9S \cdot 2Na$

STRUCTURE:

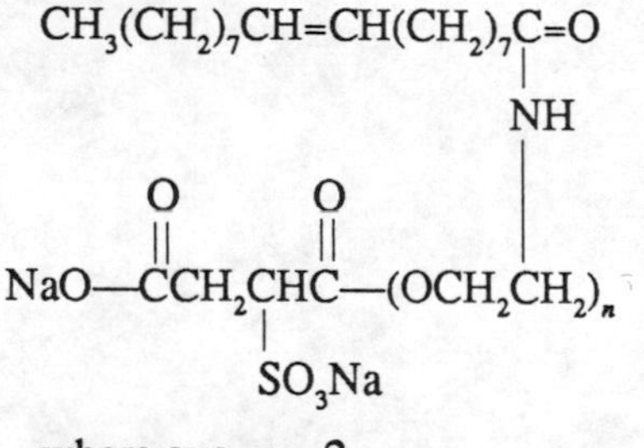

$CH_3(CH_2)_7CH{=}CH(CH_2)_7C{=}O$ — NH — $(CH_2CH_2O)_n$ — $C(=O)CH(SO_3Na)CH_2C(=O)$—ONa

$NaO{-}\overset{O}{\overset{||}{C}}CH_2\underset{SO_3Na}{CH}\overset{O}{\overset{||}{C}}{-}(OCH_2CH_2)_n{-}NH{-}C(=O)(CH_2)_7CH{=}CH(CH_2)_7CH_3$

where avg. $n = 2$

CAS No.:

56388-43-3; 68227-80-5

TRADENAME EQUIVALENTS:

Cyclopol SBG-280 [Cyclo]
Emcol 4161L [Witco/Organics]
Mackanate OD [McIntyre]
Monamate OPA-30, OPA-100 [Mona]
Standapol SH-100, SH-135 Special [Henkel]

CATEGORY:

Dispersant, wetting agent, foaming agent, detergent, emulsifier, surfactant, anti-irritant

Disodium oleamido PEG-2 sulfosuccinate (cont'd.)

APPLICATIONS:

Bath products: bubble bath (Emcol 4161L; Monamate OPA-30, OPA-100)

Cleansers: (Emcol 4161L); skin cleanser (Monamate OPA-30); soap bars (Monamate OPA-100)

Cosmetic industry preparations: (Emcol 4161L); personal care products (Emcol 4161L); shampoos (Emcol 4161L; Mackanate OD; Monamate OPA-30, OPA-100; Standapol SH-100, SH-135 Special); toiletries (Emcol 4161L)

PROPERTIES:

Form:

Liquid (Cyclopol SBG-280; Mackanate OD; Monamate OPA-30; Standapol SH-100, SH-135 Special)

Clear liquid (Emcol 4161L)

Powder (Monamate OPA-100)

Color:

Light amber (Standapol SH-100, SH-135 Special)

Light yellow (Emcol 4161L; Monamate OPA-30)

Yellow (Cyclopol SBG-280)

Odor:

Mild, pleasant (Standapol SH-100, SH-135 Special)

Composition:

29–31% active (Cyclopol SBG-280; Standapol SH-100)

30% conc. (Mackanate OD)

30% solids (Monamate OPA-30)

34–36% active (Standapol SH-135 Special)

38% solids (Emcol 4161L)

100% conc. (Monamate OPA-100)

Solubility:

Insol. in alcohols (Emcol 4161L)

Insol. in carbon tetrachloride (Emcol 4161L)

Insol. in perchloroethylene (Emcol 4161L)

Insol. in Stoddard solvent (Emcol 4161L)

Ionic Nature:

Anionic (Cyclopol SBG-280; Emcol 4161L; Mackanate OD; Monamate OPA-30, OPA-100)

Sp.gr.:

1.06 (Monamate OPA-30)

1.10 (25/4 C) (Emcol 4161L)

Density:

8.85 lb/gal (Monamate OPA-30)

Flash Pt.:

93 C (PMCC) (Emcol 4161L)

Cloud Pt.:

< 0 C (Standapol SH-100)

Disodium oleamido PEG-2 sulfosuccinate (cont'd.)

10 C max. (Standapol SH-135 Special)

Acid No.:

6.0 (Emcol 4161L)

Stability:

Stable at pH 4–8; good heat stability at its normal pH (Monamate OPA-30)

pH:

5.6 (Monamate OPA-30)
6.0–7.0 (5% sol'n.) (Standapol SH-100, SH-135 Special)
6.5 (3% aq. sol'n.) (Emcol 4161L)

Surface Tension:

32.6 dynes/cm (0.5% solids) (Emcol 4161L)

Biodegradable: (Monamate OPA-30)

TOXICITY/HANDLING:

Nonirritating (Mackanate OD; Monamate OPA-30, OPA-100; Standapol SH-100, SH-135 Special)
Nonirritating to the eyes and skin (Cyclopol SBG 280)
Low level of skin and eye irritation (Emcol 4161L)

STORAGE/HANDLING:

Store in tightly closed containers in a cool, dry area (Emcol 4161L)

Disodium stearyl sulfosuccinamate (CTFA)

SYNONYMS:

Butanedioic acid, 4-(octadecylamino)-4-oxo-2-sulfo, disodium salt
Butanedioic acid, sulfo-, monooctadecyl ester, disodium salt
Butanoic acid, 4-(octadecylamino)-4-oxo-2-sulfo-, disodium salt
Disodium N-octadecyl sulfosuccinamate
Disodium salt of a stearyl amide of sulfosuccinic acid
4-(Octadecylamino)-4-oxo-2-sulfobutanedioic acid, disodium salt
Sulfobutanedioic acid, monooctadecyl ester, disodium salt

EMPIRICAL FORMULA:

$C_{22}H_{43}NO_6S \cdot 2Na$

STRUCTURE:

$$CH_3(CH_2)_{16}CH_2—NH—\overset{O}{\overset{\|}{C}}\underset{SO_3Na}{\underset{|}{C}}HCH_2\overset{O}{\overset{\|}{C}}—ONa$$

CAS No.:

14481-60-8; 26446-37-7

Disodium stearyl sulfosuccinamate (*cont'd.*)

TRADENAME EQUIVALENTS:

Aerosol 18, 19 [American Cyanamid]
Alkasurf SS-TA [Alkaril]
Lankropol ODS/LS, ODS/PT [Lankro]
Rewopol SBF18 [Rewo]

CATEGORY:

Emulsifier, dispersant, foaming agent, wetting agent, surfactant, raw material

APPLICATIONS:

Cleansers: hand cleanser (Aerosol 19)
Cosmetic industry preparations: (Aerosol 18)
Farm products: agricultural foams (Aerosol 18)
Household detergents: (Aerosol 18); carpet & upholstery shampoos (Aerosol 19); heavy-duty cleaner (Alkasurf SS-TA; Rewopol SBF18); syndet soaps (Rewopol SBF18)
Industrial applications: carpet backing (Lankropol ODS/LS, ODS/PT); construction (Aerosol 19); foaming latexes (Aerosol 18, 19; Lankropol ODS/LS, ODS/PT); plastics (Aerosol 19); polishes and waxes (Alkasurf SS-TA); polymers/polymerization (Aerosol 18, 19; Alkasurf SS-TA); rubber latex (Alkasurf SS-TA)
Industrial cleaners: (Aerosol 18); cement, brick, tile cleaners (Alkasurf SS-TA)

PROPERTIES:

Form:

Clear liquid (Aerosol 19)
Hazy liquid (Lankropol ODS/LS)
Creamy paste (Aerosol 18)
Paste (Alkasurf SS-TA)
Flake (Rewopol SBF18)

Color:

Cream (Alkasurf SS-TA)
Light amber (Lankropol ODS/LS)
Gardner 10 (Aerosol 18, 19)

Odor:

Low (Alkasurf SS-TA)
Mild (Lankropol ODS/LS)

Composition:

35% active (Aerosol 18, 19; Lankropol ODS/LS, ODS/PT)
35% active in water (Alkasurf SS-TA)
95% (Rewopol SBF18)

Solubility:

Sol. in water: good sol. (Aerosol 18, 19); sol. @ 20 C (Lankropol ODS/LS); poor sol. (Alkasurf SS-TA)

Ionic Nature:

Anionic (Aerosol 18; Lankropol ODS/LS, ODS/PT; Rewopol SBF18)

Sp.gr.:
≈ 1.07 (Aerosol 18, 19)
1.082 (20 C) (Lankropol ODS/LS)
Density:
≈ 8.9 lb/gal (Aerosol 18, 19)
Visc.:
12 cs (20 C) (Lankropol ODS/LS)
F.P.:
Solid paste < 21 C (Aerosol 18)
Pour Pt.:
–5 C (Lankropol ODS/LS)
Flash Pt.:
> 200 F (COC) (Lankropol ODS/LS)
Stability:
Good in acid and alkaline media (Aerosol 18, 19)
pH:
7.0–9.0 (1% aq.) (Lankropol ODS/LS)
Surface Tension:
41 dynes/cm (Aerosol 18, 19)
Biodegradable: (Aerosol 18, 19)
TOXICITY/HANDLING:
Skin irritant on prolonged contact with conc. form; protective goggles and gloves should be worn; spillages may be slippery; in a fire may produce toxic fumes of sulfur dioxide (Lankropol ODS/LS, ODS/PT)

Disodium tallowiminodipropionate (CTFA)

SYNONYMS:
N-(2-Carboxyethyl)-N-(tallow acyl)-β-alanine
Disodium N-tallow-β iminodipropionate
Disodium *n*-tallow-3,3´-iminodipropionate
STRUCTURE:

$$R{-}N\begin{cases} CH_2CH_2\overset{\displaystyle O}{\overset{\|}{C}}{-}ONa \\ CH_2CH_2\underset{\displaystyle O}{\underset{\|}{C}}{-}ONa \end{cases}$$

where R represents the tallow radical

Disodium tallowiminodipropionate (cont'd.)

CAS No.

61791-56-8

TRADENAME EQUIVALENTS:

Deriphat 154 [Henkel]
Mirataine T2C [Miranol]
Monateric TDB-35 [Mona]

CATEGORY:

Surfactant, wetting agent, detergent, solubilizer, hydrotrope, lubricant, antistat

APPLICATIONS:

Cosmetic industry preparations: conditioners (Mirataine T2C); personal care products (Mirataine T2C)
Household detergents: heavy-duty cleaner (Monateric TDB-35)
Industrial applications: textile/leather processing (Mirataine T2C)
Industrial cleaners: acid cleaners (Mirataine T2C); wax stripper (Monateric TDB-35)

PROPERTIES:

Form:

Liquid (Mirataine T2C)
Clear to hazy, thin liquid (Monateric TDB-35)
Powder (Deriphat 154)

Color:

White (Deriphat 154)
Yellow (Monateric TDB-35)

Composition:

35% active (Monateric TDB-35)
35% active in water (Mirataine T2C)
98% solids (Deriphat 154)

Solubility:

Sol. in strong acids and alkalies and other ionic systems (Deriphat 154)
Clear sol. (30% solids) in 1% NaOH (Mirataine T2C)
Completely sol. (@ 1%) in 25% NaOH or KOH sol'ns. (Monateric TDB-35)

Ionic Nature:

Amphoteric

Sp.gr.:

1.03 (Monateric TDB-35)

Density:

8.6 lb/gal (Monateric TDB-35)

Bulk Density:

2 lb/gal (Deriphat 154)

pH:

11 (1% sol'n.) (Deriphat 154)
11.25 (1% sol'n.) (Monateric TDB-35)
11.5 (Mirataine T2C)

Surface Tension:

31.6 dynes/cm (1% aq.) (Mirataine T2C)

Biodegradable: (Mirataine T2C); highly (Deriphat 154)

TOXICITY/HANDLING:

Moderately irritating to eyes; mild and nonsensitizing to skin; relatively harmless by ingestion; eye protection and protective clothing recommended during handling (Deriphat 154)

STORAGE/HANDLING:

Store in closed containers above 7 C (Deriphat 154)

STD. PKGS.:

90 lb net fiber drums (Deriphat 154)

Ethyl acetate (CTFA)

SYNONYMS:

Acetic acid, ethyl ester
Acetic ester
Acetic ether
Vinegar naphtha

EMPIRICAL FORMULA:

$C_4H_8O_2$

STRUCTURE:

$$CH_3C(=O)—OCH_2CH_3$$

CAS No.:

141-78-6

TRADENAME EQUIVALENTS:

Generically sold by:
Union Oil, Union Carbide

CATEGORY:

Solvent

APPLICATIONS:

Food applications: synthetic fruit essences (generic)

Industrial applications: adhesives/cements (generic); coatings (generic); latexes (generic); organic synthesis (generic); plastics (generic); resins (generic); rubber (generic)

Pharmaceutical applications: (generic)

PROPERTIES:

Form:

Clear liquid (generic)

Color:

Colorless (generic)

Odor:

Fragrant (generic)

Solubility:

Sol. in alcohols (generic)
Sol. in chloroform (generic)
Sol. in ether (generic)
Slightly sol. in water (generic)

Sp.gr.:
0.882–0.904 (generic)
Density:
7.4 lb/gal (generic)
F.P.:
–83.6 C (generic)
B.P.:
77 C (generic)
Flash Pt.:
–4.4 C (generic)
TOXICITY/HANDLING:
Toxic by inhalation and skin absorption; irritant to skin and eyes (generic)
STORAGE/HANDLING:
Flammable—dangerous fire and explosion risk (generic)

Ethyl alcohol

SYNONYMS:
Alcohol (CTFA)
EtOH
Ethanol, undenatured
Ethyl alcohol, undenatured
Grain alcohol
EMPIRICAL FORMULA:
C_2H_6O
STRUCTURE:
CH_3CH_2OH
CAS No.:
64-17-5
TRADENAME EQUIVALENTS:
Punctilious Pure Ethyl Alcohol [U.S. Industrial Chem.]
Generically sold by:
Ashland, Shell, Union Oil, Union Carbide
CATEGORY:
Solvent, extraction medium
APPLICATIONS:
Automotive industry applications: antifreeze (generic)
Cosmetic industry preparations: (generic)
Food applications: beverages (generic)
Household detergents: (generic)

Ethyl alcohol (cont'd.)

Industrial applications: chemical mfg. (generic); coatings (generic); dyes and pigments (generic); elastomers/rubbers (generic) ; explosives (generic); fats and oils (generic); latexes (generic); petroleum industry (generic); resins (generic)
Pharmaceutical applications: (generic)

PROPERTIES:

Form:
Liquid (generic)

Color:
Colorless (generic)

Odor:
Ethereal, vinous odor (generic)

Taste:
Pungent (generic)

Solubility:
Miscible with acetone (generic 100%)
Miscible with chloroform (generic 100%; Punctilious Pure Ethyl Alcohol)
Miscible with ether (generic 100%; Punctilious Pure Ethyl Alcohol)
Sol. in methanol (Punctilious Pure Ethyl Alcohol); miscible (generic 100%)
Miscible with water (generic 100%; Punctilious Pure Ethyl Alcohol)

Sp.gr.:
≈ 0.794 (generic)
0.816 (15.5 C) (generic 95%)

Density:
6.6 lb/gal (generic)

Visc.:
0.0141 poise (20 C) (generic 95%)

F.P.:
–117.3 C (generic 100%)
–114 C (generic 95%)

B.P.:
70 C (generic 95%)
78.3 C (generic 100%)

Flash Pt.:
12.7 C (generic 95%)

Ref. Index:
1.3651 (15 C) (generic 95%)

Surface Tension:
22.3 dynes/cm (20 C) (generic 95%)

TOXICITY/HANDLING:
Depressant drug (generic)

STORAGE/HANDLING:
Flammable; dangerous fire risk (generic)

Ethylene dichloride (CTFA)

SYNONYMS:
Dichloroethane
sym-Dichloroethane
Dutch oil
Ethane, 1,2-dichloro-
Ethylene chloride

EMPIRICAL FORMULA:
$C_2H_4Cl_2$

STRUCTURE:
$Cl—CH_2CH_2—Cl$

CAS No.:
107-06-2

TRADENAME EQUIVALENTS:
Generically sold by:
Ashland, Dow, Union Oil

CATEGORY:
Solvent, diluent, fumigant

APPLICATIONS:
Degreasers: metal degreasing (generic)
Household detergents: soaps (generic)
Industrial applications: adhesives/cements (generic); ore flotation (generic); organic synthesis (generic); paint removers (generic); petroleum industry (generic); plastics (generic); rubber (generic)

PROPERTIES:

Form:
Oily liquid (generic)

Color:
Colorless (generic)

Odor:
Chloroform-like (generic)

Taste:
Sweet (generic)

Solubility:
Miscible with most common solvents (generic)
Slightly sol. in water (generic)

Sp.gr.:
1.2554 (20/4 C) (generic)

Density:
10.4 lb/gal (generic)

F.P.:
–35.5 C (generic)

Ethylene dichloride (*cont'd.*)

B.P.:
83.5 C (generic)
Flash Pt.:
56 C (generic)
Stability:
Stable to water, alkalies, acids, or active chemicals; resistant to oxidation (generic)
Ref. Index:
1.444 (generic)
TOXICITY/HANDLING:
Toxic by ingestion, inhalation, and skin absorption; strong irritant to eyes and skin; carcinogenic (generic)
STORAGE/HANDLING:
Flammable; dangerous fire risk (generic)

Ethylene glycol monobutyl ether

SYNONYMS:
Butoxyethanol (CTFA)
2-Butoxyethanol
Ethanol, 2-butoxy-
Ethylene glycol butyl ether
Monobutyl ethylene glycol ether
EMPIRICAL FORMULA:
$C_6H_{14}O_2$
STRUCTURE:
$C_4H_9OCH_2CH_2OH$
CAS No.:
111-76-2
TRADENAME EQUIVALENTS:
Butyl Cellosolve [Union Carbide]
Butyl Oxitol [Shell]
Dowanol EB [Dow]
Ektasolve EB [Eastman]
Glycol Ether EB [Ashland, Shell]
Poly-Solv EB [Olin]
CATEGORY:
Solvent, coupling agent, retarder, solubilizer, plasticizer, extraction solvent, dispersant
APPLICATIONS:
Cosmetic industry preparations: (Poly-Solv EB); nail polish (Poly-Solv EB)
Farm products: agricultural oils/sprays (Poly-Solv EB); insect repellent (Poly-Solv EB)

Household detergents: hard surface cleaner (Poly-Solv EB)
Industrial applications: adhesives (Poly-Solv EB); brake fluids (Poly-Solv EB); dyes and pigments (Butyl Oxitol; Poly-Solv EB); lubricating/cutting oils (Butyl Oxitol); paints and lacquers (Butyl Cellosolve; Butyl Oxitol; Ektasolve EB; Glycol Ether EB; Poly-Solv EB); plastics (Poly-Solv EB); polishes and waxes (Butyl Oxitol; Poly-Solv EB); printing inks (Poly-Solv EB); rubber (Butyl Cellosolve; Glycol Ether EB); textile/leather processing (Poly-Solv EB)
Industrial cleaners: (Butyl Oxitol)

PROPERTIES:

Form:
Liquid (Butyl Cellosolve; Butyl Oxitol; Ektasolve EB; Glycol Ether EB; Poly-Solv EB)

Color:
Colorless (Butyl Cellosolve; Butyl Oxitol; Ektasolve EB; Glycol Ether EB)
APHA 10 max. (Poly-Solv EB)

Odor:
Mild, agreeable (Butyl Cellosolve; Glycol Ether EB)
Faint (Ektasolve EB)
Mild (Butyl Oxitol)
Mild, characteristic (Poly-Solv EB)

Composition:
99+% conc. (Butyl Oxitol; Poly-Solv EB)
> 99% volatiles by vol. (Ektasolve EB)

Solubility:
Sol. in many organic liquids (Dowanol EB); completely misc. with most organic solvents (Butyl Oxitol); misc. (Poly-Solv EB)
Sol. in water (Dowanol EB); completely sol. (Poly-Solv EB); completely sol. @ 20 C (Ektasolve EB); completely misc. (Butyl Oxitol)

M.W.:
118.17 (Ektasolve EB; Poly-Solv EB)

Sp.gr.:
0.898–0.901 (Butyl Oxitol)
0.900–0.905 (Butyl Cellosolve; Glycol Ether EB)
0.902 (20/20 C) (Ektasolve EB)
0.903 (10/10 C) Poly-Solv EB)

Density:
0.90 kg/l (20 C) (Ektasolve EB)
7.51 lb/gal (Butyl Cellosolve; Glycol Ether EB)
7.52 lb/gal (20 C) (Poly-Solv EB)

Visc.:
3.4 cP (20 C) (Poly-Solv EB)

F.P.:
–75 C (Ektasolve EB)

Ethylene glycol monobutyl ether (cont'd.)

–70 C (Poly-Solv EB)

B.P.:

None below 163 C; 96% or more below 172 C; none above 174 C (Butyl Cellosolve)
168–173 C (Glycol Ether EB)
169.0 C min. (initial, 760 mm) (Ektasolve EB)
171 C (760 mm) (Poly-Solv EB)

Flash Pt.:

62 C (TCC) (Ektasolve EB)
63 C (TCC) (Poly-Solv EB)
138 F (Glycol Ether EB)
165 F (Butyl Cellosolve)

Fire Pt.:

70 C (Ektasolve EB)

Ref. Index:

1.4193 (20 C) (Ektasolve EB; Poly-Solv EB)

Coefficient of Linear Expansion:

0.00092 (20 C) (Poly-Solv EB)

Specific Heat:

2.438 joules/g-°C (20 C) (Poly-Solv EB)

TOXICITY/HANDLING:

Toxic (Butyl Cellosolve; Glycol Ether EB)
Readily absorbed through the skin in toxic amounts; inhalation of high concs. of vapor results in respiratory and eye irritation, narcosis, hematuria, and damage to liver and kidneys; use with adequate ventilation (Ektasolve EB)
Moderate eye irritant; mild primary skin irritant; repreated/prolonged exposure to low levels by inhalation or skin absorption may cause systemic toxicity by accumulation (Poly-Solv EB)

STORAGE/HANDLING:

Combustible—keep away from heat and open flame (Ektasolve EB)

STD. PKGS.:

55-gal (410 lb net) drums, tank cars, tank trucks (Poly-Solv EB)

Ethylene glycol monoethyl ether

SYNONYMS:

Ethoxyethanol (CTFA)
Ethanol, 2-ethoxy-
Ethylene glycol ethyl ether

EMPIRICAL FORMULA:

$C_4H_{10}O_2$

STRUCTURE:
$CH_3CH_2OCH_2CH_2OH$
CAS No.:
110-80-5
TRADENAME EQUIVALENTS:
Cellosolve [Union Carbide]
Ektasolve EE [Eastman]
Oxitol [Shell]
Poly-Solv EE [Olin]
CATEGORY:
Solvent, retarder solvent, coupling agent, solubilizer, plasticizer, extraction solvent
APPLICATIONS:
Automotive applications: brake fluids (Poly-Solv EE)
Aviation industry: fuels (Cellosolve)
Cosmetic industry preparations: (Poly-Solv EE); nail polish (Poly-Solv EE)
Farm products: agricultural sprays (Poly-Solv EE); insect repellents (Poly-Solv EE)
Household detergents: hard surface cleaner (Poly-Solv EE)
Industrial applications: adhesives (Poly-Solv EE); coatings (Ektasolve EE; Oxitol); dyes and pigments (Cellosolve; Ektasolve EE; Oxitol; Poly-Solv EE); hydraulic fluids (Ektasolve EE); paint/lacquer mfg. (Cellosolve; Ektasolve EE; Oxitol); plastics (Ektasolve EE; Poly-Solv EE); polishes and waxes (Poly-Solv EE); printing inks (Ektasolve EE; Poly-Solv EE); resins (Cellosolve; Ektasolve EE; Oxitol); soluble oils (Cellosolve); textile/leather dyeing and printing (Cellosolve; Ektasolve EE; Oxitol; Poly-Solv EE)
Industrial cleaners: glass cleaners (Ektasolve EE); metal cleaners (Ektasolve EE); solvent cleaners (Cellosolve; Oxitol); varnish remover (Cellosolve; Ektasolve EE; Oxitol)
PROPERTIES:
Form:
Liquid (Cellosolve; Ektasolve EE; Oxitol; Poly-Solv EE)
Color:
Colorless (Cellosolve; Ektasolve EE; Oxitol)
APHA 10 max. (Poly-Solv EE)
Odor:
Practically odorless (Cellosolve)
Mild (Oxitol)
Mild, agreeable (Ektasolve EE)
Mild, characteristic (Poly-Solv EE)
Composition:
100% conc. (Oxitol; Poly-Solv EE)
100% volatiles by vol. (Ektasolve EE)
Solubility:
Miscible with hydrocarbons (Cellosolve)

Ethylene glycol monoethyl ether (cont'd.)

Completely miscible with most organic solvents (Oxitol; Poly-Solv EE)
Completely sol. in water (Poly-Solv EE); completely sol. @ 20 C (Ektasolve EE); completely miscible (Oxitol); miscible (Cellosolve)

M.W.:
90.12 (Ektasolve EE; Poly-Solv EE)

Sp.gr.:
0.9256–0.9286 (Oxitol)
0.930–0.932 (20/20 C) (Ektasolve EE)
0.931 (20/20 C) (Poly-Solv EE)
0.9311 (20/20 C) (Cellosolve)

Density:
7.74 lb/gal (20 C) (Cellosolve)
7.75 lb/gal (Ektasolve EE)
7.76 lb/gal (Poly-Solv EE)

Visc.:
2.1 cP (20 C) (Poly-Solv EE)

F.P.:
–94 C (Ektasolve EE)
–70 C (Poly-Solv EE)

B.P.:
134 C min. (initial, 760 mm) (Ektasolve EE)
135 C (760 mm) (Poly-Solv EE)
135.6 C (Cellosolve)

Flash Pt.:
43 C (TCC) (Ektasolve EE)
45 C (TCC) (Poly-Solv EE)
48.9 C (Cellosolve)

Fire Pt.:
48 C (Ektasolve EE)

Stability:
Stable (Oxitol)

Ref. Index:
1.4060 (Cellosolve)
1.4076 (20 C) (Poly-Solv EE)
1.4080 (20 C) (Ektasolve EE)

Coefficient of Linear Expansion:
0.00097 (20 C) (Poly-Solv EE)

Specific Heat:
2.321 J/g-°C (20 C) (Poly-Solv EE)

TOXICITY/HANDLING:
Harmful if inhaled or absorbed through the skin; may cause adverse reproductive effects; avoid breathing vapors (Ektasolve EE)
Mild eye and primary skin irritant; repeated/prolonged exposure to low levels by

inhalation or skin absorption may cause systemic toxicity by accumulation (Poly-Solv EE)

Toxic by skin absorption (Cellosolve)

STORAGE/HANDLING:

Combustible—keep away from heat and open flames (Ektasolve EE)

Combustible—moderate fire risk (Cellosolve)

STD. PKGS.:

55-gal (420 lb net) drums, tank cars, tank trucks (Poly-Solv EE)

Ethylene glycol monoethyl ether acetate

SYNONYMS:

Acetate ester of ethylene glycol monoethyl ether
Ethanol, 2-ethoxy-, acetate
Ethoxyethanol acetate (CTFA)
2-Ethoxyethanol acetate
2-Ethoxyethyl acetate
Ethylene glycol monoethyl ether acetylated

EMPIRICAL FORMULA:

$C_6H_{12}O_3$

STRUCTURE:

$$CH_3\overset{\displaystyle O}{\overset{\|}{C}}—OCH_2CH_2OCH_2CH_3$$

CAS No.:

111-15-9

TRADENAME EQUIVALENTS:

Cellosolve Acetate [Union Carbide]
Ektasolve EE Acetate [Eastman]

CATEGORY:

Solvent

APPLICATIONS:

Industrial applications: cellulosics (Ektasolve EE Acetate); gums (Ektasolve EE Acetate); oils (Cellosolve Acetate; Ektasolve EE Acetate); paint/lacquer mfg. (Cellosolve Acetate); plastics (Ektasolve EE Acetate); resins (Cellosolve Acetate; Ektasolve EE Acetate); textile/leather processing (Cellosolve Acetate); wood stains (Cellosolve Acetate)

Industrial cleaners: varnish remover (Cellosolve Acetate)

Ethylene glycol monoethyl ether acetate (*cont'd.*)

PROPERTIES:

Form:

Liquid (Cellosolve Acetate)
Clear liquid (Ektasolve EE Acetate)

Color:

Colorless (Cellosolve Acetate; Ektasolve EE Acetate)

Odor:

Pleasant, ester-like (Cellosolve Acetate)
Mild, ester-like (Ektasolve EE Acetate)

Composition:

> 99% volatiles by volume (Ektasolve EE Acetate)

Solubility:

Miscible with aromatic hydrocarbons (Cellosolve Acetate)
Sol. 23.9% in water @ 20 C (Ektasolve EE Acetate); slightly miscible with water (Cellosolve Acetate)

Sp.gr.:

0.9748 (20/20 C) (Cellosolve Acetate)
0.975 (20/20 C) (Ektasolve EE Acetate)

Density:

8.1 lb/gal (20 C) (Cellosolve Acetate)
8.11 lb/gal (Ektasolve EE Acetate)

Visc.:

1.32 cp (20 C) (Cellosolve Acetate)

F.P.:

–61.7 C (Cellosolve Acetate)

B.P.:

156 C (Ektasolve EE Acetate)
156.3 C (Cellosolve Acetate)

Flash Pt.:

48.9 C (Cellosolve Acetate)
54 C (TCC) (Ektasolve EE Acetate)

Ref. Index:

1.4030 (Cellosolve Acetate)

TOXICITY/HANDLING:

Toxic by ingestion and skin absorption (Cellosolve Acetate)
May be harmful if inhaled or absorbed through the skin; may cause adverse reproductive effects; avoid breathing vapor (Ektasolve EE Acetate)

STORAGE/HANDLING:

Combustible (Cellosolve Acetate)
Keep container closed; combustible—keep away from heat and open flame (Ektasolve EE Acetate)

Ethylene glycol monophenyl ether

SYNONYMS:

Ethanol, 2-phenoxy-
Ethylene glycol phenyl ether
Phenoxyethanol (CTFA)
2-Phenoxyethanol
Phenoxytol

EMPIRICAL FORMULA:

$C_8H_{10}O_2$

STRUCTURE:

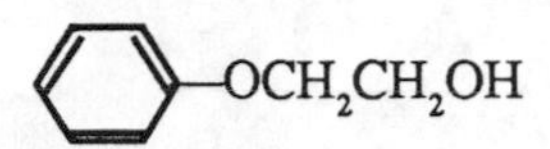

CAS No.:

122-99-6

TRADENAME EQUIVALENTS:

Dowanol EPH [Dow]
Emery 6705 [Emery]
Phenyl Cellosolve [Union Carbide]

CATEGORY:

Solvent, fixative, bactericidal agent

APPLICATIONS:

Cosmetic industry applications: perfumery (Phenyl Cellosolve)
Farm products: insect repellent (Phenyl Cellosolve)
Industrial applications: cellulosics (Emery 6705); dyes and pigments (Phenyl Cellosolve); organic synthesis (Phenyl Cellosolve); plasticizers (Phenyl Cellosolve); printing inks (Emery 6705; Phenyl Cellosolve); resins (Emery 6705; Phenyl Cellosolve)
Industrial cleaners: finish/coatings remover (Emery 6705); germicides (Phenyl Cellosolve)
Pharmaceuticals: (Phenyl Cellosolve)

PROPERTIES:

Form:

Liquid (Emery 6705; Phenyl Cellosolve)

Color:

Colorless (Phenyl Cellosolve)
Gardner 2 (Emery 6705)

Odor:

Faint aromatic (Phenyl Cellosolve)

Solubility:

Disp. in butyl stearate (@ 5%) (Emery 6705)
Insol. in glycerol trioleate (@ 5%) (Emery 6705)
Insol. in min. oil (@ 5%) (Emery 6705)
Sol. in many organic liquids (Dowanol EPH)

Ethylene glycol monophenyl ether (cont'd.)

Sol. in Stoddard (@ 5%) (Emery 6705)
Sol. in water (Dowanol EPH); partly sol. (Phenyl Cellosolve); insol. @ 5% (Emery 6705)
Sol. in xylene (@ 5%) (Emery 6705)

Sp.gr.:
1.094 (20/20 C) (Phenyl Cellosolve)

Density:
9.1 lb/gal (Emery 6705)
9.2 lb/gal (Phenyl Cellosolve)

Visc.:
10 cSt (100 F) (Emery 6705)

F.P.:
14 C (Phenyl Cellosolve)

B.P.:
244.9 C (Phenyl Cellosolve)

Pour Pt.:
13 C (Emery 6705)

Flash Pt.:
250 F (Emery 6705)
121 C (Phenyl Cellosolve)

Stability:
Stable in presence of acids and alkalies (Phenyl Cellosolve)

STORAGE/HANDLING:
Combustible (Phenyl Cellosolve)

Ethyl hydroxymethyl oleyl oxazoline (CTFA)

SYNONYMS:
4-Ethyl-2-(8-heptadecenyl)-4,5-dihydro-4-oxazolemethanol
4-Oxazolemethanol, 4-ethyl-2-(8-heptadecenyl)-4,5-dihydro-

EMPIRICAL FORMULA:
$C_{23}H_{43}NO_2$

STRUCTURE:

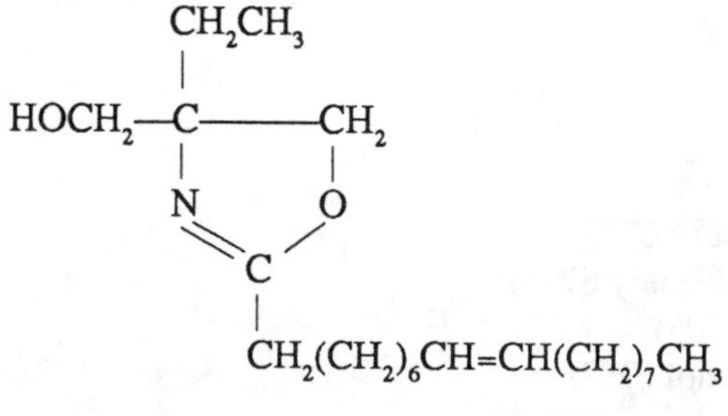

```
          CH2CH3
          |
HOCH2—C————CH2
          |          |
          N          O
            \\    /
              C
              |
              CH2(CH2)6CH=CH(CH2)7CH3
```

CAS No.:
68140-98-7
TRADENAME EQUIVALENTS:
Alkaterge E [IMC]
CATEGORY:
Detergent, emulsifier, dispersant, wetting agent, antifoam agent, antioxidant
APPLICATIONS:
Industrial applications: dyes and pigments (Alkaterge E); industrial processing (Alkaterge E); paper mfg. (Alkaterge E); textile/leather processing (Alkaterge E)
Industrial cleaners: metal processing surfactants (Alkaterge E)
PROPERTIES:
Form:
Clear liquid (Alkaterge E)
Color:
Gardner 15 max. (Alkaterge E)
Composition:
79% active (Alkaterge E)
Solubility:
Sol. in most organic liquids (Alkaterge E)
Slightly sol. in water (Alkaterge E)
Sp.gr.:
0.9 (Alkaterge E)
Density:
7.74 lb/gal (Alkaterge E)
Visc.:
155 cp (Alkaterge E)
F.P.:
–31 C (Alkaterge E)
Flash Pt.:
> 212 F (Alkaterge E)
Surface Tension:
40 dynes/cm (0.001% aq. sol'n.) (Alkaterge E)
STD. PKGS.:
Cans, drums (Alkaterge E)

Etidronic acid (CTFA)

SYNONYMS:
Hydroxyethane diphosphonic acid
(1-Hydroxyethylidene) bisphosphonic acid
Phosphonic acid, (1-hydroxyethylidene) bis-

Etidronic acid *(cont'd.)*

EMPIRICAL FORMULA:
$C_2H_8O_7P_2$
STRUCTURE:

CAS No.:
2809-21-4
TRADENAME EQUIVALENTS:
Fostex P [Henkel]
CATEGORY:
Scale inhibitor, sequestering agent
APPLICATIONS:
Industrial applications: water treatment (Fostex P)
PROPERTIES:
Form:
Liquid (Fostex P)
Composition:
59% active (Fostex P)
Ionic Nature:
Anionic (Fostex P)

Glyceryl monocaprate

SYNONYMS:

Decanoic acid, monoester with 1,2,3-propanetriol

Glyceryl caprate (CTFA)

EMPIRICAL FORMULA:

$C_{13}H_{26}O_4$

STRUCTURE:

$$CH_3(CH_2)_8C(=O){-}OCH_2CH(OH)CH_2OH$$

CAS No.:

26402-22-2

TRADENAME EQUIVALENTS:

Imwitor 310 [Dynamit-Nobel]

CATEGORY:

Solubilizer, carrier

APPLICATIONS:

Pharmaceutical applications: lipophilic drugs (Imwitor 310)

PROPERTIES:

Form:

Crystalline mass (Imwitor 310)

Color:

White (Imwitor 310)

Glyceryl monocaprylate

SYNONYMS:

Glyceryl caprylate (CTFA)

Octanoic acid, monoester with 1,2,3-propanetriol

EMPIRICAL FORMULA:

$C_{11}H_{22}O_4$

Glyceryl monocaprylate (cont' d.)

STRUCTURE:

$$CH_3(CH_2)_6\overset{\overset{\displaystyle O}{\|}}{C}—OCH_2\underset{\underset{\displaystyle OH}{|}}{C}HCH_2OH$$

CAS No.:
26402-26-6
TRADENAME EQUIVALENTS:
Imwitor 308 [Dynamit-Nobel]
CATEGORY:
Solubilizer, carrier, vehicle
APPLICATIONS:
Pharmaceutical applications: lipophilic drugs (Imwitor 308); capsulated drugs (Imwitor 308)
PROPERTIES:
Form:
Crystalline mass (Imwitor 308)
Color:
White (Imwitor 308)
M.P.:
30 C (Imwitor 308)

Heptane (CTFA)

SYNONYMS:
Dipropylmethane
n-Heptane
EMPIRICAL FORMULA:
C_7H_{16}
STRUCTURE:
$CH_3(CH_2)_5CH_3$
CAS No.:
142-82-5
TRADENAME EQUIVALENTS:
Amsco Heptane [Union Oil]
Commercial Heptane [Phillips]
Generically sold by:
Ashland, Union Oil
CATEGORY:
Solvent
APPLICATIONS:
Industrial applications: adhesives (generic; Commercial Heptane); laboratory reagents (generic); octane rating determination (generic); organic synthesis (generic); rubber (generic; Commercial Heptane)
Pharmaceutical applications: anesthetic (generic)
PROPERTIES:
Form:
Liquid (generic; Commercial Heptane)
Clear liquid (Amsco Heptane)
Color:
Colorless (generic; Amsco Heptane; Commercial Heptane)
Odor:
Faint, typical (generic; Commercial Heptane)
Light aliphatic (Amsco Heptane)
Composition:
26.5 vol.% *n*-heptane (Amsco Heptane)
Solubility:
Sol. in alcohols (generic)
Sol. in chloroform (generic)
Sol. in ether (generic)

Heptane (cont'd.)

Miscible with most common organic solvents (Amsco Heptane)
Insol. in water (generic)

Sp.gr.:
0.699 (D891) (Amsco Heptane)
0.706–0.727 (generic)

Density:
5.82 lb/gal (Amsco Heptane)

Visc.:
0.61 cSt (20 C, D446) (Amsco Heptane)

F.P.:
–90.595 C (generic)

B.P.:
92.6 C (initial) (Amsco Heptane)
98.428 C (generic)
195–210 F (Commercial Heptane)

Flash Pt.:
–3.89 C (CC) (generic)
< 20 F (TCC, D56) (Amsco Heptane)

Ref. Index:
1.38764 (20 C) (generic)
1.3906 (20 C, D1218) (Amsco Heptane)

TOXICITY/HANDLING:
Toxic by inhalation; 400 ppm TLV in air (generic)
400 ppm TLV, 500 ppm STEL (*n*-heptane) (Amsco Heptane)

STORAGE/HANDLING:
Volatile; highly flammable; dangerous fire risk (generic; Commercial Heptane)
DOT flammable liquid (Amsco Heptane)

STD. PKGS.:
5-gal cans, 55-gal drums, tank trucks, tank cars, barges (Amsco Heptane)
Drum, tank car, tank truck (Commercial Heptane)
Drum, tank car, tank truck, barge (generic)

Hexaglyceryl distearate

SYNONYMS:
Hexaglycerol distearate
Polyglyceryl-6 distearate (CTFA)

CAS No.:
977067-15-4

TRADENAME EQUIVALENTS:
Aldo HGDS [Glyco]
Caprol 6G2S [Capital City]
Drewmulse 6-2-S [PVO Int'l.]
Drewpol 6-2-S [PVO Int'l.]
Hodag SVO-629 [Hodag] (veg. grade)
Polyaldo HGDS [Lonza]

CATEGORY:
Emulsifier, solubilizer, dispersant

APPLICATIONS:
Cosmetic industry preparations: (Drewmulse 6-2-S; Polyaldo HGDS); creams and lotions (Drewmulse 6-2-S)
Food applications: food additives (Drewpol 6-2-S); food emulsifying (Caprol 6G2S; Drewpol 6-2-S; Hodag SVO-629; Polyaldo HGDS)
Pharmaceutical applications: (Drewmulse 6-2-S)

PROPERTIES:

Form:
Solid (Aldo HGDS; Caprol 6G2S; Drewmulse 6-2-S)
Beads (Drewpol 6-2-S; Polyaldo HGDS)
Flakes (Hodag SVO-629)

Color:
Amber (Drewpol 6-2-S)
Tan (Aldo HGDS)
Gardner 13 (Drewmulse 6-2-S)

Composition:
100% conc. (Caprol 6G2S; Drewpol 6-2-S; Hodag SVO-629; Polyaldo HGDS)

Solubility:
Sol. hot in isopropanol (Drewmulse 6-2-S)
Sol. hot in min. oil (Drewmulse 6-2-S)
Sol. hot in peanut oil (Drewmulse 6-2-S)
Disp. in propylene glycol (Drewmulse 6-2-S)

Ionic Nature:
Nonionic (Caprol 6G2S; Drewmulse 6-2-S; Hodag SVO-629; Polyaldo HGDS)

M.P.:
53–57 C (Aldo HGDS)

HLB:
4.0 (Aldo HGDS; Drewpol 6-2-S; Hodag SVO-629)
7.0 (Polyaldo HGDS)
8.0 (Drewmulse 6-2-S)
8.5 (Caprol 6G2S)

Saponification No.:
100–130 (Drewpol 6-2-S)
105–125 (Aldo HGDS; Drewmulse 6-2-S)

Hydrogenated castor oil (CTFA)

SYNONYMS:

Castor oil, hydrogenated
Glyceryl tri(12-hydroxystearate)

STRUCTURE:

$C_3H_5(OOCC_{17}H_{34}OH)_3$

CAS No.:

8001-78-3

TRADENAME EQUIVALENTS:

Castorwax, MP-70, MP-80 [CasChem]
Cenwax G [Union Camp]

CATEGORY:

Dispersant, release agent, suspending aid, lubricant, intermediate, wax modifier

APPLICATIONS:

Cosmetic industry preparations: (Castorwax MP-80); creams and lotions (Castorwax, MP-70); toiletries (Castorwax MP-80)
Industrial applications: coatings (Cenwax G); hot-melt adhesives (Castorwax MP-80); paper coatings (Cenwax G)
Pharmaceutical applications: antiperspirant/deodorant (Castorwax MP-80)

PROPERTIES:

Form:

Flake (Castorwax MP-80; Cenwax G)

Color:

White (Castorwax MP-80)
White to pale yellow (Cenwax G)

Composition:

100% active (Castorwax MP-80)

Solubility:

Sol. in organic cosmetic oils (Castorwax, MP-70)

M.P.:

70 C (Castorwax MP-70)
80 C (Castorwax MP-80)
86 C (Castorwax)
87 C (Cenwax G)

Flash Pt.:

320 C (COC) (Cenwax G)

Fire Pt.:

338 C (Cenwax G)

Acid No.:

3.0 (Cenwax G)

Iodine No.:

2.5 (Cenwax G)

Saponification No.:

180 (Cenwax G)

Isobutyl acetate (CTFA)

SYNONYMS:

Acetic acid, 2-methylpropyl ester
2-Methylpropyl acetate

EMPIRICAL FORMULA:

$C_6H_{12}O_2$

STRUCTURE:

$$CH_3C(=O)\text{—}OCH_2CH(CH_3)CH_3$$

CAS No.:

110-19-0

TRADENAME EQUIVALENTS:

Generically sold by:
Ashland, Eastman, Union Carbide, Union Oil

CATEGORY:

Solvent, flavoring agent

APPLICATIONS:

Cosmetic industry preparations: perfumery (generic)
Food applications: flavoring (generic)
Industrial applications: adhesives/sealants (generic); lacquers (generic); latexes (generic); nitrocellulose (generic); rubbers (generic); thinners (generic)

PROPERTIES:

Form:

Liquid (generic)

Color:

Colorless (generic)

Odor:

Characteristic, nonresidual fruit-like odor (generic)

Solubility:

Sol. in alcohols (generic)
Sol. in ether (generic)
Sol. in hydrocarbons (generic)
Partly sol. in water (generic)

Sp.gr.:

0.8685 (15 C) (generic)

Isobutyl acetate *(cont'd.)*

Density:
7.23 lb/gal (generic)
F.P.:
–99 C (generic)
B.P.:
112–119 C (generic)
Flash Pt.:
17.7 C (CC) (generic)
Ref. Index:
1.392 (generic)
TOXICITY/HANDLING:
TLV 150 ppm in air (generic)
STORAGE/HANDLING:
Flammable; dangerous fire risk (generic)
STD. PKGS.:
Drum, tank truck, tank car (generic)

Isopropyl acetate (CTFA)

SYNONYMS:
Acetic acid, 1-methylethyl ester
1-Methylethyl acetate
EMPIRICAL FORMULA:
$C_5H_{10}O_2$
STRUCTURE:

$$CH_3\overset{\overset{\displaystyle O}{\|}}{C}—OCH(CH_3)_2$$

CAS No.:
108-21-4
TRADENAME EQUIVALENTS:
Generically sold by:
Ashland, Eastman, Union Carbide, Union Oil
CATEGORY:
Solvent
APPLICATIONS:
Cosmetic industry preparations: perfumery (generic)
Industrial applications: adhesives/cements (generic); organic synthesis (generic); paints/lacquers (generic); printing inks (generic); resin gums (generic); rubber (generic)

PROPERTIES:
Form:
Liquid (generic)
Color:
Colorless (generic)
Odor:
Aromatic (generic)
Solubility:
Miscible with most organic solvents (generic)
2.9% sol. in water (generic)
Sp.gr.:
0.8690 (25/4 C) (generic)
Density:
7.17 lb/gal (20 C) (generic)
Visc.:
0.49 cp (generic)
F.P.:
–73.4 C (generic)
B.P.:
89.4 C (generic)
Flash Pt.:
4.4 C (generic)
Acidity:
0.02% max. (as acetic acid) (generic)
Ref. Index:
1.378 (20 C) (generic)
Specific Heat:
0.46 cal/g (generic)
TOXICITY/HANDLING:
TLV 250 ppm in air (generic)
STORAGE/HANDLING:
Flammable; dangerous fire risk (generic)
STD. PKGS.:
Drums, tank trucks, tank cars (generic)

Isopropyl alcohol (CTFA)

SYNONYMS:
Dimethylcarbinol
IPA
Isopropanol

Isopropyl alcohol (cont'd.)

SYNONYMS *(cont'd.):*

2-Propanol
sec-Propyl alcohol

EMPIRICAL FORMULA:

C_3H_8O

STRUCTURE:

$CH_3—CH—CH_3$
||
OH

CAS No.:

67-63-0

TRADENAME EQUIVALENTS:

Generically sold by:
Ashland, Eastman, Shell, Union Carbide, Union Oil

CATEGORY:

Solvent, deicing agent, dehydrating agent, preservative, denaturant

APPLICATIONS:

Cosmetic industry preparations: perfumery/essential oils (generic)
Industrial applications: adhesives/cements (generic); cellulosics (generic); chemical mfg. (generic); coatings (generic); extraction processes (generic); fuels (generic); gums (generic); lacquers (generic); resins (generic); rubber (generic)

PROPERTIES:

Form:
Liquid (generic)
Color:
Colorless (generic)
Odor:
Pleasant (generic)
Solubility:
Sol. in alcohols (generic)
Sol. in ether (generic)
Sol. in water (generic)
Sp.gr.:
0.7863 (20/20 C) (generic)
Density:
6.54 lb/gal (generic)
Visc.:
2.1 cp (generic)
F.P.:
–86 C (generic)
B.P.:
81–84 C (generic)

Flash Pt.:
11.7 C (TOC) (generic)
Ref. Index:
1.3756 (20 C) (generic)
Specific Heat:
0.65 cal/g (generic)
TOXICITY/HANDLING:
Toxic by ingestion and inhalation; TLV 400 ppm in air (generic)
STORAGE/HANDLING:
Flammable; dangerous fire risk (generic)
STD. PKGS.:
Drum, tank truck, tank car (generic)

Isostearyl benzoate (CTFA)

SYNONYMS:
Benzoic acid, isostearyl ester
EMPIRICAL FORMULA:
$C_{25}H_{42}O_2$
STRUCTURE:

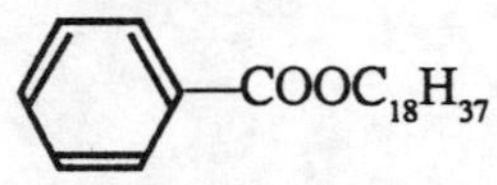

TRADENAME EQUIVALENTS:
Finsolv SB [Finetex]
CATEGORY:
Solubilizer
APPLICATIONS:
Cosmetic industry preparations: (Finsolv SB); perfumery (Finsolv SB)
PROPERTIES:
Form:
Liquid (Finsolv SB)
Solubility:
Sol. in organic solvents (Finsolv SB)
Insol. in water (Finsolv SB)
TOXICITY/HANDLING:
Nonirritating to eyes and skin (Finsolv SB)

Lecithin (CTFA)

SYNONYMS:

Mixture of diglycerides of stearic, palmitic and oleic acids, linked to the choline ester of phosphoric acid

CAS No.:

8002-43-5

TRADENAME EQUIVALENTS:

Actiflo 68, 70 [Central Soya]
Alcolec 439-C [American Lecithin]
Alcolec 495, 619-B, 621, 628G, 634, 638, 650, 658, 662 [American Lecithin]
Alcolec 4135 [American Lecithin]
Alcolec Extra A [American Lecithin]
Alcolec Granules [American Lecithin]
Alcolec HS-3 [American Lecithin]
Alcolec Powder [American Lecithin]
Alcolec RCX-1 [American Lecithin]
Alcolec S [American Lecithin]
Alcolec Z-3 [American Lecithin] (hydroxylated)
Centrocap Series [Central Soya]
Centrol Series [Central Soya]
Centrolene Series [Central Soya] (hydroxylated)
Centrolex Series [Central Soya]
Centromix Series [Central Soya]
Centrophase 31 [Central Soya/Chemurgy]
Centrophase HR [Central Soya] (heat-resistant)
Centrophil Series [Central Soya]
Clearate Special Extra, WDF [W.A. Cleary]
Emulbesto [Lucas Meyer GmbH]
Emulfluid [Lucas Meyer GmbH]
Lexin K [American Lecithin]
Soyalec DBF, DBP, SBF, SBP, UBF, WDF-FG [Canada Packers]
Troykyd Lecithin W.D. [Troy]

CATEGORY:

Emulsifier, dispersant, suspending agent, wetting agent, penetrant, release agent, stabilizer, conditioner, softening agent, lubricant, emollient, anticorrosive agent

APPLICATIONS:

Cosmetic industry preparations: (Actiflo 68, 70; Alcolec 4135, Granules, Powder; Centrol Series; Centrolene Series; Centrolex Series; Centromix Series; Centrophil

Series)

Farm products: animal feeds (Emulbesto); pesticides (Centrol Series; Centromix Series)

Food applications: (Actiflo 68, 70; Centrol Series; Centrolene Series; Centrolex Series; Centromix Series; Centrophil Series; Emulfluid); food additives (Alcolec Granules); food emulsifying (Alcolec 495, 619-B, 628G, 650, 662, Extra A, RCX-1, S, Z-3; Clearate Special Extra, WDF; Lexin K; Soyalec DBF, DBP, SBF, SBP, UBF, WDF-FG)

Industrial applications: adhesives (Centromix Series); coatings (Centrol Series; Centrolene Series; Centromix Series); construction (Centrol Series; Centrolene Series; Centromix Series; Centrophil Series); dyes and pigments (Troykyd Lecithin W.D.); glass and ceramics processing (Centrol Series; Centrolene Series; Centromix Series); magnetic media (Centrophase 31); metalworking (Centrol Series; Centrolene Series; Centromix Series; Centrophil Series); paint mfg. (Alcolec 439-C; Troykyd Lecithin W.D.); paper mfg. (Centrol Series; Centrolene Series; Centromix Series); petroleum industry (Centrol Series; Centrolene Series); plastics (Actiflo 68, 70; Centrol Series; Centrolene Series; Centrolex Series; Centromix Series; Centrophil Series); printing (Centrol Series; Centrolene Series; Centromix Series); rubber (Actiflo 68, 70; Centrol Series; Centrolene Series; Centrolex Series; Centromix Series; Centrophil Series); textile/leather processing (Actiflo 68, 70; Centrol Series; Centrolene Series; Centromix Series)

Pharmaceutical applications: (Alcolec Granules, Powder; Centrol Series; Centrolene Series; Centrolex Series; Centrophil Series)

PROPERTIES:

Form:

Liquid (Actiflo 68, 70; Alcolec 439-C, 495, 619-B, 621, 628G, 634, 638, 650, 658, 662, 4135, Extra A, HS-3, RCX-1, S, Z-3; Centrocap Series; Centrol Series; Centrolene Series; Centromix Series; Centrophase 31; Centrophil Series; Clearate Special Extra, WDF; Emulbesto; Emulfluid; Soyalec DBF, SBF, UBF, WDF-FG)

Viscous liquid (Troykyd Lecithin W.D.)

Paste (Centromix Series; Soyalec DBP, SBP)

Semisolid (Lexin K)

Solid (Centrophil Series)

Beads (Alcolec Granules)

Granular (Centrolex Series)

Powder (Alcolec Powder)

Color:

Amber (Alcolec Granules; Troykyd Lecithin W.D.)

Odor:

Bland (Alcolec Granules)

Low (Lexin K)

Composition:

52–55% conc. (Soyalec WDF-FG)

60% conc. (Clearate WDF)
60–65% conc. (Emulbesto)
62% active (Centrophase 31)
62–66% conc. (Soyalec DBF, SBF, UBF)
66–72% active (Actiflo 68, 70)
67–72% conc. (Soyalec DBP, SBP)
95% solids (Troykyd Lecithin W.D.)
95+% active (Alcolec Granules)
97% conc. (Alcolec Powder)
100% conc. (Alcolec 439-C, 495, 650, 658, 4135, Extra A; Centrocap Series; Control Series; Centrolene Series; Centrolex Series; Centromix Series; Centrophil Series; Emulfluid)

Solubility:
Sol. in alcohols (Alcolec 658)
Sol. in cyclohexanone (Centrophase 31)
Partly sol. in MEK (Centrophase 31)
Partly sol. in MIBK (Centrophase 31)
Sol. in THF (Centrophase 31)
Sol. in toluene (Centrophase 31)
Sol. in water (Alcolec 634, 638, HS-3); disp. (Alcolec 439-C; Emulfluid; Troykyd Lecithin W.D.)

Ionic Nature:
Amphoteric (Actiflo 68, 70; Centrocap Series; Centrol Series; Centrolene Series; Centrolex Series; Centromix Series; Centrophase 31; Centrophil Series; Emulfluid)
Anionic (Alcolec 638, HS-3)
Nonionic (Alcolec 439-C, 495, 619-B, 621, 628G, 634, 650, 658, 662, 4135, Extra A, Granules, Powder, RCX-1, S, Z-3; Clearate Special Extra, WDF; Emulbesto)

Sp.gr.:
0.5 (Alcolec Granules)
1.02–1.05 (80 F) (Troykyd Lecithin W.D.)

Density:
8.5–8.8 lb/gal (80 F) (Troykyd Lecithin W.D.)

HLB:
8.0 (Clearate WDF)

pH:
7.3–8.2 (5% disp.) (Troykyd Lecithin W.D.)

STD. PKGS.:
200 lb drums (Alcolec Granules)
55-gal (400 lb net) drums, 15-gal (100 lb net) drums, 5-gal (40 lb net) pails (Troykyd Lecithin W.D.)

Methylene chloride (CTFA)

SYNONYMS:
Dichloromethane
Methane, dichloro-
Methylene dichloride
EMPIRICAL FORMULA:
CH_2Cl_2
STRUCTURE:
CH_2Cl_2
CAS No.:
75-09-2
TRADENAME EQUIVALENTS:
Aerothene MM [Dow] (inhibited)
Stauffer MC+ [Stauffer]
Generically sold by:
Ashland, Dow, Ethyl Corp., Stauffer, Union Oil
CATEGORY:
Solvent, blowing agent, aerosol propellant
APPLICATIONS:
Cosmetic industry preparations: hair sprays (Aerothene MM)
Degreasers: (generic)
Farm products: insecticides/pesticides (Aerothene MM)
Industrial applications: adhesives (Aerothene MM); aerosols (Aerothene MM; generic); electronics (Aerothene MM); extraction processes (Aerothene MM; generic); paints and coatings (Aerothene MM); paint removers (generic); plastics (generic); waxes (generic)
Industrial cleaners: solvent cleaners (Stauffer MC+)
PROPERTIES:
Form:
Liquid (generic)
Clear liquid (Stauffer MC+)
Color:
Colorless (generic)
APHA 10 max. (Stauffer MC+)
Odor:
Penetrating, ether-like (generic)
Composition:
10 ppm max. nonvolatiles (Stauffer MC+)

Methylene chloride (cont'd.)

Solubility:
Sol. in alcohols (generic)
Sol. in ether (generic)
Sol. in most organic solvents (Aerothene MM)
Slightly sol. in water (Aerothene MM; generic)
M.W.:
84.93 (Stauffer MC+)
Sp.gr.:
1.312 (Stauffer MC+)
1.335 (14/5 C) (generic)
Density:
10.92 lb/gal (Stauffer MC+)
11.07 lb/gal (20 C) (generic)
Visc.:
0.430 cp (20 C) (generic)
F.P.:
–97 C (generic)
B.P.:
40.1 C (generic)
Flash Pt.:
Nonflammable (generic)
Ref. Index:
1.4244 (20 C) (generic)
TOXICITY/HANDLING:
Toxic, carcinogenic (generic)
STD. PKGS.:
55-gal (600 lb net) drums; bulk in tank trucks, 400-, 8000-, 10,000-, and 15,000-gal net tank cars, and barges (Stauffer MC+)

Methyl ethyl ketone

SYNONYMS:
2-Butanone
Ethyl methyl ketone
MEK (CTFA)
EMPIRICAL FORMULA:
C_4H_8O
STRUCTURE:

$$CH_3-\overset{\overset{\displaystyle O}{\|}}{C}-CH_2CH_3$$

CAS No.:
78-93-3
TRADENAME EQUIVALENTS:
Generically sold by:
Ashland, Eastman, Harwick, Shell, Union Carbide, Union Oil
CATEGORY:
Solvent
APPLICATIONS:
Industrial applications: adhesives/cements (generic); lubricating/cutting oils (generic); nitrocellulose coatings (generic); organic synthesis (generic); plastics (generic); rubber (generic)
Industrial cleaners: (generic); paint removers (generic)
PROPERTIES:
Form:
Liquid (generic)
Color:
Colorless (generic)
Odor:
Acetone-like (generic)
Solubility:
Sol. in alcohols (generic)
Sol. in benzene (generic)
Sol. in ether (generic)
Misc. with oils (generic)
2.6% sol. in water (generic)
M.W.:
72.10 (generic)
Sp.gr.:
0.7997 (25/4 C) (generic)
Density:
6.71 lb/gal (20 C) (generic)
Visc.:
0.401 cps (generic)
F.P.:
–86.4 C (generic)
B.P.:
79.6 C (generic)
Flash Pt.:
–4.4 C (TOC) (generic)
Ref. Index:
1.379 (20 C) (generic)
Specific Heat:
0.549 cal/g (generic)

Methyl ethyl ketone *(cont'd.)*

TOXICITY/HANDLING:
TLV 200 ppm in air (generic)
STORAGE/HANDLING:
Flammable; dangerous fire risk (generic)

Methyl hydroxystearate (CTFA)

SYNONYMS:
12-Hydroxyoctadecanoic acid, methyl ester
Methyl 12-hydroxyoctadecanoate
Methyl 12-hydroxystearate
Octadecanoic acid, 12-hydroxy-, methyl ester
EMPIRICAL FORMULA:
$C_{19}H_{38}O_3$
STRUCTURE:

$$CH_3(CH_2)_5CH(OH)(CH_2)_{10}C(=O)—OCH_3$$

CAS No.:
141-23-1
TRADENAME EQUIVALENTS:
Cenwax ME [Union Camp]
CATEGORY:
Dispersant, lubricant, wetting agent, mold release agent
APPLICATIONS:
Cosmetic industry preparations: (Cenwax ME); lipsticks (Cenwax ME); makeup dyes and pigments (Cenwax ME)
Industrial applications: mold release (Cenwax ME)
PROPERTIES:
Form:
Flake (Cenwax ME)
Color:
White to pale yellow (Cenwax ME)
M.P.:
52 C (Cenwax ME)
Flash Pt.:
216 C (COC) (Cenwax ME)
Fire Pt.:
229 C (COC) (Cenwax ME)

Acid No.:
6.0 (Cenwax ME)
Iodine No.:
3.0 (Cenwax ME)
Saponification No.:
178 (Cenwax ME)

Methyl isobutyl ketone

SYNONYMS:
Hexone
Isopropylacetone
4-Methyl-2-pentanone
MIBK (CTFA)
2-Pentanone, 4-methyl-
EMPIRICAL FORMULA:
$C_6H_{12}O$
STRUCTURE:

$$CH_3C(=O){-}CH_2CH(CH_3)CH_3$$

CAS No.:
108-10-1
TRADENAME EQUIVALENTS:
Generically sold by:
Ashland, Eastman, Shell, Union Carbide, Union Oil
CATEGORY:
Solvent, intermediate, denaturant
APPLICATIONS:
Industrial applications: adhesives/cements (generic); chemical mfg. (generic); extraction processes (generic); nitrocellulosics (generic); organic synthesis (generic); paints/varnishes/lacquers (generic); rubber (generic)
PROPERTIES:
Form:
Liquid (generic)
Color:
Colorless (generic)
Odor:
Pleasant, characteristic (generic)

Methyl isobutyl ketone (cont'd.)

Solubility:
Misc. with most organic solvents (generic)
Slightly sol. in water (generic)
M.W.:
100.2 (generic)
Sp.gr.:
0.8042 (20/20 C) (generic)
Density:
6.68 lb/gal (20 C) (generic)
F.P.:
–85 C (generic)
B.P.:
115.8 C (generic)
Flash Pt.:
22.7 C (generic)
Ref. Index:
1.3959 (20 C) (generic)
TOXICITY/HANDLING:
Avoid ingestion and inhalation; absorbed by skin; TLV 50 ppm in air (generic)
STORAGE/HANDLING:
Flammable; dangerous fire risk; explosive limits in air 1.4–7.5% (generic)
STD. PKGS.:
Drum, tank truck, tank car (generic)

Monoisopropanolamine dodecylbenzenesulfonate

SYNONYMS:
Benzenesulfonic acid, dodecyl-, compd. with 1-amino-2-propanol (1:1)
Dodecylbenzenesulfonic acid, compd. with 1-amino-2-propanol (1:1)
MIPA-dodecylbenzenesulfonate (CTFA)
EMPIRICAL FORMULA:
$C_{18}H_{30}O_3S \cdot C_3H_9NO$
STRUCTURE:
$CH_2(CH_2)_{10}CH_3$ (attached to benzene ring, para to)
$SO_3H \cdot NH_2CH_2CHCH_3$ (with OH on CH)

Monoisopropanolamine dodecylbenzenesulfonate *(cont'd.)*

CAS No.:
42504-46-1; 54590-52-2
TRADENAME EQUIVALENTS:
Hetsulf IPA [Heterene]
CATEGORY:
Dispersant, wetting agent, emulsifier
PROPERTIES:
Form:
Liquid (Hetsulf IPA)
Ionic Nature:
Anionic (Hetsulf IPA)
pH:
5.0–6.5 (5% aq. sol'n.) (Hetsulf IPA)
Biodegradable: (Hetsulf IPA)

Octyldodecanol

SYNONYMS:

1-Dodecanol, 2-octyl-
2-Octyl dodecanol

EMPIRICAL FORMULA:

$C_{20}H_{42}O$

STRUCTURE:

$CH_3(CH_2)_9CHCH_2OH$
|
$CH_3(CH_2)_6CH_2$

CAS No.:

5333-42-6

TRADENAME EQUIVALENTS:

Eutanol G [Henkel]
Primarol 1208 [Henkel]
Standamul G [Henkel]

CATEGORY:

Dispersant, solubilizer, carrier, intermediate, coupling agent, wetting agent

APPLICATIONS:

Cosmetic industry preparations: (Eutanol G; Standamul G); baby oils (Standamul G); toiletries (Standamul G)

Household detergents: (Primarol 1208)

Industrial applications: chemical mfg. (Primarol 1208); dyes and pigments (Primarol 1208; Standamul G); lubricating/cutting oils (Primarol 1208); paper coatings (Primarol 1208); plastics (Primarol 1208); polymers (Primarol 1208); printing inks (Primarol 1208); rubber (Primarol 1208); solvents (Primarol 1208); textile/leather processing (Primarol 1208)

Industrial cleaners: metal processing (Primarol 1208); sanitary chemicals (Primarol 1208)

Pharmaceutical applications: (Eutanol G)

PROPERTIES:

Form:

Liquid (Eutanol G)
Clear liquid (Primarol 1208; Standamul G)

Color:

Water-white (Primarol 1208; Standamul G)

Composition:
100% conc. (Eutanol G)
Ionic Nature:
Nonionic (Eutanol G)
Cloud Pt.:
≈ –50 C (Standamul G)
< –20 C (Primarol 1208)
Acid No.:
0.5 max. (Standamul G)
< 1.0 (Primarol 1208)
Iodine No.:
8.0 max. (Standamul G)
< 9.0 (Primarol 1208)
Saponification No.:
2.0–6.0 (Primarol 1208)
5.0 max. (Standamul G)
Hydroxyl No.:
170–185 (Primarol 1208; Standamul G)
Storage Stability:
Stable under normal storage conditions (Primarol 1208)
Indefinite shelf life (Standamul G)
TOXICITY/HANDLING:
Eye protection and gloves are recommended (Primarol 1208)
STORAGE/HANDLING:
Store in closed containers to avoid absorption of moisture or contamination (Primarol 1208)
STD. PKGS.:
55-gal (385 lb) drums, bulk, tank wagons, rail cars (Primarol 1208)

Pareth-91-6 (CTFA)

SYNONYMS:
C_{9-11} linear primary alcohol 6 mole ethoxylate

CAS No.:
68439-46-3 (generic)

TRADENAME EQUIVALENTS:
Neodol 91-6 [Shell]

CATEGORY:
Dispersant, wetting agent, surfactant

APPLICATIONS:
Household detergents: hard surface cleaner (Neodol 91-6); laundry detergent (Neodol 91-6); liquid detergents (generic)
Industrial applications: pulp/paper processing (Neodol 91-6); textile/leather processing (Neodol 91-6)
Industrial cleaners: janitorial cleaners (Neodol 91-6); maintenance cleaners (Neodol 91-6); transportation cleaners (Neodol 91-6)

PROPERTIES:

Form:
Liquid (Neodol 91-6)

Color:
Practically colorless (Neodol 91-6)

Composition:
100% active (Neodol 91-6)

Solubility:
Sol. in water (generic)

Ionic Nature:
Nonionic (Neodol 91-6)

M.W.:
424 avg. (Neodol 91-6)

Sp.gr.:
0.991 (Neodol 91-6)

Density:
8.24 lb/gal (Neodol 91-6)

Visc.:
23 cs (100 F) (Neodol 91-6)

M.P.:
6–9 C (Neodol 91-6)

Pour Pt.:
7 C (Neodol 91-6)
Flash Pt.:
334 F (PMCC) (Neodol 91-6)
Cloud Pt.:
52 C (1% sol'n.) (Neodol 91-6)
HLB:
12.5 (Neodol 91-6)
Hydroxyl No.:
132 (Neodol 91-6)
pH:
6.0 (1% aq.) (Neodol 91-6)
Biodegradable: Rapidly biodegradable (Neodol 91-6)
TOXICITY/HANDLING:
Severe eye irritant and skin irritant on prolonged contact with concentrated form (Neodol 91-6)

Pentasodium aminotrimethylene phosphonate (CTFA)

SYNONYMS:
Aminotri (methylenephosphonic acid), pentasodium salt
Pentasodium amino tris (methylene phosphonate)
Pentasodium [nitrilotris (methylene)] tris phosphonate
Phosphonic acid, [nitrilotris (methylene)] tris-, pentasodium salt
EMPIRICAL FORMULA:
$C_3H_{12}NO_9P_3 \cdot 5Na$
STRUCTURE:

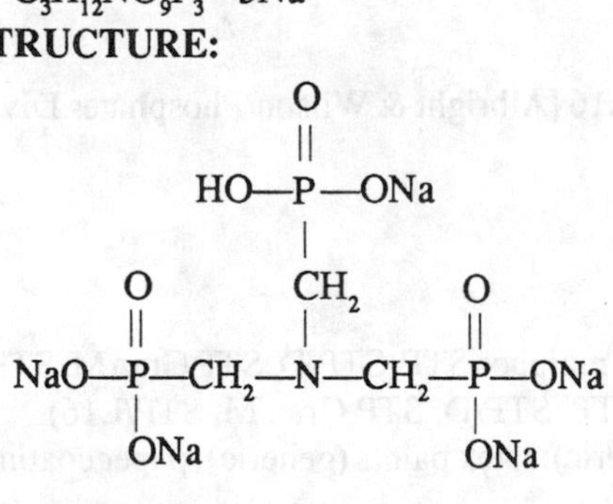

```
              O
              ||
         HO—P—ONa
              |
   O         CH2         O
   ||         |          ||
NaO—P—CH2—N—CH2—P—ONa
    |                    |
   ONa                  ONa
```

CAS No.:
2235-43-0
TRADENAME EQUIVALENTS:
Dequest 2006 [Monsanto]

Pentasodium aminotrimethylene phosphonate (cont'd.)

CATEGORY:
Dispersant, scale and corrosion inhibitor
APPLICATIONS:
Industrial applications: aqueous industrial processing (Dequest 2006); metalworking (Dequest 2006); paper mfg. (Dequest 2006); textile/leather processing (Dequest 2006)
Industrial cleaners: (Dequest 2006)
PROPERTIES:
Form:
Liquid (Dequest 2006)
Color:
Straw (Dequest 2006)
Composition:
40% active in water (Dequest 2006)

Pentasodium triphosphate (CTFA)

SYNONYMS:
Sodium tripolyphosphate
Triphosphoric acid, pentasodium salt
EMPIRICAL FORMULA:
$H_5O_{10}P_3 \cdot 5Na$
STRUCTURE:
$Na_5P_3O_{10}$
CAS No.:
7758-29-4
TRADENAME EQUIVALENTS:
Empiphos STP, STP/D, STP Gran M, STP/L16 [Albright & Wilson/Phosphates Div.]
Generically sold by: Monsanto
CATEGORY:
Dispersant, deflocculant
APPLICATIONS:
Household detergents: heavy-duty cleaner (Empiphos STP, STP/D, STP Gran M, STP/L16); powdered detergents (Empiphos STP, STP/D, STP Gran M, STP/L16)
Industrial applications: cement slurries (generic); latex paints (generic); paper coating (generic); petroleum industry (generic)
PROPERTIES:
Form:
Free-flowing granules (generic)
Free-flowing powder (generic; Empiphos STP, STP/D)

Free-flowing granular powder (Empiphos STP Gran M, STP/L16)

Color:

White (generic; Empiphos STP, STP/D, STP Gran M, STP/L16)

Composition:

94% $Na_5P_3O_{10}$; 56.7% P_2O_5 (Empiphos STP, STP/D, STP Gran M, STP/L16)

Bulk Density:

0.5 g/cm^3 (as packed) (Empiphos STP Gran M, STP/L16)
0.8 g/cm^3 (as packed) (Empiphos STP, STP/D)

pH:

9.8 ± 0.3 (Empiphos STP, STP/D, STP Gran M, STP/L16)

Perchloroethylene

SYNONYMS:

Tetrachloroethylene

STRUCTURE:

$Cl_2C{:}CCl_2$

CAS No.:

127-18-4

TRADENAME EQUIVALENTS:

Perk [Stauffer Chem.] (dry cleaning grade, industrial grade)
Perkare [PPG] (activated; with detergent and antistat)
PPG Perchlor [PPG] (dry cleaning grade, degreasing/general solvent grade)
PPG Perchlor HD [PPG] (heavy-duty grade)
Generically sold by:
Ashland, Ethyl Corp., Union Oil

CATEGORY:

Solvent, chemical intermediate, carrier solvent

APPLICATIONS:

Degreasers: (generic; Perk); vapor degreasing (PPG Perchlor; PPG Perchlor HD)
Industrial applications: chemical mfg. (generic); metals (generic; Perk); paper deinking (Perk); silicone products (Perk); solder flux removal (generic—Ethyl); textile/leather processing (Perk); waxes, targs, gums, oils (Perk)
Industrial cleaners: bleaching (Perk); cold cleaning (PPG Perchlor); drycleaning compositions (generic; Perkare; PPG Perchlor); metal cleaning (PPG Perchlor)

PROPERTIES:

Form:

Liquid (generic; Perk)
Clear liquid (Perkare; PPG Perchlor)

Color:

Colorless (generic; Perk)

Perchloroethylene (cont'd.)

Water-white (PPG Perchlor)
APHA 18–20 (Perkare)

Odor:
Ether-like (generic)
Pleasant (Perkare)

Composition:
99.96% active (Perk industrial grade)
99.97% active (Perk dry cleaning grade)

Solubility:
Miscible with alcohols (generic; Perk)
Miscible with ether (generic; Perk)
Miscible with oils (generic; Perk)
Completely miscible with most organic liquids (PPG Perchlor)
Very slightly sol. in water (Perk); insol. in water (generic); sol. 0.015 g/100 g water (PPG Perchlor)

M.W.:
165.85 (PPG Perchlor)

Sp.gr.:
1.614 (20/20 C) (Perkare)
1.619 (Perk)
1.623–1.628 (20/20 C) (PPG Perchlor)
1.625 (20/20 C) (generic)

Density:
13.4 lb/gal (20 C) (Perkare)
13.46 lb/gal (26 C) (generic)
13.47 lb/gal (Perk)
13.57 lb/gal (20 C) (PPG Perchlor)

Visc.:
0.88 cps (20 C) (PPG Perchlor)

F.P.:
–22.4 C (generic; Perk)
–22.3 C (PPG Perchlor)

B.P.:
120–122.0 C (760 mm Hg) (Perk)
121 C (generic)
121.1 C (PPG Perchlor)

Flash Pt.:
None (generic; PPG Perchlor)

Fire Pt.:
None (PPG Perchlor)

Stability:
Extremely stable; resistant to hydrolysis (generic)
Nonexplosive; good stability (Perk)

Ref. Index:
1.5029 (generic)
1.5053 (PPG Perchlor)
pH:
Close to neutral (Perkare)
Specific Heat:
0.205 Cal/(g)(C) (20 C) (PPG Perchlor)
Dielectric Constant:
2.365 (1000 cps) (PPG Perchlor)
TOXICITY/HANDLING:
Irritant to eyes and skin; TLV: 50 ppm in air (generic)
Relatively low toxicity (Perk)
Central nervous system depressant and irritant; acute inhalation overexposure can cause irritation of the respiratory tract, dizziness, nausea, headache, loss of coordination, unconsciousness, and even death in confined or poorly ventilated areas; chronic inhalation overexposure can cause liver and kidney damage; contact with eyes can cause pain and irritation; prolonged/repeated skin contact can cause irritation or dermatitis; if exposed to open flames, can decompose to form toxic and corrosive acid fumes (Perkare; PPG Perchlor)
STORAGE/HANDLING:
Nonflammable (generic; Perk)
STD. PKGS.:
700-lb net drums, 1000-gal min. tank trucks, 4000-, 8000-, 10,000-, and 16,000-gal tank trucks (Perk)
55-gal drums, tank trucks, tank cars (Perkare)

POE (5) castor oil

SYNONYMS:
PEG-5 castor oil (CTFA)
PEG (5) castor oil
CAS No.:
61791-12-6 (generic)
TRADENAME EQUIVALENTS:
Alkasurf CO-5 [Alkaril]
Chemax CO-5 [Chemax]
Surfactol 318 [NL Industries]
Trylox CO-5 [Emery]
CATEGORY:
Emulsifier, coemulsifier, dispersant, solubilizer, antifoaming agent, stabilizer, emollient

POE (5) castor oil (cont'd.)

APPLICATIONS:

Cosmetic industry preparations: (Surfactol 318)

Farm products: herbicides (Surfactol 318); insecticides/pesticides (Surfactol 318)

Household applications: (Surfactol 318)

Industrial applications: chlorinated solvents (Alkasurf CO-5; Chemax CO-5; Surfactol 318); dyes and pigments (Alkasurf CO-5; Surfactol 318; Trylox CO-5); lubricating/cutting oils (Surfactol 318); paint mfg. (Alkasurf CO-5; Surfactol 318; Trylox CO-5); paper mfg. (Surfactol 318); textile/leather processing (Alkasurf CO-5; Surfactol 318; Trylox CO-5)

PROPERTIES:

Form:

Liquid (Alkasurf CO-5; Chemax CO-5; Surfactol 318; Trylox CO-5)

Color:

Gardner 3 (Surfactol 318)

Gardner 4 (Alkasurf CO-5; Trylox CO-5)

Odor:

Faint (Surfactol 318)

Composition:

100% active (Trylox CO-5)

100% conc. (Surfactol 318)

Solubility:

Disp. in aromatic solvent (@ 10%) (Alkasurf CO-5)

Sol. in butyl acetate (Surfactol 318)

Disp. in butyl stearate (@ 5%) (Trylox CO-5)

Sol. in glycerol trioleate (@ 5%) (Trylox CO-5)

Sol. in MEK (Surfactol 318)

Disp. in min. spirits (@ 10%) (Alkasurf CO-5)

Sol. in oil (Chemax CO-5)

Disp. in perchloroethylene (@ 10%) (Alkasurf CO-5); (@ 5%) (Trylox CO-5)

Disp. in Stoddard solvent (@ 5%) (Trylox CO-5)

Sol. in toluene (Surfactol 318)

Disp. in water (@ 10%) (Alkasurf CO-5); (@ 5%) (Trylox CO-5)

Ionic Nature:

Nonionic (Surfactol 318; Trylox CO-5)

Sp.gr.:

0.984 (Surfactol 318)

Density:

0.98 g/ml (Alkasurf CO-5)

8.2 lb/gal (Trylox CO-5)

Visc.:

690 cps (Surfactol 318)

800 cs (Trylox CO-5)

Cloud Pt.:
< 25 C (Trylox CO-5)

HLB:
3.6 (Surfactol 318)
3.8 (Alkasurf CO-5; Chemax CO-5; Trylox CO-5)

Saponification No.:
145–150 (Alkasurf CO-5)

STORAGE/HANDLING:
Keep containers closed (Surfactol 318)

STD. PKGS.:
55-gal drums (Surfactol 318)

POE (10) castor oil

SYNONYMS:
PEG-10 castor oil (CTFA)
PEG 500 castor oil

CAS No.:
61791-12-6 (generic)

TRADENAME EQUIVALENTS:
Alkasurf CO-10 [Alkaril]
Etocas 10 [Croda Chem. Ltd.]
Incrocas 10 [Croda Surfactants]
Nikkol CO-10 [Nikko]
Trylox CO-10 [Emery]

CATEGORY:
Emulsifier, dispersant, solubilizer, hydrotrope, stabilizer, emollient, superfatting agent, lubricant, antistat, softener, leveling agent, detergent

APPLICATIONS:
Bath products: bubble bath (Etocas 10)
Cosmetic industry preparations: (Etocas 10; Incrocas 10; Nikkol CO-10); perfumery (Etocas 10); shampoos (Etocas 10); skin preparations (Etocas 10; Incrocas 10)
Farm products: herbicides (Etocas 10; Incrocas 10); insecticides/pesticides (Etocas 10; Incrocas 10)
Household detergents: (Etocas 10; Incrocas 10)
Industrial applications: chlorinated solvents (Alkasurf CO-10); dyes and pigments (Alkasurf CO-10); metalworking (Etocas 10; Incrocas 10); paint mfg. (Alkasurf CO-10); polymers/polymerization (Etocas 10; Incrocas 10); textile/leather processing (Alkasurf CO-10; Etocas 10; Incrocas 10)
Pharmaceutical applications: (Nikkol CO-10)

POE (10) castor oil (cont'd.)

PROPERTIES:

Form:

Liquid (Alkasurf CO-10; Etocas 10; Incrocas 10; Nikkol CO-10; Trylox CO-10)

Color:

Pale yellow (Etocas 10; Incrocas 10)
Gardner 4 (Alkasurf CO-10; Trylox CO-10)

Composition:

97% active (Etocas 10)
100% active (Incrocas 10)
100% conc. (Nikkol CO-10)

Solubility:

Partly sol. in arachis oil (Etocas 10); (@ 10%) (Incrocas 10)
Disp. in aromatic solvent (@ 10%) (Alkasurf CO-10)
Partly sol. in butyl stearate (Etocas 10); (@ 10%) (Incrocas 10); disp. (@ 5%) (Trylox CO-10)
Sol. in ethanol (Etocas 10); (@ 10%) (Incrocas 10)
Sol. in glycerol trioleate (@ 5%) (Trylox CO-10)
Sol. in MEK (Etocas 10); (@ 10%) (Incrocas 10)
Disp. in min. spirits (@ 10%) (Alkasurf CO-10)
Sol. in naphtha (Etocas 10); (@ 10%) (Incrocas 10)
Sol. in oil (Trylox CO-10)
Sol. in oleic acid (Etocas 10); (@ 10%) (Incrocas 10)
Sol. in oleyl alcohol (Etocas 10); (@ 10%) (Incrocas 10)
Disp. in perchloroethylene (@ 10%) (Alkasurf CO-10)
Disp. in Stoddard solvent (@ 5%) (Trylox CO-10)
Sol. in trichlorethylene (Etocas 10); (@ 10%) (Incrocas 10)
Disp. in water (Etocas 10); (@ 10%) (Alkasurf CO-10; Incrocas 10)
Sol. in xylene (@ 5%) (Trylox CO-10)

Ionic Nature:

Nonionic (Etocas 10; Incrocas 10; Trylox CO-10)

Density:

1.00 g/ml (Alkasurf CO-10)
8.3 lb/gal (Trylox CO-10)

Visc.:

332 cSt (100 F) (Trylox CO-10)

Pour Pt.:

–5 C (Trylox CO-10)

Flash Pt.:

560 F (Trylox CO-10)

Cloud Pt.:

20 C (1% in 10% brine) (Etocas 10; Incrocas 10)
45–49 C (10% in 25% butyl Carbitol) (Alkasurf CO-10)

HLB:
6.3 (Etocas 10; Incrocas 10)
6.4 (Trylox CO-10)
6.5 (Nikkol CO-10)
7.2 (Alkasurf CO-10)
Acid No.:
1.0 max. (Etocas 10; Incrocas 10)
Iodine No.:
55–60 (Etocas 10; Incrocas 10)
Saponification No.:
120–130 (Etocas 10; Incrocas 10)
121–126 (Alkasurf CO-10)
Hydroxyl No.:
120–140 (Etocas 10; Incrocas 10)
pH:
6.0–7.5 (Etocas 10); (3% aq.) (Incrocas 10)

POE (25) castor oil

SYNONYMS:
PEG-25 castor oil (CTFA)
PEG (25) castor oil
CAS No.:
61791-12-6 (generic)
RD No.: 977063-56-1
TRADENAME EQUIVALENTS:
Alkasurf CO-25 [Alkaril]
Chemax CO-25 [Chemax]
Hetoxide C-25 [Heterene]
Trylox CO-25 [Emery]
CATEGORY:
Emulsifier, solubilizer, dispersant, emollient, viscosity control agent, lubricant, scouring agent, dyeing assistant, dye carrier, intermediate
APPLICATIONS:
Cosmetic industry preparations: (Alkasurf CO-25; Hetoxide C-25); perfumery (Hetoxide C-25)
Household detergents: (Hetoxide C-25)
Industrial applications: dyes and pigments (Chemax CO-25; Hetoxide C-25); lubricating/cutting oils (Trylox CO-25); metalworking (Hetoxide C-25); textile/leather processing (Alkasurf CO-25; Chemax CO-25; Hetoxide C-25; Trylox CO-25)

POE (25) castor oil (cont'd.)

PROPERTIES:

Form:
Liquid (Alkasurf CO-25; Chemax CO-25; Trylox CO-25); (@ 30 C) (Hetoxide C-25)

Color:
Yellow (Alkasurf CO-25)
Gardner 4 (Trylox CO-25)

Odor:
Typical (Alkasurf CO-25)

Composition:
100% conc. (Alkasurf CO-25)

Solubility:
Disp. in butyl stearate (@ 5%) (Trylox CO-25)
Disp. in glycerol trioleate (@ 5%) (Trylox CO-25)
Sol. in isopropanol (Hetoxide C-25)
Sol. in min. oil (Hetoxide C-25)
Disp. in Stoddard solvent (@ 5%) (Trylox CO-25)
Sol. in water (Chemax CO-25); disp. (@ 5%) (Trylox CO-25)
Disp. in xylene (@ 5%) (Trylox CO-25)

Ionic Nature:
Nonionic (Hetoxide C-25; Trylox CO-25)

Density:
8.6 lb/gal (Trylox CO-25)

Visc.:
396 cSt (100 F) (Trylox CO-25)

Pour Pt.:
–5 C (Trylox CO-25)

Flash Pt.:
565 F (Trylox CO-25)

Cloud Pt.:
66 C (1% saline) (Trylox CO-25)

HLB:
10.5 (Chemax CO-25)
10.8 (Trylox CO-25)
15.6 (Hetoxide C-25)

Acid No.:
1.0 (Hetoxide C-25)

Saponification No.:
74–82 (Hetoxide C-25)

SYNONYMS:

PEG-30 castor oil (CTFA)
PEG (30) castor oil

CAS No.:

61791-12-6 (generic)
RD No.: 977059-28-1

TRADENAME EQUIVALENTS:

Alkasurf CO-30 [Alkaril]
Chemax CO-30 [Chemax]
Emulphor EL-620 [GAF]
Etocas 30 [Croda]
Hetoxide C-30 [Heterene]
Incrocas 30 [Croda Surfactants]
Merpoxen RO300 [Kempen]
Trylox CO-30 [Emery]

CATEGORY:

Emulsifier, dispersant, solubilizer, degreaser, lubricant, emollient, superfatting agent, antistat, wetting agent, surfactant, viscosity control agent, scouring agent, dyeing assistant, dye carrier, leveling agent, softener, intermediate, detergent, stabilizer

APPLICATIONS:

Automobile cleaners: wax polishes (Emulphor EL-620)
Cosmetic industry preparations: (Emulphor EL-620; Hetoxide C-30; Incrocas 30); perfumery (Hetoxide C-30; Incrocas 30); skin preparations (Incrocas 30)
Farm products: herbicides (Emulphor EL-620; Incrocas 30); insecticides/pesticides (Emulphor EL-620; Incrocas 30)
Household detergents: (Hetoxide C-30; Incrocas 30); heavy-duty cleaner (Incrocas 30)
Industrial applications: dyes and pigments (Alkasurf CO-30; Chemax CO-30; Emulphor EL-620; Trylox CO-30); glass processing (Emulphor EL-620); magnetic media (Emulphor EL-620); metalworking (Etocas 30; Hetoxide C-30; Incrocas 30); paint mfg. (Emulphor EL-620; Trylox CO-30); paper mfg. (Emulphor EL-620; Trylox CO-30); polishes and waxes (Emulphor EL-620); polymers/polymerization (Incrocas 30); textile/leather processing (Alkasurf CO-30; Chemax CO-30; Emulphor EL-620; Etocas 30; Hetoxide C-30; Incrocas 30; Merpoxen RO300; Trylox CO-30); urethane foams (Emulphor EL-620; Trylox CO-30)
Industrial cleaners: lime soap dispersant (Incrocas 30)
Pharmaceutical applications: (Emulphor EL-620)

PROPERTIES:

Form:

Liquid (Alkasurf CO-30; Chemax CO-30; Etocas 30; Hetoxide C-30; Incrocas 30; Trylox CO-30)
Clear liquid (Emulphor EL-620)
Wax (Merpoxen RO300)

POE (30) castor oil (cont'd.)

Color:
White (Merpoxen RO300)
Pale yellow (Incrocas 30)
Light brown (Emulphor EL-620)
Gardner 2 (Trylox CO-30)
Gardner 4 (Alkasurf CO-30)
Gardner 5 max. (Hetoxide C-30)

Odor:
Mild oily (Emulphor EL-620)

Composition:
100% active (Emulphor EL-620; Incrocas 30; Merpoxen RO300; Trylox CO-30)

Solubility:
Sol. in acetone (Emulphor EL-620)
Partly sol. in arachis oil (@ 10%) (Incrocas 30)
Sol. in aromatic solvent (@ 10%) (Alkasurf CO-30)
Sol. in butyl Cellosolve (Emulphor EL-620)
Partly sol. in butyl stearate (@ 10%) (Incrocas 30); disp. (@ 5%) (Trylox CO-30)
Sol. in carbon tetrachloride (Emulphor EL-620)
Sol. in ethanol (Emulphor EL-620); (@ 10%) (Incrocas 30)
Sol. in ether (Emulphor EL-620)
Disp. in glycerol trioleate (@ 5%) (Trylox CO-30)
Sol. in isopropanol (Hetoxide C-30)
Sol. in MEK (@ 10%) (Incrocas 30)
Sol. in methanol (Emulphor EL-620)
Disp. in min. oil (Hetoxide C-30); disp. (@ 10%) (Alkasurf CO-30)
Sol. in naphtha (@ 10%) (Incrocas 30)
Sol. in oleic acid (@ 10%) (Incrocas 30)
Sol. in oleyl alcohol (@ 10%) (Incrocas 30)
Sol. in perchloroethylene (@ 10%) (Alkasurf CO-30); disp. (@ 5%) (Trylox CO-30)
Disp. in Stoddard solvent (@ 5%) (Trylox CO-30)
Sol. in toluene (Emulphor EL-620)
Sol. in trichlorethylene (@ 10%) (Incrocas 30)
Sol. in veg. oil (Emulphor EL-620)
Sol. in water (Chemax CO-30; Emulphor EL-620); (@ 10%) (Alkasurf CO-30; Incrocas 30); (@ 5%) (Trylox CO-30); disp. (Hetoxide C-30)
Sol. in xylene (Emulphor EL-620)

Ionic Nature:
Nonionic (Emulphor EL-620; Hetoxide C-30; Incrocas 30; Trylox CO-30)

Sp. Gr.:
1.04–1.05 (Emulphor EL-620)

Density:
1.04 g/ml (Alkasurf CO-30)
8.7–8.8 lb/gal (Emulphor EL-620)

8.8 lb/gal (Trylox CO-30)

Visc.:

600–1000 cps (Emulphor EL-620)
720 cs (Trylox CO-30)

Flash Pt.:

291–295 C (Emulphor EL-620)

Fire Pt.:

322–326 C (Emulphor EL-620)

Cloud Pt.:

35–40 C (1% in 10% brine) (Incrocas 30)
40–45 C (1% aq.) (Emulphor EL-620)
48–61 C (1% in 5% sodium chloride) (Alkasurf CO-30)
> 100 C (Trylox CO-30)

HLB:

11.7 (Alkasurf CO-30; Chemax CO-30; Hetoxide C-30; Trylox CO-30)
12.5 (Incrocas 30)

Acid No.:

2.0 max. (Hetoxide C-30; Incrocas 30)

Iodine No.:

30–40 (Incrocas 30)

Saponification No.:

65–75 (Hetoxide C-30)
72–82 (Incrocas 30)
73–78 (Alkasurf CO-30)

Hydroxyl No.:

80–100 (Incrocas 30)

Stability:

Stable to metallic ions and weak alkali; partially stable to 1% acid (Emulphor EL-620)

pH:

6.0–7.5 (3% aq.) (Incrocas 30)

Surface Tension:

41 dynes/cm (0.1% sol'n.) (Emulphor EL-620)
41.5 dynes/cm (0.1% DW) (Incrocas 30)

STD. PKGS.:

200-kg net iron drums (Merpoxen RO300)

POE (40) castor oil

SYNONYMS:

PEG-40 castor oil (CTFA)
PEG 2000 castor oil

POE (40) castor oil (*cont'd.*)

CAS No.:

61791-12-6 (generic)
RD No.: 977055-00-7

TRADENAME EQUIVALENTS:

Alkasurf CO-40M [Alkaril]
Chemax CO-40 [Chemax]
Emulphor EL-719 [GAF]
Etocas 40 [Croda]
Hetoxide C40 [Heterene]
Merpoxen RO400 [Kempen]
Nikkol CO-40TX [Nikko]
Surfactol-365 [NL Ind.]
T-Det C-40 [Thompson-Hayward]
Trylox CO-40 [Emery]

CATEGORY:

Emulsifier, coemulsifier, dispersant, solubilizer, hydrotrope, degreaser, lubricant, antistat, emollient, wetting agent, surfactant, scouring agent, viscosity control agent, dyeing assistant, dye carrier, leveling agent, defoamer, stabilizer, penetrant, intermediate, softener

APPLICATIONS:

Automobile cleaners: polish (Emulphor EL-719)

Cosmetic industry preparations: (Emulphor EL-719; Hetoxide C40; Nikkol CO-40TX; Surfactol-365); creams and lotions (Hetoxide C40); perfumery (Hetoxide C40; Surfactol-365)

Farm products: herbicides (Emulphor EL-719; Surfactol-365); insecticides/pesticides (Emulphor EL-719; Surfactol-365)

Household detergents: (Hetoxide C40; Surfactol-365)

Industrial applications: dyes and pigments (Alkasurf CO-40M; Chemax CO-40; Emulphor EL-719; Hetoxide C40; Surfactol-365; T-Det C-40; Trylox CO-40); glass processing (Emulphor EL-719); lubricating/cutting oils (Surfactol-365; T-Det C-40); magnetic media (Emulphor EL-719); metalworking (Etocas 40; Hetoxide C40; T-Det C-40); paint mfg. (Emulphor EL-719; Surfactol-365; T-Det C-40); paper mfg. (Emulphor EL-719; T-Det C-40); polishes and waxes (Emulphor EL-719; T-Det C-40); printing inks (T-Det C-40); rubber (T-Det C-40); textile/leather processing (Alkasurf CO-40M; Chemax CO-40; Emulphor EL-719; Etocas 40; Hetoxide C40; Merpoxen RO400; Surfactol-365; T-Det C-40; Trylox CO-40); urethane foams (Emulphor EL-719)

Pharmaceutical applications: (Emulphor EL-719; Nikkol CO-40TX)

PROPERTIES:

Form:

Liquid (Alkasurf CO-40M; Chemax CO-40; Etocas 40; Nikkol CO-40TX; Surfactol-365; T-Det C-40; Trylox CO-40)

Clear liquid (Emulphor EL-719)
Paste @ 30 C (Hetoxide C40)
Wax (Merpoxen RO400)

Color:
White (Merpoxen RO400)
Yellow (Emulphor EL-719; T-Det C-40)
Gardner 2 (Trylox CO-40)
Gardner 3 (Surfactol-365)
Gardner 4 (Alkasurf CO-40M)

Odor:
Oily (Emulphor EL-719)
Mild, oily (T-Det C-40)

Composition:
96% active min. (Emulphor EL-719)
97% conc. (Etocas 40)
99.5% active min. (T-Det C-40)
100% active (Merpoxen RO400; Trylox CO-40)
100% conc. (Nikkol CO-40TX; Surfactol-365)

Solubility:
Sol. in acetone (Emulphor EL-719; T-Det C-40)
Sol. in aromatic solvent (@ 10%) (Alkasurf CO-40M)
Sol. in butyl acetate (Surfactol-365)
Sol. in butyl Cellosolve (Emulphor EL-719; T-Det C-40)
Disp. in butyl stearate (@ 5%) (Trylox CO-40)
Sol. in carbon tetrachloride (Emulphor EL-719; T-Det C-40)
Sol. in ethanol (Emulphor EL-719; Surfactol-365; T-Det C-40)
Sol. in ether (Emulphor EL-719; T-Det C-40)
Sol. in ethylene glycol (T-Det C-40)
Disp. in glycerol trioleate (@ 5%) (Trylox CO-40)
Sol. in isopropanol (Hetoxide C40)
Sol. in MEK (Surfactol-365)
Sol. in methanol (Emulphor EL-719; T-Det C-40)
Sol. in perchloroethylene (@ 10%) (Alkasurf CO-40M); disp. (@ 5%) (Trylox CO-40)
Disp. in Stoddard solvent (@ 5%) (Trylox CO-40)
Sol. in toluene (Emulphor EL-719; Surfactol-365)
Sol. in veg. oil (Emulphor EL-719; T-Det C-40)
Sol. in water (Emulphor EL-719; Hetoxide C40; Surfactol-365; T-Det C-40); sol. (@ 10%) (Alkasurf CO-40M); (@ 5%) (Trylox CO-40)
Sol. in xylene (Emulphor EL-719; T-Det C-40)

Ionic Nature:
Nonionic (Emulphor EL-719; Etocas 40; Hetoxide C40; Surfactol-365; T-Det C-40; Trylox CO-40)

POE (40) castor oil (cont'd.)

Sp.gr.:
≈ 1.05 (T-Det C-40)
1.054 (Surfactol-365)
1.06–1.07 (Emulphor EL-719)

Density:
1.04 g/ml (Alkasurf CO-40M)
8.7 lb/gal (T-Det C-40)
8.8 lb/gal (Trylox CO-40)
8.9–9.0 lb/gal (Emulphor EL-719)

Visc.:
500 cps (Surfactol-365)
500–800 cps (Emulphor EL-719)
650 cs (Trylox CO-40)

Pour Pt.:
≈ 60 F (T-Det C-40)

Flash Pt.:
275–279 C (Emulphor EL-719)
> 200 F (TOC) (T-Det C-40)

Fire Pt.:
328–332 C (Emulphor EL-719)

Cloud Pt.:
65–71 C (1% in 5% sodium chloride) (Alkasurf CO-40M)
83–88 C (1% aq.) (Emulphor EL-719)
> 100 C (Trylox CO-40)

HLB:
12.5 (Nikkol CO-40TX)
12.9 (Chemax CO-40; Trylox CO-40)
13.0 (Alkasurf CO-40M; Surfactol-365)
14.2 (T-Det C-40)
16.9 (Hetoxide C40)

Acid No.:
1.0 (Hetoxide C40)

Saponification No.:
55–65 (Hetoxide C40)
58–63 (Alkasurf CO-40M)

Stability:
Stable to metallic ions, weak alkali; partially stable to 1% acid (Emulphor EL-719)
Moderately stable to acids, bases, and salts (T-Det C-40)

pH:
≈ 6.0 (1%) (T-Det C-40)

Surface Tension:
38 dynes/cm (0.1%) (Emulphor EL-719)

STORAGE/HANDLING:

Keep from contact with oxidizing materials; keep containers closed (Surfactol-365)

STD. PKGS.:

200 kg net iron drums (Merpoxen RO400)

55-gal drums (Surfactol-365)

470-lb net steel closed-head drum (T-Det C-40)

POE (100) castor oil

SYNONYMS:

PEG-100 castor oil (CTFA)

PEG (100) castor oil

CAS No.:

61791-12-6 (generic)

RD No.: 977063-24-3

TRADENAME EQUIVALENTS:

Etocas 100 [Croda Surfactants]

Incrocas 100 [Croda Surfactants]

CATEGORY:

Solubilizer, dispersant, emulsifier, lubricant, emollient, superfatting agent, antistat, softener, detergent

APPLICATIONS:

Bath products: bubble bath (Etocas 100)

Cosmetic industry preparations: (Incrocas 100); perfumery (Etocas 100); shampoos (Etocas 100); skin care preparations (Etocas 100; Incrocas 100)

Farm products: herbicides (Etocas 100; Incrocas 100); insecticides/pesticides (Etocas 100; Incrocas 100)

Household detergents: (Etocas 100; Incrocas 100)

Industrial applications: dye leveling (Etocas 100; Incrocas 100); emulsion polymerization (Etocas 100; Incrocas 100); industrial processing (Incrocas 100); metalworking (Etocas 100; Incrocas 100); textile/leather processing (Etocas 100; Incrocas 100)

Industrial cleaners: lime soap dispersant (Incrocas 100)

PROPERTIES:

Form:

Waxy solid (Etocas 100; Incrocas 100)

Color:

Pale yellow (Etocas 100; Incrocas 100)

Composition:

97% active (Etocas 100)

POE (100) castor oil (cont'd.)

100% active (Incrocas 100)

Solubility:

Sol. in ethanol (Etocas 100); (@ 10%) (Incrocas 100)
Sol. in MEK (Etocas 100); (@ 10%) (Incrocas 100)
Sol. in naphtha (Etocas 100); (@ 140%) (Incrocas 100)
Sol. in oleic acid (Etocas 100); (@ 10%) (Incrocas 100)
Partly sol. in oleyl alcohol (Etocas 100); (@ 10%) (Incrocas 100)
Sol. in trichlorethylene (Etocas 100); (@ 10%) (Incrocas 100)
Sol. in water (Etocas 100); (@ 10%) (Incrocas 100)

Ionic Nature:

Nonionic (Etocas 100; Incrocas 100)

Drop Pt.:

42 C (Incrocas 100)

Cloud Pt.:

66 C (Etocas 100); (1% in 10% brine) (Incrocas 100)

HLB:

16.5 (Etocas 100; Incrocas 100)

Acid No.:

1.0 max. (Etocas 100; Incrocas 100)

Iodine No.:

14–18 (Etocas 100; Incrocas 100)

Saponification No.:

25–35 (Etocas 100; Incrocas 100)

Hydroxyl No.:

35–45 (Etocas 100; Incrocas 100)

pH:

6.0–7.5 (Etocas 100); (3% aq.) (Incrocas 100)

Surface Tension:

41.6 dynes/cm (0.1% deionized water) (Etocas 100; Incrocas 100)

POE (2) cetyl/stearyl ether

SYNONYMS:

Ceteareth-2 (CTFA)
PEG-2 cetyl/stearyl ether
PEG 100 cetyl/stearyl ether

STRUCTURE:

$R(OCH_2CH_2)_nOH$

where R represents a blend of cetyl and stearyl radicals and avg. $n = 2$

TRADENAME EQUIVALENTS:
Macol CSA-2 [Mazer]
CATEGORY:
Dispersant, solubilizer, coupling agent, detergent, wetting agent, emulsifier
APPLICATIONS:
Cosmetic industry preparations: (Macol CSA-2)
Household detergents: (Macol CSA-2)
Industrial applications: (Macol CSA-2); lubricating/cutting oils (Macol CSA-2); metalworking (Macol CSA-2); textile/leather processing (Macol CSA-2)
PROPERTIES:
Form:
Solid (Macol CSA-2)
Color:
White (Macol CSA-2)
Solubility:
Sol. in isopropanol (@ 1%) (Macol CSA-2)
Insol. in min. oil (@ 1%) (Macol CSA-2)
Insol. in water (@ 1%) (Macol CSA-2)
Ionic Nature:
Nonionic (Macol CSA-2)
M.P.:
39 C (Macol CSA-2)
HLB:
5.1 (Macol CSA-2)
Hydroxyl No.:
160 (Macol CSA-2)
Stability:
Very stable to hydrolysis by strong acids and alkalis (Macol CSA-2)

POE (3) cetyl/stearyl ether

SYNONYMS:
Ceteareth-3 (CTFA)
PEG-3 cetyl/stearyl ether
PEG (3) cetyl/stearyl ether
POE (3) ceto stearyl ether
STRUCTURE:
$R(OCH_2CH_2)_nOH$
where R represents a blend of cetyl and stearyl radicals and
avg. $n = 3$

POE (3) cetyl/stearyl ether (cont'd.)

CAS No.:
RD No.: 977054-58-2

TRADENAME EQUIVALENTS:
Hostacerin T-3 [American Hoechst]
Volpo CS3 [Croda Chem. Ltd.]

CATEGORY:
Emulsifier, dispersant, solubilizer, ointment base, superfatting agent, surfactant, wetting agent, gelling agent, scouring agent

APPLICATIONS:
Cleansers: (Hostacerin T-3)
Cosmetic industry preparations: (Volpo CS3); creams and lotions (Hostacerin T-3); personal care products (Hostacerin T-3); shampoos (Hostacerin T-3)
Industrial applications: (Volpo CS3)
Pharmaceutical applications: ointments (Hostacerin T-3)

PROPERTIES:

Form:
Soft, waxy solid (Volpo CS3)
Soft, waxlike substance (Hostacerin T-3)

Color:
White (Hostacerin T-3)
Off-white (Volpo CS3)

Composition:
97% conc. (Volpo CS3)
100% active (Hostacerin T-3)

Solubility:
Partly sol. in arachis oil (Volpo CS3)
Partly sol. in butyl stearate (Volpo CS3)
Sol. in ethanol (Volpo CS3)
Sol. warm in fatty alcohols (Hostacerin T-3)
Sol. warm in all hydrocarbons (Hostacerin T-3)
Sol. in kerosene (Volpo CS3)
Almost sol. in min. oil (Volpo CS3)
Sol. in oleic acid (Volpo CS3)
Sol. in oleyl alcohol (Volpo CS3)
Sol. in trichloroethylene (Volpo CS3)

Ionic Nature:
Nonionic (Hostacerin T-3)

Sp.gr.:
0.905 (50 C) (Hostacerin T-3)

Visc.:
15 ± 3 cps (50 C) (Hostacerin T-3)

Flash Pt.:
≈ 220 C (Hostacerin T-3)

Cloud Pt.:
≈ 54 C (in butyl diglycol) (Hostacerin T-3)
HLB:
7.3 (Volpo CS3)
Acid No.:
1.0 max. (Hostacerin T-3; Volpo CS3)
Iodine No.:
2.0 max. (Volpo CS3)
Saponification No.:
1.0 max. (Hostacerin T-3)
Hydroxyl No.:
145–155 (Volpo CS3)
Stability:
Stable to alkalis and strong mineral acids (Volpo CS3)
Ref. Index:
≈ 1.446 (Hostacerin T-3)
pH:
6.0–7.5 (3%) (Volpo CS3)
Biodegradable: (Volpo CS3)

POE (4) cetyl/stearyl ether

SYNONYMS:
Ceteareth-4 (CTFA)
PEG-4 cetyl/stearyl ether
PEG 200 cetyl/stearyl ether
POE (4) cetearyl ether
POE (4) cetyl/stearyl alcohol
CAS No.:
RD No.: 977054-59-3
STRUCTURE:
$R(OCH_2CH_2)_nOH$
where R represents a blend of cetyl and stearyl radicals and avg. $n = 4$
TRADENAME EQUIVALENTS:
Lipocol SC-4 [Lipo]
Macol CSA-4 [Mazer]
Siponic E-2 [Alcolac]
CATEGORY:
Surfactant, dispersant, antistat, emulsifier, defoamer, detergent, wetting agent, solubilizer, coupling agent, conditioning agent, lubricant, emollient, leveling agent,

POE (4) cetyl/stearyl ether (cont'd.)

dyeing assistant

APPLICATIONS:

Cosmetic industry preparations: (Macol CSA-4); creams and lotions (Lipocol SC-4; Siponic E-2); shampoos (Siponic E-2)

Household detergents: (Macol CSA-4)

Industrial applications: (Macol CSA-4); dyes and pigments (Lipocol SC-4); lubricating/cutting oils (Macol CSA-4); metalworking (Macol CSA-4); polymers/polymerization (Siponic E-2); textile/leather processing (Macol CSA-4; Siponic E-2)

Pharmaceutical applications: antiperspirant/deodorant (Lipocol SC-4); depilatories (Lipocol SC-4)

PROPERTIES:

Form:

Solid (Lipocol SC-4; Macol CSA-4)

Color:

White (Macol CSA-4)

Composition:

100% active (Siponic E-2)
100% conc. (Lipocol SC-4)

Solubility:

Sol. in isopropanol (@ 1%) (Macol CSA-4)
Insol. in min. oil (@ 1%) (Macol CSA-4)
Insol. in water (@ 1%) (Macol CSA-4)

Ionic Nature:

Nonionic (Lipocol SC-4; Macol CSA-4; Siponic E-2)

M.P.:

38 C (Macol CSA-4)

HLB:

7.9 (Macol CSA-4)
8.0 (Lipocol SC-4; Siponic E-2)

Hydroxyl No.:

128 (Macol CSA-4)

Stability:

Acid and alkaline stable (Lipocol SC-4)
Very stable to hydrolysis by strong acids and alkalis (Macol CSA-4)

POE (6) cetyl/stearyl ether

SYNONYMS:

Ceteareth-6 (CTFA)
PEG-6 cetyl/stearyl ether
PEG 300 cetyl/stearyl ether
POE (6) cetyl/stearyl alcohol

POE (6) cetyl/stearyl ether (cont'd.)

STRUCTURE:
$R(OCH_2CH_2)_nOH$
where R represents a blend of cetyl and stearyl radicals and avg. $n = 6$

CAS No.:
RD No.: 977054-61-7

TRADENAME EQUIVALENTS:
Siponic E-3 [Alcolac]

CATEGORY:
Conditioning agent, dispersant, solubilizer, lubricant, emollient, emulsifier, leveling agent, dyeing assistant, surfactant

APPLICATIONS:
Cosmetic industry preparations: (Siponic E-3); creams and lotions (Siponic E-3); shampoos (Siponic E-3)
Industrial applications: (Siponic E-3); dyes and pigments (Siponic E-3); textile/leather processing (Siponic E-3)
Industrial cleaners: metal processing surfactants (Siponic E-3); textile cleaning (Siponic E-3)

PROPERTIES:

Form:
Solid (Siponic E-3)

Composition:
100% active (Siponic E-3)

Ionic Nature:
Nonionic (Siponic E-3)

HLB:
10.1 (Siponic E-3)

POE (10) cetyl/stearyl ether

SYNONYMS:
Ceteareth-10 (CTFA)
PEG-10 cetyl/stearyl ether
PEG 500 cetyl/stearyl ether
POE 10 ceto stearyl ether
POE (10) cetyl/stearyl alcohol
Stearyl-cetyl alcohol ethoxylate (10 moles EO)

CAS No.:
RD No.: 977054-63-9

POE (10) cetyl/stearyl ether (cont'd.)

STRUCTURE:

$R(OCH_2CH_2)_nOH$

where R represents a blend of cetyl and stearyl radicals and avg. $n = 10$

TRADENAME EQUIVALENTS:

Macol CSA-10 [Mazer]
Siponic E-5 [Alcolac]
Volpo CS10 [Croda Chem. Ltd.]

CATEGORY:

Dispersant, solubilizer, coupling agent, detergent, wetting agent, emulsifier, lubricant, emollient, dye assistant, leveling agent, conditioning agent, surfactant, gelling agent, scouring agent

APPLICATIONS:

Cosmetic industry preparations: (Macol CSA-10; Volpo CS10); shampoos (Siponic E-5)
Household detergents: (Macol CSA-10)
Industrial applications: (Macol CSA-10; Volpo CS10); lubricating/cutting oils (Macol CSA-10); metalworking (Macol CSA-10); paper mfg. (Siponic E-5); polishes and waxes (Siponic E-5); polymers/polymerization (Siponic E-5); rubber (Siponic E-5); textile/leather processing (Siponic E-5)
Pharmaceutical applications: (Siponic E-5)

PROPERTIES:

Form:

Solid (Macol CSA-10; Volpo CS10)
Wax (Siponic E-5)

Color:

White (Macol CSA-10)

Composition:

97% conc. (Volpo CS10)
100% active (Siponic E-5)

Solubility:

Sol. in isopropanol (@ 1%) (Macol CSA-10)
Insol. in min. oil (@ 1%) (Macol CSA-10)
Disp. in water (@ 1%) (Macol CSA-10)

Ionic Nature:

Nonionic (Macol CSA-10; Siponic E-5; Volpo CS10)

M.P.:

39 C (Macol CSA-10)

Cloud Pt.:

73 C (1% NaCl) (Siponic E-5)

HLB:

12.4 (Siponic E-5)
12.6 (Macol CSA-10)

Hydroxyl No.:
80 (Macol CSA-10)
Stability:
Very stable to hydrolysis by strong acids and strong alkalis (Macol CSA-10)
pH:
6.5 (1% sol'n.) (Siponic E-5)

POE (12) cetyl/stearyl ether

SYNONYMS:
Ceteareth-12 (CTFA)
Cetyl-stearyl alcohol with 12 mol EO
PEG-12 cetyl/stearyl ether
PEG 600 cetyl/stearyl ether
CAS No.:
RD No.: 977054-64-0
STRUCTURE:
$R(OCH_2CH_2)_nOH$
where R represents a blend of cetyl and stearyl radicals and
avg. $n = 12$
TRADENAME EQUIVALENTS:
Eumulgin B-1 [Henkel]
Incropol CS-12 [Croda Surfactants]
Standamul B-1 [Henkel]
CATEGORY:
Emulsifier, solubilizer, coupling agent, lubricant, detergent, surfactant, antistat
APPLICATIONS:
Bath products: (Standamul B-1)
Cosmetic industry preparations: (Incropol CS-12); creams and ointments (Eumulgin B-1; Standamul B-1); perfumery (Standamul B-1); skin preparations (Standamul B-1)
Household detergents: (Incropol CS-12; Standamul B-1)
Industrial applications: (Incropol CS-12); textile/leather processing (Incropol CS-12)
Pharmaceutical applications: antiperspirant/deodorant (Standamul B-1); sunscreens (Standamul B-1)
PROPERTIES:
Form:
Solid (Incropol CS-12)
Waxy solid (Eumulgin B-1; Standamul B-1)
Color:
Gardner 1 max. (Incropol CS-12)

POE (12) cetyl/stearyl ether (cont'd.)

Composition:
99.5–100.0% solids (Standamul B-1)
100% active (Incropol CS-12)
100% conc. (Eumulgin B-1)
Ionic Nature:
Nonionic (Eumulgin B-1; Incropol CS-12; Standamul B-1)
Sp.gr.:
0.97 (70 C) (Standamul B-1)
Solidification Pt.:
35 C (Standamul B-1)
HLB:
12.0 (Standamul B-1)
Hydroxyl No.:
66–76 (Incropol CS-12)
69–74 (Standamul B-1)
pH:
5.5–7.5 (3%) (Incropol CS-12)
Biodegradable: (Incropol CS-12)
TOXICITY/HANDLING:
Although nontoxic, ingestion should be avoided (Standamul B-1)
STORAGE/HANDLING:
Store for prolonged periods in a cool, dry place in sealed containers at temps. below 30 C (Standamul B-1)
STD. PKGS.:
450-lb net steel drums, 110-lb net fiber drums (Standamul B-1)

POE (15) cetyl/stearyl ether

SYNONYMS:
Ceteareth-15 (CTFA)
PEG-15 cetyl/stearyl ether
PEG (15) cetyl/stearyl ether
POE 15 ceto stearyl ether
POE (15) cetyl/stearyl alcohol
STRUCTURE:
$R(OCH_2CH_2)_nOH$
where R represents a blend of cetyl and stearyl radicals and avg. $n = 15$
TRADENAME EQUIVALENTS:
Hetoxol CS-15, CSA-15 [Heterene]
Lipal 15CSA [PVO Int'l.]

TRADENAME EQUIVALENTS *(cont'd.)*:
Lipocol SC-15 [Lipo]
Macol CSA-15 [Mazer]
Siponic E7 [Alcolac]
Volpo CS15 [Croda Chem. Ltd.]

CATEGORY:
Surfactant, emulsifier, dispersant, solubilizer, coupling agent, wetting agent, gelling agent, detergent, scouring agent, lubricant, leveling agent, intermediate, defoamer, conditioning agent, antistat

APPLICATIONS:
Bath products: bath oils (Hetoxol CS-15, CSA-15)
Cleansers: germicidal skin cleanser (Lipal 15CSA)
Cosmetic industry preparations: (Macol CSA-15; Siponic E7; Volpo CS15); creams and lotions (Hetoxol CS-15, CSA-15; Lipal 15CSA; Lipocol SC-15); hair preparations (Lipal 15CSA); makeup (Lipal 15CSA); shampoos (Hetoxol CS-15, CSA-15); shaving preparations (Lipal 15CSA)
Household detergents: (Hetoxol CS-15, CSA-15; Macol CSA-15)
Industrial applications: (Macol CSA-15; Siponic E7; Volpo CS15); dyes and pigments (Hetoxol CS-15, CSA-15; Lipocol SC-15; Siponic E7); lubricating/cutting oils (Macol CSA-15); metalworking (Macol CSA-15); polymers/polymerization (Siponic E7); textile/leather processing (Hetoxol CS-15, CSA-15; Macol CSA-15)
Industrial cleaners: metal processing surfactants (Siponic E7); textile cleaning (Hetoxol CS-15, CSA-15; Siponic E7)
Pharmaceutical applications: antiperspirant/deodorant (Lipal 15CSA; Lipocol SC-15); depilatories (Lipocol SC-15); ointments (Lipal 15CSA)

PROPERTIES:

Form:
Solid (Hetoxol CS-15, CSA-15; Lipal 15CSA; Macol CSA-15; Siponic E7)
Waxy solid (Volpo CS15)
Solid wax (Lipocol SC-15)

Color:
White (Lipocol SC-15; Macol CSA-15)
Off-white (Volpo CS15)
Gardner 1 (Lipal 15CSA)
Gardner 1 max. (Hetoxol CS-15, CSA-15)

Composition:
97% conc. (Volpo CS15)
100% active (Lipal 15CSA; Lipocol SC-15)
100% conc. (Siponic E7)

Solubility:
Partly sol. in arachis oil (Volpo CS15)
Sol. in ethanol (Volpo CS15)
Sol. in isopropanol (Hetoxol CS-15, CSA-15; Lipal 15CSA); sol. (@ 1%) (Macol

POE (15) cetyl/stearyl ether (cont'd.)

CSA-15)

Insol. in min. oil (Hetoxol CS-15, CSA-15); (@ 1%) (Macol CSA-15)

Sol. in oleic acid (Volpo CS15)

Almost sol. in oleyl alcohol (Volpo CS15)

Sol. in propylene glycol (Lipal 15CSA)

Sol. in trichloroethylene (Volpo CS15)

Sol. in water (Hetoxol CS-15, CSA-15; Lipal 15CSA); sol. (@ 1%) (Macol CSA-15); almost sol. (Volpo CS15)

Ionic Nature:

Nonionic (Hetoxol CS-15, CSA-15; Lipal 15CSA; Lipocol SC-15; Macol CSA-15; Siponic E7; Volpo CS15)

M.P.:

40 C (Macol CSA-15)

Cloud Pt.:

67–70 C (1% sol'n.) (Lipal 15CSA)

94 C (1% aq.) (Volpo CS15)

HLB:

14.2 (Hetoxol CS-15, CSA-15)

14.3 (Macol CSA-15; Siponic E7)

14.3 ± 1 (Lipocol SC-15)

14.4 ± 1 (Lipal 15CSA)

14.6 (Volpo CS15)

Acid No.:

1.0 max. (Hetoxol CSA-15; Volpo CS15)

2.0 max. (Hetoxol CS-15; Lipocol SC-15)

Iodine No.:

2.0 max. (Volpo CS15)

Hydroxyl No.:

50–65 (Lipocol SC-15)

56–68 (Lipal 15CSA)

59–69 (Hetoxol CSA-15)

61 (Macol CSA-15)

65–73 (Hetoxol CS-15)

65–75 (Volpo CS15)

Stability:

Stable over a wide pH range (Lipal 15CSA)

Acid and alkali stable (Lipocol SC-15)

Stable to alkalis and strong mineral acids (Volpo CS15)

pH:

6.0–7.5 (3%) (Volpo CS15)

Surface Tension:

35.5 dynes/cm (0.1% aq.) (Volpo CS15)

Biodegradable: (Volpo CS15)

SYNONYMS:
Ceteareth-20 (CTFA)
Cetearyl alcohol (EO 20) polyethylene glycol ether
Cetyl-stearyl alcohol + EO 20 mole
PEG-20 cetyl/stearyl ether
PEG 1000 cetyl/stearyl ether
POE 20 ceto stearyl ether

STRUCTURE:
$R(OCH_2CH_2)_nOH$
where R represents a blend of cetyl and stearyl radicals and avg. $n = 20$

CAS No.:
RD No.: 977054-66-2

TRADENAME EQUIVALENTS:
Empilan KM20 [Albright & Wilson/Marchon]
Eumulgin B2 [Henkel]
Hetoxol CS20 [Heterene]
Incropol CS-20 [Croda Surfactants]
Lipocol SC-20 [Lipo]
Macol CSA-20 [Mazer]
Siponic E-10 [Alcolac]
Standamul B-2 [Henkel]
Volpo CS20 [Croda Chem. Ltd.]

CATEGORY:
Emulsifier, coupling agent, solubilizer, dispersant, defoamer, detergent, scouring agent, leveling agent, intermediate, dye assistant, lubricant, emollient, antistat, wetting agent, surfactant, conditioning agent, gelling agent

APPLICATIONS:
Automobile cleaners: (Standamul B-2)
Bath products: bath oils (Hetoxol CS20)
Cosmetic industry preparations: (Incropol CS-20; Macol CSA-20; Volpo CS20); creams, lotions, ointments (Eumulgin B2; Hetoxol CS20; Lipocol SC-20; Macol CSA-20; Standamul B-2); hair preparations (Standamul B-2); perfumery (Standamul B-2); shampoos (Hetoxol CS20; Macol CSA-20); skin preparations (Standamul B-2)
Household detergents: (Empilan KM20; Hetoxol CS20; Incropol CS-20; Macol CSA-20); floor cleaners (Standamul B-2); heavy-duty cleaner (Empilan KM20)
Industrial applications: (Incropol CS-20; Macol CSA-20; Volpo CS20); dyes and pigments (Hetoxol CS20; Lipocol SC-20); lubricating/cutting oils (Macol CSA-20); metalworking (Macol CSA-20); paper mfg. (Standamul B-2); polishes and waxes (Standamul B-2); polymers/polymerization (Macol CSA-20); textile/leather processing (Hetoxol CS20; Incropol CS-20; Macol CSA-20; Macol CSA-20; Standamul B-2)

POE (20) cetyl/stearyl ether (cont'd.)

Industrial cleaners: leather cleaning (Standamul B-2)
Pharmaceutical applications: antiperspirant/deodorant (Lipocol SC-20; Standamul B-2); depilatories (Lipocol SC-20)

PROPERTIES:

Form:
Solid (Hetoxol CS20; Incropol CS-20; Lipocol SC-20; Macol CSA-20; Volpo CS20)
Waxy flake/block (Empilan KM20)
Waxy solid (Eumulgin B2; Standamul B-2)

Color:
Colorless to pale straw (Empilan KM20)
White (Macol CSA-20)
Gardner 1 max. (Incropol CS-20)

Composition:
97% conc. (Volpo CS20)
99.7–100.0% solids (Standamul B-2)
100% active (Incropol CS-20; Macol CSA-20)
100% conc. (Eumulgin B2; Lipocol SC-20)

Solubility:
Sol. in isopropanol (Hetoxol CS20); sol. (@ 1%) (Macol CSA-20)
Insol. in min. oil (@ 1%) (Macol CSA-20)
Sol. in water (Hetoxol CS20); (@ 1%) (Macol CSA-20)

Ionic Nature:
Nonionic (Empilan KM20; Eumulgin B2; Hetoxol CS20; Incropol CS-20; Lipocol SC-20; Macol CSA-20; Macol CSA-20; Standamul B-2; Volpo CS20)

Sp.gr.:
1.0 (70 C) (Standamul B-2)

M.P.:
40 C (Macol CSA-20)

Setting Pt.:
≈ 38 C (Empilan KM20)

Solidification Pt.:
40 C (Standamul B-2)

HLB:
14.0 (Standamul B-2)
15.3 (Macol CSA-20)
15.4 (Hetoxol CS20; Lipocol SC-20; Macol CSA-20)

Hydroxyl No.:
46–52 (Incropol CS-20)
48–54 (Standamul B-2)
49 (Empilan KM20)
50–70 (Hetoxol CS20)
52 (Macol CSA-20)

pH:
5.5–7.5 (3%) (Incropol CS-20)
Biodegradable: (Incropol CS-20)
TOXICITY/HANDLING:
Although nontoxic, ingestion should be avoided (Standamul B-2)
STORAGE/HANDLING:
Store for prolonged periods in a cool, dry place in sealed containers at temps. below 30 C (Standamul B-2)
STD. PKGS.:
450-lb net steel drums; 110-lb net fiber drums (Standamul B-2)

POE (4) decyl ether phosphate

SYNONYMS:
Deceth-4 phosphate (CTFA)
PEG-4 decyl ether phosphate
PEG 200 decyl ether phosphate
CAS No.:
9004-80-2 (generic)
TRADENAME EQUIVALENTS:
Gafac RA-600 [GAF]
CATEGORY:
Dispersant, detergent, wetting agent, coupling agent
APPLICATIONS:
Household detergents: (Gafac RA-600); hard surface cleaner (Gafac RA-600)
Industrial applications: textile/leather processing (Gafac RA-600)
Industrial cleaners: (Gafac RA-600)
PROPERTIES:
Form:
Clear to slightly hazy viscous liquid (Gafac RA-600)
Composition:
98.5% active (Gafac RA-600)
Solubility:
Sol. in butyl Cellosolve (Gafac RA-600)
Sol. in ethanol (Gafac RA-600)
Sol. in kerosene (Gafac RA-600)
Disp. in min. oil (Gafac RA-600)
Sol. in perochloroethylene (Gafac RA-600)
Sol. in water (Gafac RA-600)
Sol. in xylene (Gafac RA-600)

POE (4) decyl ether phosphate (cont'd.)

Ionic Nature:
Anionic (Gafac RA-600)
Sp.gr.:
1.06–1.08 (Gafac RA-600)
Density:
8.9 lb/gal (Gafac RA-600)
Pour Pt.:
< 0 C (Gafac RA-600)
Acid No.:
100–115 (Gafac RA-600)
Stability:
Stable to alkaline and neutral media (Gafac RA-600)
pH:
< 2.5 (10% sol'n.) (Gafac RA-600)
TOXICITY/HANDLING:
Protect skin and eyes from contact (Gafac RA-600)

POE (4) dilaurate

SYNONYMS:
PEG-4 dilaurate (CTFA)
PEG 200 dilaurate
STRUCTURE:

$$CH_3(CH_2)_{10}\overset{O}{\overset{\|}{C}}\text{—}(OCH_2CH_2)_nO\text{—}\overset{O}{\overset{\|}{C}}(CH_2)_{10}CH_3$$

where avg. $n = 4$
CAS No.:
9005-02-1 (generic)
RD No.: 977055-02-9
TRADENAME EQUIVALENTS:
Alkamuls 200-DL [Alkaril]
Cithrol 2DL [Croda Chem. Ltd.]
Emerest 2622 [Emery]
Emerest 2704 [Emery]
Hodag 22L [Hodag]
Kessco PEG 200 Dilaurate [Armak]
Lipopeg 2 DL [Lipo]
Mapeg 200DL [Mazer]
Pegosperse 200DL [Glyco]
Scher PEG 200 Dilaurate [Scher]

CATEGORY:

Emulsifier, coemulsifier, surfactant, solubilizer, dispersant, spreading agent, emollient, thickener, wetting agent, softener, lubricant, defoamer, cosolvent, plasticizer, stabilizer, viscosity control agent, mold release agent

APPLICATIONS:

Bath products: bath oils (Kessco PEG 200 Dilaurate; Lipopeg 2DL; Scher PEG 200 Dilaurate)

Cosmetic industry preparations: (Cithrol 2DL; Hodag 22L; Kessco PEG 200 Dilaurate; Mapeg 200DL); body oils (Pegosperse 200DL); creams and lotions (Scher PEG 200 Dilaurate); hair preparations (Kessco PEG 200 Dilaurate; Pegosperse 200DL); makeup (Pegosperse 200DL); perfumery (Kessco PEG 200 Dilaurate; Scher PEG 200 Dilaurate); shampoos (Kessco PEG 200 Dilaurate)

Farm products: (Kessco PEG 200 Dilaurate)

Food applications: (Kessco PEG 200 Dilaurate)

Industrial applications: (Alkamuls 200-DL; Cithrol 2DL; Emerest 2622); metalworking lubricants (Mapeg 200DL); paper mfg. (Alkamuls 200-DL; Emerest 2622); plastics (Kessco PEG 200 Dilaurate); textile/leather processing (Alkamuls 200-DL; Emerest 2622; Mapeg 200DL)

Pharmaceutical applications: (Hodag 22L; Kessco PEG 200 Dilaurate; Mapeg 200DL)

PROPERTIES:

Form:

Liquid (Emerest 2622, 2704; Hodag 22L; Kessco PEG 200 Dilaurate; Lipopeg 2DL; Pegosperse 200DL)

Clear liquid (Mapeg 200DL; Scher PEG 200 Dilaurate)

Liquid to paste (Alkamuls 200-DL)

Paste (Cithrol 2DL)

Color:

Light straw (Scher PEG 200 Dilaurate)

Light yellow (Kessco PEG 200 Dilaurate; Pegosperse 200DL)

Yellow (Alkamuls 200-DL; Mapeg 200DL)

Gardner 2 (Emerest 2622)

Odor:

Mild, typical (Scher PEG 200 Dilaurate)

Composition:

97% conc. (Cithrol 2DL)

100% active (Emerest 2622)

100% conc. (Emerest 2704; Hodag 22L; Lipopeg 2DL; Mapeg 200DL)

Solubility:

Sol. in acetate esters (Scher PEG 200 Dilaurate)

Sol. in acetone (Kessco PEG 200 Dilaurate)

Sol. in alcohols (Scher PEG 200 Dilaurate)

Sol. in aliphatic hydrocarbons (Scher PEG 200 Dilaurate)

Sol. in aromatic hydrocarbons (Scher PEG 200 Dilaurate)

POE (4) dilaurate (cont'd.)

Sol. in aromatic solvent (@ 10%) (Alkamuls 200-DL)
Sol. in butyl stearate (Emerest 2622)
Sol. in carbon tetrachloride (Kessco PEG 200 Dilaurate)
Sol. in ethanol (Pegosperse 200DL)
Sol. in ethyl acetate (Kessco PEG 200 Dilaurate)
Sol. in glycerol trioleate (Emerest 2622)
Sol. in glycol ethers (Scher PEG 200 Dilaurate)
Sol. in glycols (Scher PEG 200 Dilaurate)
Sol. in isopropanol (Kessco PEG 200 Dilaurate; Mapeg 200DL)
Sol. in isopropyl myristate (Kessco PEG 200 Dilaurate)
Sol. in ketones (Scher PEG 200 Dilaurate)
Sol. in min. oils (Emerest 2622; Mapeg 200DL; Pegosperse 200DL); sol. (@ 10%) (Alkamuls 200-DL); partly sol. (Scher PEG 200 Dilaurate)
Sol. in min. spirits (@ 10%) (Alkamuls 200-DL)
Sol. in perchloroethylene (Emerest 2622); (@ 10%) (Alkamuls 200-DL)
Sol. in soybean oil (Mapeg 200DL)
Sol. in Stoddard solvent (Emerest 2622)
Sol. in toluol (Kessco PEG 200 Dilaurate; Mapeg 200DL)
Sol. in veg. oil (Pegosperse 200DL); partly sol. (Scher PEG 200 Dilaurate)
Disp. in water (Emerest 2622; Kessco PEG 200 Dilaurate; Mapeg 200DL; Scher PEG 200 Dilaurate); disp. cold (Pegosperse 200DL); disp. (@ 10%) (Alkamuls 200-DL)
Sol. in white oil (Kessco PEG 200 Dilaurate)

Ionic Nature:
Nonionic (Cithrol 2DL; Emerest 2622, 2704; Hodag 22L; Kessco PEG 200 Dilaurate; Lipopeg 2DL; Mapeg 200DL; Pegosperse 200DL)

M.W.:
584 (Scher PEG 200 Dilaurate)

Sp.gr.:
0.95 (Mapeg 200DL; Scher PEG 200 Dilaurate)
0.951 (Kessco PEG 200 Dilaurate)
0.96 (Pegosperse 200DL)

Density:
0.96 g/ml (Alkamuls 200-DL)
7.9 lb/gal (Kessco PEG 200 Dilaurate)
8.0 lb/gal (Emerest 2622)

Visc.:
35 cs (Emerest 2622)

F.P.:
< 9 C (Kessco PEG 200 Dilaurate)

M.P.:
10 C (Mapeg 200DL)

Solidification Pt.:
3 C (Pegosperse 200DL)

Flash Pt.:
> 170 C (OC) (Scher PEG 200 Dilaurate)
460 F (COC) (Kessco PEG 200 Dilaurate)
Fire Pt.:
510 F (Kessco PEG 200 Dilaurate)
Cloud Pt.:
13 C (Scher PEG 200 Dilaurate)
< 25 C (Emerest 2622)
HLB:
5.9 (Kessco PEG 200 Dilaurate)
6.2 (Emerest 2704)
6.6 (Hodag 22L)
6.8 (Emerest 2622; Mapeg 200DL; Scher PEG 200 Dilaurate)
7 ± 1 (Pegosperse 200DL)
7.4 (Alkamuls 200-DL)
Acid No.:
5.0 max. (Pegosperse 200DL)
10.0 max. (Kessco PEG 200 Dilaurate; Mapeg 200DL; Scher PEG 200 Dilaurate)
Iodine No.:
8.5 max. (Pegosperse 200DL)
12.0 max. (Mapeg 200DL)
12.5 max. (Kessco PEG 200 Dilaurate; Scher PEG 200 Dilaurate)
Saponification No.:
170–185 (Pegosperse 200DL)
170–190 (Scher PEG 200 Dilaurate)
176–186 (Kessco PEG 200 Dilaurate)
176–192 (Mapeg 200DL)
190–200 (Alkamuls 200-DL)
Stability:
Slight crystallization may occur on standing (Scher PEG 200 Dilaurate)
Ref. Index:
1.4519 (Scher PEG 200 Dilaurate)

POE (6) dilaurate

SYNONYMS:
PEG-6 dilaurate (CTFA)
PEG 300 dilaurate

POE (6) dilaurate (cont'd.)

STRUCTURE:

$$CH_3(CH_2)_{10}\overset{\overset{O}{\|}}{C}—(OCH_2CH_2)_nO—\overset{\overset{O}{\|}}{C}(CH_2)_{10}CH_3$$

where avg. $n = 6$

CAS No.:

9005-02-1 (generic)

RD No.: 977055-03-0

TRADENAME EQUIVALENTS:

Kessco PEG 300 Dilaurate [Armak]

CATEGORY:

Surfactant, solubilizer, thickener

APPLICATIONS:

Bath products: bath oils (Kessco PEG 300 Dilaurate)

Cosmetic industry preparations: (Kessco PEG 300 Dilaurate); hair preparations (Kessco PEG 300 Dilaurate); perfumery (Kessco PEG 300 Dilaurate); shampoos (Kessco PEG 300 Dilaurate)

Farm products: (Kessco PEG 300 Dilaurate)

Food applications: (Kessco PEG 300 Dilaurate)

Industrial applications: (Kessco PEG 300 Dilaurate); plastics (Kessco PEG 300 Dilaurate)

Pharmaceutical applications: (Kessco PEG 300 Dilaurate)

PROPERTIES:

Form:

Liquid (Kessco PEG 300 Dilaurate)

Color:

Light yellow (Kessco PEG 300 Dilaurate)

Solubility:

Sol. in acetone (Kessco PEG 300 Dilaurate)

Sol. in carbon tetrachloride (Kessco PEG 300 Dilaurate)

Sol. in ethyl acetate (Kessco PEG 300 Dilaurate)

Sol. in isopropanol (Kessco PEG 300 Dilaurate)

Sol. in isopropyl myristate (Kessco PEG 300 Dilaurate)

Partly sol. in kerosene (Kessco PEG 300 Dilaurate)

Sol. in naphtha (Kessco PEG 300 Dilaurate)

Sol. in toluol (Kessco PEG 300 Dilaurate)

Disp. in water (Kessco PEG 300 Dilaurate)

Partly sol. in white oil (Kessco PEG 300 Dilaurate)

Ionic Nature:

Nonionic (Kessco PEG 300 Dilaurate)

Sp.gr.:

0.975 (Kessco PEG 300 Dilaurate)

POE (6) dilaurate (cont'd.)

Density:
8.1 lb/gal (Kessco PEG 300 Dilaurate)
F.P.:
< 13 C (Kessco PEG 300 Dilaurate)
Flash Pt.:
475 F (COC) (Kessco PEG 300 Dilaurate)
Fire Pt.:
550 F (Kessco PEG 300 Dilaurate)
HLB:
9.8 (Kessco PEG 300 Dilaurate)
Acid No.:
10.0 max. (Kessco PEG 300 Dilaurate)
Iodine No.:
10.0 max. (Kessco PEG 300 Dilaurate)
Saponification No.:
148–158 (Kessco PEG 300 Dilaurate)
STD. PKGS.:
450-lb steel drums (Kessco PEG 300 Dilaurate)

POE (8) dilaurate

SYNONYMS:
PEG-8 dilaurate (CTFA)
PEG 400 dilaurate
STRUCTURE:

$$CH_3(CH_2)_{10}\overset{\overset{\displaystyle O}{\|}}{C}—(OCH_2CH_2)_nO—\overset{\overset{\displaystyle O}{\|}}{C}(CH_2)_{10}CH_3$$

where avg. $n = 8$
CAS No.:
9005-02-1 (generic)
RD No.: 977055-04-1
TRADENAME EQUIVALENTS:
Alkamuls 400-DL [Alkaril]
Cithrol 4DL [Croda]
CPH-79-N [C.P. Hall]
Emerest 2652 [Emery]
Emerest 2706 [Emery]
Industrol DL-9 [BASF Wyandotte]
Kessco PEG 400 Dilaurate [Armak]
Lipal 400DL [PVO Int'l.]

POE (8) dilaurate (cont'd.)

TRADENAME EQUIVALENTS *(cont'd.)*:

Lipopeg 4DL [Lipo]
Mapeg 400DL [Mazer]
Pegosperse 400DL [Glyco]
Scher PEG 400 Dilaurate [Scher]
Varonic 400DL [Sherex]

CATEGORY:

Dispersant, emulsifier, lubricant, release agent, softener, solubilizer, surfactant, wetting agent, defoamer

APPLICATIONS:

Bath products: bath oils (Kessco PEG-400 Dilaurate; Lipal 400DL; Lipopeg 4DL; Scher PEG 400 Dilaurate)
Cosmetic industry preparations (Alkamuls 400DL; Cithrol 4DL; Emerest 2706; Kessco PEG-400 Dilaurate; Scher PEG 400 Dilaurate); hair rinses (Lipal 400DL); makeup (Lipal 400DL); perfumery (Scher PEG 400 Dilaurate); shampoos (Kessco PEG-400 Dilaurate; Lipal 400DL); shaving preparations (Lipal 400DL)
Farm products (Alkamuls 400DL; Kessco PEG-400 Dilaurate)
Food applications (Kessco PEG-400 Dilaurate)
Industrial applications: (Cithrol 4DL; Emerest 2706); dyes and pigments (CPH-79-N); metalworking (Alkamuls 400DL); paper mfg.(Emerest 2652); plastics (Kessco PEG-400 Dilaurate); textile/ leather processing (Emerest 2652, 2706)
Pharmaceutical applications (Kessco PEG-400 Dilaurate)

PROPERTIES:

Form:

Liquid (Cithrol 4DL; CPH-79-N; Emerest 2652, 2706; Industrol DL9; Kessco PEG 400 Dilaurate; Lipal 400DL; Lipopeg 4-DL; Mapeg 400DL; Pegosperse 400DL)
Clear liquid (Scher PEG 400 Dilaurate)
Liquid to paste (Alkamuls 400DL)

Color:

Light yellow (Kessco PEG 400 Dilaurate; Mapeg 400 DL)
Yellow (Alkamuls 400DL; Lipopeg 40DL; Pegosperse 400DL)
Light straw (Scher PEG 400 Dilaurate)
Gardner 2 (Emerest 2652)
Gardner 3 (Industrol DL9)
Gardner 4 (Lipal 400DL)

Composition:

97% conc. (Cithrol 4DL)
100% active (Alkamuls 400-DL; Emerest 2652, 2706; Industrol DL-9; Lipal 400DL; Varonic 400DL)
100% conc. (Mapeg 400DL; Pegosperse 400DL)

Solubility:

Sol. in acetate esters (Scher PEG 400 Dilaurate)
Sol. in acetone (Kessco PEG 400 Dilaurate; Pegosperse 400DL)

Sol. in alcohols (Scher PEG 400 Dilaurate)
Sol. in aromatic and aliphatic hydrocarbons (Scher PEG 400 Dilaurate)
Sol. in aromatic solvents (Alkamuls 400DL)
Sol. in butyl stearate (Emerest 2652)
Sol. in carbon tetrachloride (Kessco PEG 400 Dilaurate)
Sol. in ethanol (Pegosperse 400DL)
Sol. in ethyl acetate (Kessco PEG 400 Dilaurate; Pegosperse 400DL)
Sol. in glycerol trioleate (Emerest 2652)
Sol. in glycols, glycol ethers (Scher PEG 400 Dilaurate)
Sol. in isopropanol (Kesco PEG 400 Dilaurate; Lipal 400DL; Mapeg 400DL)
Sol. in isopropyl myristate (Kessco PEG 400 Dilaurate)
Partly sol. in kerosene (Kessco PEG 300 Dilaurate, Kessco PEG 400 Dilaurate)
Sol. in ketones (Scher PEG 400 Dilaurate)
Sol. in methanol (Pegosperse 400DL)
Sol. in min. oil (Alkamuls 400DL; Emerest 2652; Lipal 400DL; Pegosperse 400DL); partly sol. (Scher PEG 400 Dilaurate)
Sol. in min. spirits (Alkamuls 400DL)
Sol. in naphtha (Kessco PEG 400 Dilaurate; Pegosperse 400DL)
Sol. in peanut oil (Kessco PEG 400 Dilaurate; Lipal 400DL)
Sol. in perchloroethylene (Alkamuls 400DL)
Sol. in soybean oil (Mapeg 40DL)
Sol. in Stoddard solvent (Emerest 2652)
Sol. in toluol (Kessco PEG 400 Dilaurate; Mapeg 400DL)
Sol. in veg. oil (Pegosperse 400DL); partly sol. (Scher PEG 400 Dilaurate)
Disp. in water (Alkamuls 400DL; Emerest 2652; Kessco PEG 400 Dilaurate; Lipal 400DL; Mapeg 400DL); insol. in water (Industrol DL9)
Sol. in xylene (Emerest 2652)

Ionic Nature:
Nonionic

M.W.:
802 avg. (Scher PEG 400 Dilaurate)

Density:
8.3 lb/gal (Kessco PEG 400 Dilaurate)
1.00 g/ml (Alkamuls 400DL)

Sp.gr.:
0.98 (Mapeg 400DL)
0.99 (Industrol DL-9; Pegosperse 400DL; Scher PEG 400 Dilaurate)

M.P.:
18 C (Mapeg 400DL)

Visc.:
38 cst @ 100 F (Emerest 2652)

Pour Pt.:
8 C (Emerest 2652; Industrol DL-9)

POE (8) dilaurate (cont'd.)

Solidification Pt.:
5–12 C (Pegosperse 400DL)

Cloud Pt.:
< 25 C (1% aq.) (Emerest 2652; Industrol DL9)
6 C (Scher PEG 400 Dilaurate)

HLB:
9.7 (CPH-79-N)
9.8 (Kessco PEG 400 Dilaurate; Mapeg 400DL)
9.9 (Emerest 2706)
10.0 (Alkamuls 40DL)
10.0 ± 0.5 (Pegosperse 400DL)
10.0 ± 1 (Lipopeg 4-DL)
10.2 (Scher PEG 400 Dilaurate)
10.4 ± 1 (Lipal 400DL)
10.8 (Emerest 2652, Industrol DL-9)

Acid No.:
< 5 (Pegosperse 40DL)
10 (Kessco PEG 400 Dilaurate; Industrol DL-9; Lipal 400DL; Lipopeg 4DL; Mapeg 400DL)
10 max. (Scher PEG 400 Dilaurate)

Iodine No.:
< 7 (Pegosperse 400DL)
8 (Lipal 400DL)
9.5 max. (Kessco PEG 400 Dilaurate)

Saponification No.:
125–140 (Lipal 400DL)
125–142 (Lipopeg 4-DL)
127–137 (Kessco PEG 400 Dilaurate; Scher PEG 400 Dilaurate)
127–140 (Pegosperse 400DL)
130–140 (Mapeg 400DL)
138–148 (Alkamuls 400DL)

Stability:
Stable over a wide pH range (Lipal 400DL)

Refractive Index:
1.4560 (Scher PEG 400 Dilaurate)

pH:
3–5 (5% aq. disp.) (Pegosperse 400DL)
6.0–7.5 (5% aq. disp.) (Industrol DL-9)

Flash Pt.:
> 170 C (OC) (Scher PEG 400 Dilaurate)
420 F (Emerest 2652)
480 F (Kessco PEG 400 Dilaurate)

Fire Pt.:
555 F (Kessco PEG 400 Dilaurate)
STD. PKGS.:
450 lb steel drums (Kessco PEG 400 Dilaurate; Industrol DL-9)

POE (12) dilaurate

SYNONYMS:
PEG-12 dilaurate (CTFA)
PEG 600 dilaurate
STRUCTURE:

$$CH_3(CH_2)_{10}\overset{O}{\overset{\|}{C}}—(OCH_2CH_2)_nO—\overset{O}{\overset{\|}{C}}(CH_2)_{10}CH_3$$

where avg. $n = 12$
CAS No.:
9005-02-1 (generic)
RD No.: 977063-27-6
TRADENAME EQUIVALENTS:
Alkamuls 600DL [Alkaril]
Cithrol 6DL [Croda Chem. Ltd.]
Kessco PEG 600 Dilaurate [Armak]
Mapeg 600DL [Mazer]
CATEGORY:
Surfactant, dispersant, solubilizer, emulsifier, thickener, wetting agent, cosolvent
APPLICATIONS:
Bath products: bath oils (Kessco PEG 600 Dilaurate)
Cosmetic industry preparations: (Alkamuls 600DL; Cithrol 6DL; Kessco PEG 600 Dilaurate; Mapeg 600DL); hair preparations (Kessco PEG 600 Dilaurate); perfumery (Kessco PEG 600 Dilaurate); shampoos (Kessco PEG 600 Dilaurate); toiletries (Alkamuls 600DL)
Farm products: (Kessco PEG 600 Dilaurate); agricultural oils/sprays (Alkamuls 600DL)
Food applications: (Kessco PEG 600 Dilaurate)
Industrial applications: (Alkamuls 600DL; Cithrol 6DL; Kessco PEG 600 Dilaurate); lubricating/cutting oils (Alkamuls 600DL; Mapeg 600DL); metalworking (Alkamuls 600DL; Mapeg 600DL); plastics (Kessco PEG 600 Dilaurate); textile/leather processing (Alkamuls 600DL; Mapeg 600DL)
Pharmaceutical applications: (Kessco PEG 600 Dilaurate; Mapeg 600DL)

POE (12) dilaurate (cont'd.)

PROPERTIES:

Form:

Liquid (Alkamuls 600DL; Cithrol 6DL)
Semisolid (Mapeg 600DL)
Soft solid (Kessco PEG 600 Dilaurate)

Color:

Cream (Kessco PEG 600 Dilaurate)
Light yellow (Mapeg 600DL)
Yellow (Alkamuls 600DL)

Composition:

97% conc. (Cithrol 6DL)
100% conc. (Alkamuls 600DL)

Solubility:

Sol. in acetone (Kessco PEG 600 Dilaurate)
Sol. in aromatic solvent (@ 10%) (Alkamuls 600DL)
Sol. in carbon tetrachloride (Kessco PEG 600 Dilaurate)
Sol. in ethyl acetate (Kessco PEG 600 Dilaurate)
Sol. in isopropanol (Kessco PEG 600 Dilaurate; Mapeg 600DL)
Sol. in isopropyl myristate (Kessco PEG 600 Dilaurate)
Partly sol. in min. oil (Mapeg 600DL); disp. (@ 10%) (Alkamuls 600DL)
Disp. in min. spirits (@ 10%) (Alkamuls 600DL)
Partly sol. in naphtha (Kessco PEG 600 Dilaurate)
Sol. in perchloroethylene (@ 10%) (Alkamuls 600DL)
Insol. in propylene glycol (Mapeg 600DL)
Sol. in soybean oil (Mapeg 600DL)
Sol. in toluol (Kessco PEG 600 Dilaurate; Mapeg 600DL)
Disp. in water (Kessco PEG 600 Dilaurate; Mapeg 600DL); (@ 10%) (Alkamuls 600DL)

Ionic Nature:

Nonionic (Alkamuls 600DL; Cithrol 6DL; Kessco PEG 600 Dilaurate)

Sp.gr.:

0.9820 (65 C) (Kessco PEG 600 Dilaurate)
0.99 (Mapeg 600DL)

Density:

1.03 g/ml (Alkamuls 600DL)

F.P.:

24 C (Kessco PEG 600 Dilaurate)

M.P.:

24 C (Mapeg 600DL)

Flash Pt.:

465 F (COC) (Kessco PEG 600 Dilaurate)

Fire Pt.:

530 F (Kessco PEG 600 Dilaurate)

HLB:
11.5 (Alkamuls 600DL)
12.2 (Kessco PEG 600 Dilaurate; Mapeg 600DL)
Acid No.:
10.0 max. (Kessco PEG 600 Dilaurate; Mapeg 600DL)
Iodine No.:
7.5 max. (Kessco PEG 600 Dilaurate; Mapeg 600DL)
Saponification No.:
102–112 (Kessco PEG 600 Dilaurate; Mapeg 600DL)
106–116 (Alkamuls 600DL)

POE (150) dinonyl phenyl ether

SYNONYMS:
Dinonylphenol-150 mole ethylene oxide adduct
Nonyl nonoxynol-150 (CTFA)
PEG-150 dinonyl phenyl ether
PEG (150) dinonyl phenyl ether
POE (150) dinonyl phenol
STRUCTURE:
$(C_9H_{19})_2C_6H_3(OCH_2CH_2)_nOH$
where avg. $n = 150$
CAS No.:
9014-93-1 (generic)
RD No.: 977069-35-4
TRADENAME EQUIVALENTS:
Chemax DNP-150 [Chemax]
Iconol DNP-150 [BASF Wyandotte]
Igepal DM-970 [GAF]
T-Det D-150 [Thompson-Hayward]
Trycol DNP-150, DNP-150/50 [Emery]
CATEGORY:
Dispersant, emulsifier, detergent, wetting agent, stabilizer, dye leveling agent
APPLICATIONS:
Cleansers: (Igepal DM-970); cleansing creams (Igepal DM-970)
Farm products: insecticides/pesticides (Igepal DM-970; Trycol DNP-150, DNP-150/50)
Household detergents: (Igepal DM-970); built detergents (Trycol DNP-150, DNP-150/50); commercial laundry compound (Chemax DNP-150; Igepal DM-970); hard surface cleaner (Chemax DNP-150; Igepal DM-970; T-Det D-150; Trycol DNP-150, DNP-150/50); heavy-duty cleaner (Trycol DNP-150, DNP-150/50);

laundry detergent (Igepal DM-970)

Industrial applications: coatings (Igepal DM-970); construction (Igepal DM-970); dyes and pigments (Igepal DM-970; Trycol DNP-150, DNP-150/50); paint mfg. (Igepal DM-970); petroleum industry (Igepal DM-970); textile/leather processing (Trycol DNP-150, DNP-150/50)

Industrial cleaners: bottle cleaners (Chemax DNP-150; Trycol DNP-150, DNP-150/ 50); dairy cleaners (Igepal DM-970; Trycol DNP-150, DNP-150/50); institutional cleaners (Igepal DM-970); marine fuel tank cleaners (Igepal DM-970); railway cleaners (Igepal DM-970); textile cleaning (Chemax DNP-150; Igepal DM-970)

PROPERTIES:

Form:

Liquid (Trycol DNP-150/50)
Cast solid (Iconol DNP-150)
Solid (Chemax DNP-150; T-Det D-150; Trycol DNP-150)
Flake (Iconol DNP-150; Igepal DM-970; Trycol DNP-150)

Color:

White (Igepal DM-970)
< Gardner 1 (Trycol DNP-150/50)
Gardner 1 (Trycol DNP-150)
Gardner 1 max. (Iconol DNP-150)

Composition:

50% active (Trycol DNP-150/50)
100% active (Iconol DNP-150; Igepal DM-970; Trycol DNP-150)
100% conc. (T-Det D-150)

Solubility:

Sol. in aromatic solvent (Igepal DM-970)
Sol. in ethanol (Igepal DM-970)
Sol. in methanol (Igepal DM-970)
Sol. in perchloroethylene (@ 5%) (Trycol DNP-150)
Sol. in water (Iconol DNP-150; Igepal DM-970); (@ 5%) (Trycol DNP-150, DNP-150/50)

Ionic Nature:

Nonionic (Iconol DNP-150; Igepal DM-970; T-Det D-150)

M.W.:

6900 (Iconol DNP-150)

Sp.gr.:

1.05 (Iconol DNP-150); (@ 80 C) (Igepal DM-970)

Density:

9.0 lb/gal (Trycol DNP-150/50)

M.P.:

55 C (Iconol DNP-150; Trycol DNP-150)

Solidification Pt.:

54 C (Igepal DM-970)

Cloud Pt.:
> 100 C (Trycol DNP-150, DNP-150/50); (1% aq.) (Iconol DNP-150; Igepal DM-970)
HLB:
19.0 (Iconol DNP-150; Igepal DM-970; T-Det D-150; Trycol DNP-150, DNP-150/50)
Stability:
Stable to hydrolysis by acids and alkalis (Trycol DNP-150, DNP-150/50)
Storage Stability:
Good (Igepal DM-970)
pH:
6.0–7.5 (5% aq.) (Iconol DNP-150)
STORAGE/HANDLING:
Contact with conc. oxidizing or reducing agents may be explosive (Igepal DM-970)
STD. PKGS.:
55-gal (250 lb net) fiber drums (Iconol DNP-150, flake)
55-gal (450 lb net) steel drums (Iconol DNP-150, cast solid)
50-lb bags (Igepal DM-970)

POE 1500 dioleate

SYNONYMS:
PEG-6-32 dioleate (CTFA)
PEG 1500 dioleate
CAS No.:
9005-07-6 (generic); 52688-97-0 (generic)
RD No.: 977065-78-3
TRADENAME EQUIVALENTS:
Pegosperse 1500 DO [Glyco]
CATEGORY:
Emulsifier, dispersant
APPLICATIONS:
Industrial applications: industrial processing (Pegosperse 1500 DO)
PROPERTIES:
Form:
Soft solid (Pegosperse 1500 DO)
Color:
Amber (Pegosperse 1500 DO)
Composition:
100% conc. (Pegosperse 1500 DO)
Solubility:
Sol. in acetone (Pegosperse 1500 DO)
Sol. in ethanol (Pegosperse 1500 DO)

POE 1500 dioleate (cont'd.)

Sol. in ethyl acetate (Pegosperse 1500 DO)
Sol. in methanol (Pegosperse 1500 DO)
Sol. in toluol (Pegosperse 1500 DO)
Sol. hot in veg. oil (Pegosperse 1500 DO)
Disp. hot in water (Pegosperse 1500 DO)

Ionic Nature:
Nonionic (Pegosperse 1500 DO)

Sp.gr.:
1.05 (Pegosperse 1500 DO)

M.P.:
30–38 C (Pegosperse 1500 DO)

HLB:
7.8 ± 0.5 (Pegosperse 1500 DO)

Acid No.:
< 11 (Pegosperse 1500 DO)

Iodine No.:
65–72 (Pegosperse 1500 DO)

Saponification No.:
104–112 (Pegosperse 1500 DO)

pH:
3.5–5.0 (5% aq. disp.) (Pegosperse 1500 DO)

POE (32) distearate

SYNONYMS:
PEG-32 distearate (CTFA)
PEG 1540 distearate

STRUCTURE:

$$CH_3(CH_2)_{16}\overset{O}{\overset{\|}{C}}—(OCH_2CH_2)_nO—\overset{O}{\overset{\|}{C}}(CH_2)_{16}CH_3$$

where avg. $n = 32$

CAS No.:
9005-08-7 (generic)
RD No.: 977055-09-6

TRADENAME EQUIVALENTS:
Kessco PEG 1540 Distearate [Armak]
Mapeg 1540 DS [Mazer]

CATEGORY:
Dispersant, emulsifier, lubricant, softener, solubilizer, defoamer, antistat, intermediate, surfactant, thickener

APPLICATIONS:

Bath products: bath oils (Kessco PEG 1540 Distearate)
Cosmetic industry preparations: (Kessco PEG 1540 Distearate; Mapeg 1540DS); cream rinses (Kessco PEG 1540 Distearate); hair preparations (Kessco PEG 1540 Distearate); perfumery (Kessco PEG 1540 Distearate); shampoos (Kessco PEG 1540 Distearate)
Farm products: (Kessco PEG 1540 Distearate)
Food applications: (Kessco PEG 1540 Distearate)
Industrial applications: metalworking (Mapeg 1540DS); plastics (Kessco PEG 1540 Distearate); textile/leather processing (Mapeg 1540DS)
Pharmaceutical applications: (Kessco PEG 1540 Distearate; Mapeg 1540DS)

PROPERTIES:

Form:
Solid (Mapeg 1540DS)
Flake (Mapeg 1540DS)
Wax (Kessco PEG 1540 Distearate)

Color:
White (Mapeg 1540DS)
Cream (Kessco PEG 1540 Distearate)

Composition:
100% conc. (Mapeg 1540DS)

Solubility:
Sol. in acetone (Kessco PEG 1540 Distearate)
Sol. in carbon tetrachloride (Kessco PEG 1540 Distearate)
Sol. in ethyl acetate (Kessco PEG 1540 Distearate)
Sol. in isopropanol (Kessco PEG 1540 Distearate; Mapeg 1540DS)
Partly sol. in peanut oil (Kessco PEG 1540 Distearate)
Sol. in propylene glycol (Mapeg 1540DS)
Sol. in toluol (Kessco PEG 1540 Distearate; Mapeg 1540DS)
Sol. in water (Kessco PEG 1540 Distearate; Mapeg 1540DS)

Ionic Nature:
Nonionic (Kessco PEG 1540 Distearate; Mapeg 1540DS)

Sp.gr.:
1.015 (65 C) (Kessco PEG 1540 Distearate)

M.P.:
45 C (Kessco PEG 1540 Distearate; Mapeg 1540DS)

Flash Pt.:
490 F (COC) (Kessco PEG 1540 Distearate)

Fire Pt.:
540 F (Kessco PEG 1540 Distearate)

HLB:
14.8 (Kessco PEG 1540 Distearate; Mapeg 1540DS)

POE (32) distearate (cont'd.)

Acid No.:
10.0 max. (Kessco PEG 1540 Distearate; Mapeg 1540DS)
Iodine No.:
0.25 max. (Kessco PEG 1540 Distearate)
1.0 max. (Mapeg 1540DS)
Saponification No.:
49–58 (Kessco PEG 1540 Distearate; Mapeg 1540DS)
pH:
5.0 (3% disp.) (Kessco PEG 1540 Distearate)

POE (12) ditallate

SYNONYMS:
PEG-12 ditallate (CTFA)
PEG 600 ditallate
STRUCTURE:

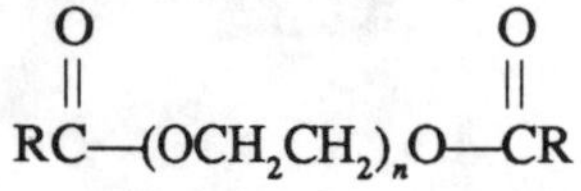

where RCO⁻ represents the tall oil fatty radicals and avg. $n = 12$
CAS No.:
61791-01-3 (generic)
RD No.: 977063-28-7
TRADENAME EQUIVALENTS:
Industrol DT-13 [BASF Wyandotte]
Mapeg 600 DOT [Mazer]
Nopalcol 6-DTW [Diamond Shamrock]
CATEGORY:
Dispersant, emulsifier, surfactant, softener, solubilizer, antistat, defoamer, intermediate, wetting agent, plasticizer, lubricant, binding agent, thickener
APPLICATIONS:
Cosmetic industry preparations: (Mapeg 600 DOT; Nopalcol 6-DTW)
Industrial applications: adhesives (Nopalcol 6-DTW); dyes and pigments (Nopalcol 6-DTW); metalworking (Mapeg 600 DOT); paper mfg. (Nopalcol 6-DTW); textile/leather processing (Mapeg 600 DOT; Nopalcol 6-DTW)
Industrial cleaners: drycleaning compositions (Nopalcol 6-DTW)
Pharmaceutical applications: (Mapeg 600 DOT)
PROPERTIES:
Form:
Liquid (Industrol DT-13; Mapeg 600 DOT)

Color:
Amber (Mapeg 600 DOT)
Gardner 4 max. (Industrol DT-13)
Composition:
100% active (Industrol DT-13; Nopalcol 6-DTW)
Solubility:
Sol. in isopropanol (Mapeg 600 DOT)
Insol. in min. oil (Mapeg 600 DOT)
Insol. in propylene glycol (Mapeg 600 DOT)
Sol. in soybean oil (Mapeg 600 DOT)
Sol. in toluol (Mapeg 600 DOT)
Disp. in water (Industrol DT-13; Mapeg 600 DOT)
Ionic Nature:
Nonionic (Industrol DT-13)
Sp.gr.:
1.00 (Industrol DT-13; Mapeg 600 DOT)
M.P.:
15 C (Mapeg 600 DOT)
Pour Pt.:
10 C (Industrol DT-13)
Cloud Pt.:
< 25 C (1% aq.) (Industrol DT-13)
HLB:
10.5 (Industrol DT-13)
13.2 (Mapeg 600 DOT)
Acid No.:
10.0 max. (Industrol DT-13; Mapeg 600 DOT)
Iodine No.:
65 max. (Mapeg 600 DOT)
Saponification No.:
85–95 (Mapeg 600 DOT)
Stability:
Stable to acid and alkali at R.T.; may hydrolyze at higher temps. or in conc. acid and alkaline sol'ns. (Nopalcol 6-DTW)
pH:
6.0–7.5 (5% aq.) (Industrol DT-13)
STD. PKGS.:
55-gal (450 lb net) steel drums (Industrol DT-13)

POE (8) ditriricinoleate

SYNONYMS:
PEG-8 ditriricinoleate (CTFA)
PEG 400 ditriricinoleate
CAS No.:
RD No.: 977055-12-1
TRADENAME EQUIVALENTS:
Pegosperse 400 DTR [Glyco]
CATEGORY:
Emulsifier, dispersant
APPLICATIONS:
Industrial applications: industrial processing (Pegosperse 400 DTR)
PROPERTIES:
Form:
Liquid (Pegosperse 400 DTR)
Color:
Amber (Pegosperse 400 DTR)
Composition:
100% conc. (Pegosperse 400 DTR)
Solubility:
Sol. in acetone (Pegosperse 400 DTR)
Partly sol. in ethanol (Pegosperse 400 DTR)
Sol. in ethyl acetate (Pegosperse 400 DTR)
Sol. in min. oils (Pegosperse 400 DTR)
Sol. in naphtha (Pegosperse 400 DTR)
Sol. in toluol (Pegosperse 400 DTR)
Sol. in veg. oil (Pegosperse 400 DTR)
Disp. in water (Pegosperse 400 DTR)
Ionic Nature:
Nonionic (Pegosperse 400 DTR)
Sp.gr.:
0.95–0.97 (Pegosperse 400 DTR)
Solidification Pt.:
< 10 C (Pegosperse 400 DTR)
HLB:
1.8 ± 0.5 (Pegosperse 400 DTR)
Acid No.:
< 12 (Pegosperse 400 DTR)
Iodine No.:
75–82 (Pegosperse 400 DTR)
Saponification No.:
156–166 (Pegosperse 400 DTR)
pH:
3.5–5.0 (5% aq. disp.) (Pegosperse 400 DTR)

POE (40) hydrogenated castor oil

SYNONYMS:

PEG-40 hydrogenated castor oil (CTFA)
PEG-40 castor oil, hydrogenated
PEG 2000 hydrogenated castor oil

CAS No.:

61788-85-0 (generic)
RD No.: 977055-15-4

TRADENAME EQUIVALENTS:

Cremophor RH40 [BASF AG]
Cremophor RH410 [BASF-Wyandotte]
Croduret 40 [Croda Chem. Ltd.]
Eumulgin HRE40 [Henkel KGaA]
Eumulgin RO-40 [Henkel/Canada]
Hetoxide HC-40 [Heterene]
Nikkol HCO-40 [Nikko]
Tagat R40 [Th. Goldschmidt AG]

CATEGORY:

Solubilizer, hydrotrope, emulsifier, dispersant, emollient, superfatting agent, detergent, antistat, viscosity control agent, lubricant, dye assistant, dye carrier, intermediate

APPLICATIONS:

Bath products: bubble bath (Croduret 40)
Cosmetic industry preparations: (Croduret 40; Eumulgin HRE40; Hetoxide HC-40; Nikkol HCO-40); perfumery (Cremophor RH40, RH410; Eumulgin RO-40; Hetoxide HC-40; Tagat R40); shampoos (Croduret 40); skin preparations (Croduret 40)
Farm products: herbicides (Croduret 40); insecticides/pesticides (Croduret 40)
Food applications: flavors (Tagat R40)
Household detergents: (Croduret 40; Hetoxide HC-40)
Industrial applications: metalworking (Croduret 40; Hetoxide HC-40); polymers/polymerization (Croduret 40); textile/leather processing (Croduret 40; Hetoxide HC-40)
Industrial cleaners: textile cleaning (Hetoxide HC-40)
Pharmaceutical applications: (Eumulgin HRE40; Nikkol HCO-40); vitamins (Croduret 40; Tagat R40)

PROPERTIES:

Form:

Liquid (Cremophor RH410; Nikkol HCO-40)
Viscous liquid (Eumulgin RO-40)
Viscous paste (Croduret 40)
Paste (Cremophor RH40)
Semisolid (Hetoxide HC-40)
Solid (Eumulgin HRE40; Tagat R40)

POE (40) hydrogenated castor oil (*cont'd.*)

Color:
White (Cremophor RH40)
Off-white (Croduret 40)
Gardner 4 max. (Hetoxide HC-40)
Composition:
90% active (Cremophor RH410)
100% active (Cremophor RH40)
100% conc. (Croduret 40; Eumulgin HRE40, RO-40; Nikkol HCO-40)
Solubility:
Sol. in benzene (Cremophor RH40, RH410)
Sol. in carbon tetrachloride (Cremophor RH40, RH410)
Sol. in chloroform (Cremophor RH40, RH410)
Sol. in ethanol (Cremophor RH40, RH410; Croduret 40)
Sol. in ethyl acetate (Cremophor RH40, RH410)
Sol. in isopropanol (Cremophor RH40, RH410; Hetoxide HC-40)
Sol. in MEK (Croduret 40)
Insol. in min. oil (Hetoxide HC-40)
Sol. in naphtha (Croduret 40)
Sol. in oleic acid (Croduret 40)
Sol. in *n* -propanol (Cremophor RH40, RH410)
Sol. in toluene (Cremophor RH40, RH410)
Sol. in trichlorethylene (Cremophor RH40, RH410; Croduret 40)
Sol. in water (Cremophor RH40, RH410; Croduret 40; Hetoxide HC-40)
Sol. in xylene (Cremophor RH40, RH410)
Ionic Nature:
Nonionic (Cremophor RH40; Croduret 40; Eumulgin HRE40, RO-40; Hetoxide HC-40; Nikkol HCO-40)
Visc.:
1500–2000 cps (Cremophor RH410)
Cloud Pt.:
62 C (Croduret 40)
HLB:
12.5 (Nikkol HCO-40)
13.0 (Croduret 40; Tagat R40)
13.1 (Hetoxide HC-40)
14–16 (Cremophor RH40, RH410; Eumulgin HRE40)
Acid No.:
< 1 (Cremophor RH40, RH410)
1.5 max. (Hetoxide HC-40)
Iodine No.:
< 1 (Cremophor RH40, RH410)
Saponification No.:
45–55 (Cremophor RH410)

50–60 (Cremophor RH40; Hetoxide HC-40)
60–65 (Croduret 40)

Hydroxyl No.:
60–80 (Cremophor RH40)

Stability:
Stable to aq. alcohol (Cremophor RH40, RH410)

Ref. Index:
1.453–1.457 (60 C) (Cremophor RH40)
1.457–1.462 (Cremophor RH410)

pH:
6.0–7.0 (10% aq.) (Cremophor RH40, RH410)

Surface Tension:
46 dynes/cm (0.1% sol'n.) (Croduret 40)

STORAGE/HANDLING:
Drums or tanks should be kept tightly closed to prevent growth of microorganisms (Cremophor RH40, RH410)

POE (50) hydrogenated castor oil

SYNONYMS:
PEG-50 hydrogenated castor oil (CTFA)
PEG (50) hydrogenated castor oil

CAS No.:
61788-85-0 (generic)

TRADENAME EQUIVALENTS:
Nikkol HCO-50 [Nikko]

CATEGORY:
Solubilizer, hydrotrope, emulsifier

APPLICATIONS:
Cosmetic industry preparations: (Nikkol HCO-50)
Pharmaceutical applications: (Nikkol HCO-50)

PROPERTIES:

Form:
Paste (Nikkol HCO-50)

Composition:
100% conc. (Nikkol HCO-50)

Ionic Nature:
Nonionic (Nikkol HCO-50)

HLB:
13.5 (Nikkol HCO-50)

POE (60) hydrogenated castor oil

SYNONYMS:

PEG-60 hydrogenated castor oil (CTFA)
PEG (60) hydrogenated castor oil

CAS No.:

61788-85-0 (generic)
RD No.: 977060-08-4

TRADENAME EQUIVALENTS:

Cremophor RH60 [BASF]
Croduret 60 [Croda Chem. Ltd.]
Eumulgin HRE60 [Henkel KGaA]
Hetoxide HC-60 [Heterene]
Nikkol HCO-60 [Nikko]
Tagat R60 [Th. Goldschmidt AG]

CATEGORY:

Solubilizer, emulsifier, emollient, superfatting agent, detergent, viscosity control agent, lubricant, dispersant, dyeing assistant, dye carrier, intermediate, antistat

APPLICATIONS:

Bath products: bubble bath (Croduret 60)
Cosmetic industry preparations: (Croduret 60; Eumulgin HRE60; Hetoxide HC-60); perfumery (Cremophor RH60; Hetoxide HC-60; Tagat R60); shampoos (Croduret 60); skin preparations (Croduret 60)
Farm products: herbicides (Croduret 60); insecticides/pesticides (Croduret 60)
Food applications: flavors (Tagat R60)
Household detergents: (Croduret 60; Hetoxide HC-60)
Industrial applications: metalworking (Croduret 60; Hetoxide HC-60); polymers/ polymerization (Croduret 60); textile/leather processing (Croduret 60; Hetoxide HC-60)
Industrial cleaners: textile cleaning (Hetoxide HC-60)
Pharmaceutical applications: (Eumulgin HRE60; Nikkol HCO-60); vitamins (Croduret 60; Tagat R60)

PROPERTIES:

Form:

Semiliquid (Nikkol HCO-60)
Paste (Cremophor RH60)
Stiff paste (Croduret 60)
Solid (Eumulgin HRE60; Hetoxide HC-60; Tagat R60)

Color:

White (Cremophor RH60)
Off-white (Croduret 60)
Gardner 4 max. (Hetoxide HC-60)

Composition:

100% active (Cremophor RH60)
100% conc. (Eumulgin HRE60; Nikkol HCO-60)

Solubility:
Sol. in benzene (Cremophor RH60)
Sol. in carbon tetrachloride (Cremophor RH60)
Sol. in chloroform (Cremophor RH60)
Sol. in ethanol (Cremophor RH60; Croduret 60)
Sol. in ethyl acetate (Cremophor RH60)
Sol. in isopropanol (Cremophor RH60; Hetoxide HC-60)
Sol. in MEK (Croduret 60)
Insol. in min. oil (Hetoxide HC-60)
Sol. in oleic acid (Croduret 60)
Sol. in *n*-propanol (Cremophor RH60)
Sol. in toluene (Cremophor RH60)
Sol. in trichlorethylene (Cremophor RH60; Croduret 60)
Sol. in water (Cremophor RH60; Croduret 60; Hetoxide HC-60)
Sol. in xylene (Cremophor RH60)

Ionic Nature:
Nonionic (Cremophor RH60; Croduret 60; Eumulgin HRE60; Hetoxide HC-60; Nikkol HCO-60)

Cloud Pt.:
71 C (Croduret 60)

HLB:
14.5 (Nikkol HCO-60)
14.7 (Croduret 60)
14.8 (Hetoxide HC-60)
15.0 (Tagat R60)
15–17 (Cremophor RH60; Eumulgin HRE60)

Acid No.:
< 1 (Cremophor RH60)
1.5 max. (Hetoxide HC-60)

Iodine No.:
< 1 (Cremophor RH60)

Saponification No.:
40–50 (Cremophor RH60)
41–51 (Hetoxide HC-60)
45–50 (Croduret 60)

Hydroxyl No.:
39–49 (Hetoxide HC-60)
50–70 (Cremophor RH60)

Stability:
Stable to aq. alcohol (Cremophor RH60)

Ref. Index:
1.453–1.457 (60 C) (Cremophor RH60)

POE (60) hydrogenated castor oil (cont'd.)

pH:
6.0–7.0 (10% aq.) (Cremophor RH60)
Surface Tension:
47.5 dynes/cm (0.1% sol'n.) (Croduret 60)
STORAGE/HANDLING:
Drums or tanks should be tightly closed to prevent growth of microorganisms (Cremophor RH60)

POE (100) hydrogenated castor oil

SYNONYMS:
PEG-100 hydrogenated castor oil (CTFA)
PEG (100) hydrogenated castor oil
CAS No.:
61788-85-0 (generic)
RD No.: 977063-25-4
TRADENAME EQUIVALENTS:
Croduret 100 [Croda Chem. Ltd.]
Nikkol HCO-100 [Nikko]
CATEGORY:
Emulsifier, solubilizer, hydrotrope, emollient, superfatting agent, detergent
APPLICATIONS:
Bath products: bubble bath (Croduret 100)
Cosmetic industry preparations: (Croduret 100; Nikkol HCO-100); shampoos (Croduret 100); skin preparations (Croduret 100)
Farm products: herbicides (Croduret 100); insecticides/pesticides (Croduret 100)
Household detergents: (Croduret 100)
Industrial applications: metalworking (Croduret 100); polymers/polymerization (Croduret 100); textile/leather processing (Croduret 100)
Pharmaceutical applications: (Nikkol HCO-100); vitamins (Croduret 100)
PROPERTIES:
Form:
Solid (Nikkol HCO-100)
Waxy solid (Croduret 100)
Color:
Off-white (Croduret 100)
Composition:
100% conc. (Nikkol HCO-100)
Solubility:
Sol. in ethanol (Croduret 100)
Sol. in MEK (Croduret 100)

Sol. in trichlorethylene (Croduret 100)
Sol. in water (Croduret 100)
Ionic Nature:
Nonionic (Croduret 100; Nikkol HCO-100)
Cloud Pt.:
65 C (Croduret 100)
HLB:
16.5 (Croduret 100; Nikkol HCO-100)
Saponification No.:
25–35 (Croduret 100)
Surface Tension:
46.2 dynes/cm (0.1% sol'n.) (Croduret 100)
TOXICITY/HANDLING:
Low toxicity; conforms to JSCI (Nikkol HCO-100)

POE (200) hydrogenated castor oil

SYNONYMS:
Hydrogenated castor oil ethoxylate (200 EO)
PEG-200 hydrogenated castor oil (CTFA)
PEG (200) hydrogenated castor oil
CAS No.:
61788-85-0 (generic)
RD No.: 977063-55-0
TRADENAME EQUIVALENTS:
Chemax HCO-200/50 [Chemax]
Croduret 200 [Croda Chem. Ltd.]
Industrol COH-200 [BASF Wyandotte]
Trylox HCO-200/50 [Emery]
CATEGORY:
Emulsifier, solubilizer, wetting agent, lubricant, antistatic agent, surfactant
PROPERTIES:
Form:
Liquid (Chemax HCO-200/50)
Clear liquid (Trylox HCO-200/50)
Solid (Croduret 200)
Wax (Industrol COH-200)
Color:
< Gardner 1 (Trylox HCO-200/50)
Gardner 2 max. (Industrol COH-200)

POE (200) hydrogenated castor oil *(cont'd.)*

Composition:
50% active (Chemax HCO-200/50)
50% active in water (Trylox HCO-200/50)
100% active (Industrol COH-200)
100% conc. (Croduret 200)

Solubility:
Sol. in water (Chemax HCO-200/50; Industrol COH-200); (@ 5%) (Trylox HCO-200/50)

Ionic Nature:
Nonionic (Croduret 200; Industrol COH-200; Trylox HCO-200/50)

M.W.:
9800 (theoret.) (Industrol COH-200)

Density:
8.8 lb/gal (Trylox HCO-200/50)

M.P.:
45 C (Industrol COH-200)

Cloud Pt.:
> 100 C (Trylox HCO-200/50)

HLB:
18.0 (Croduret 200)
18.1 (Chemax HCO-200/50; Industrol COH-200; Trylox HCO-200/50)

Acid No.:
2.0 max. (Industrol COH-200)

pH:
6.0–7.5 (5% aq.) (Industrol COH-200)

STD. PKGS.:
55-gal (450 lb net) steel drums (Industrol COH-200)

POE (50) hydrogenated tallow amide

SYNONYMS:
PEG (50) hydrogenated tallow amide
PEG-50 tallow amide (CTFA)

STRUCTURE:

$$RC(=O)\text{—NH—}(CH_2CH_2O)_nH$$

where RCO⁻ represents the tallow acid radical and avg. $n = 50$

CAS No.:
8051-63-6

TRADENAME EQUIVALENTS:

Ethomid HT/60 [Armak]
Schercomid HT-60 [Scher]

CATEGORY:

Dispersant, emulsifier, detergent, leveling agent, foaming agent, stabilizer, thickener

APPLICATIONS:

Cosmetic industry preparations: shampoos (Ethomid HT/60); shaving preparations (Ethomid HT/60)

Industrial applications: silicone finishing agents (Ethomid HT/60); sizing lubricants (Ethomid HT/60); textile/leather processing (Ethomid HT/60)

PROPERTIES:

Form:

Hard solid (Ethomid HT/60)
Hard waxy solid (Schercomid HT-60)

Color:

Gardner 7 max. (molten) (Schercomid HT-60)
Gardner 11 max. (Ethomid HT/60)

Odor:

Ammoniacal (Schercomid HT-60)

Composition:

100% active (Schercomid HT-60)

Solubility:

Sol. in alcohols (Schercomid HT-60)
Sol. in some aromatic hydrocarbons (Schercomid HT-60)
Sol. in benzene (< 80 C) (Ethomid HT/60)
Sol. in carbon tetrachloride (< 75 C) (Ethomid HT/60)
Sol. in some chlorinated hydrocarbons (Schercomid HT-60)
Sol. in dioxane (< 75 C) (Ethomid HT/60)
Insol. in natural fats (Schercomid HT-60)
Sol. in glycol ethers (Schercomid HT-60)
Sol. in glycols (Schercomid HT-60)
Sol. in isopropanol (< 75 C) (Ethomid HT/60)
Insol. in min. oil (Schercomid HT-60)
Sol. in polyols (Schercomid HT-60)
Sol. in triols (Schercomid HT-60)
Insol. in veg. oil (Schercomid HT-60)
Sol. in water (Schercomid HT-60); sol. (< 75 C) (Ethomid HT/60)

Ionic Nature:

Nonionic (Ethomid HT/60; Schercomid HT-60)

Sp.gr.:

1.064 (60 C) (Schercomid HT-60)
1.14 (Ethomid HT/60)

POE (50) hydrogenated tallow amide (cont'd.)

Density:
9.6 lb/gal (Schercomid HT-60)
M.P.:
50–55 C (Schercomid HT-60)
Flash Pt.:
> 180 C (OC) (Schercomid HT-60)
Hydroxyl No.:
40–50 (Ethomid HT/60)
40–60 (Schercomid HT-60)
Stability:
Stable in acid or alkaline sol'ns. (Ethomid HT/60)
Storage Stability:
1 yr min. shelf life in closed containers (Schercomid HT-60)
pH:
Neutral (Ethomid HT/60)
9.0–10.0 (10% aq. sol'n.) (Schercomid HT-60)
Surface Tension:
47 dynes/cm (0.1%) (Ethomid HT/60)

POE (20) isocetyl ether

SYNONYMS:
Isoceteth-20 (CTFA)
PEG-20 isocetyl ether
PEG 1000 isocetyl ether
POE (20) isohexadecyl alcohol
POE (20) isohexadecyl ether
STRUCTURE:
$C_{16}H_{33}(OCH_2CH_2)_nOH$
where avg. $n = 20$
TRADENAME EQUIVALENTS:
Arlasolve 200 [ICI United States]
CATEGORY:
Emulsifier, solubilizer, surfactant
APPLICATIONS:
Cosmetic industry preparations: (Arlasolve 200)
PROPERTIES:
Form:
Soft waxy solid (Arlasolve 200)
Color:
White (Arlasolve 200)

Composition:
100% active (Arlasolve 200)
Solubility:
Sol. in ethanol (Arlasolve 200)
Sol. in propylene glycol (Arlasolve 200)
Sol. in water (Arlasolve 200)
Ionic Nature:
Nonionic (Arlasolve 200)
M.W.:
Sp.gr.:
≈ 1.0 (Arlasolve 200)
Flash Pt.:
> 230 F (Arlasolve 200)
Fire Pt.:
> 230 F (Arlasolve 200)
HLB:
15.7 (Arlasolve 200)
TOXICITY/HANDLING:
Eye irritant (Arlasolve 200)

POE (2) lauryl ether

SYNONYMS:
Laureth-2 (CTFA)
PEG-2 lauryl ether
PEG 100 lauryl ether
EMPIRICAL FORMULA:
$C_{16}H_{34}O_3$
STRUCTURE:
$CH_3(CH_2)_{10}CH_2(OCH_2CH_2)_nOH$
where avg. $n = 2$
CAS No.:
3055-93-4; 9002-92-0 (generic)
RD No.: 977062-02-4
TRADENAME EQUIVALENTS:
Dehydol LS2 [Henkel KGaA]
Empilan KB2 [Albright & Wilson/Marchon]
Incropol L-2 [Croda]
Nikkol BL-2 [Nikko]

POE (2) lauryl ether (cont'd.)

CATEGORY:

Solubilizer, emulsifier, intermediate, foam booster, superfatting agent, lubricant, detergent, coupling agent, antistat

APPLICATIONS:

Cosmetic industry preparations: (Incropol L-2)

Household detergents: (Dehydol LS2; Empilan KB2; Incropol L-2); cold cleaners (Dehydol LS2); detergent base/intermediate (Dehydol LS2; Empilan KB2); dishwashing (Dehydol LS2)

Industrial applications: (Incropol L-2); industrial processing (Dehydol LS2; Empilan KB2)

PROPERTIES:

Form:

Liquid (Dehydol LS2; Empilan KB2; Incropol L-2; Nikkol BL-2)

Color:

Almost colorless (Empilan KB2)

Gardner 1 max. (Incropol L-2)

Composition:

100% active (Empilan KB2; Incropol L-2; Nikkol BL-2)

Solubility:

Insol. in cold water (Empilan KB2)

Ionic Nature:

Nonionic (Empilan KB2; Incropol L-2; Nikkol BL-2)

Sp.gr.:

0.9 (Empilan KB2)

Visc.:

25 cs (Empilan KB2)

HLB:

9.5 (Nikkol BL-2)

Hydroxyl No.:

198–205 (Incropol L-2)

Stability:

Stable to acids and alkalies (Empilan KB2)

pH:

5.5–7.3 (3%) (Incropol L-2)

Biodegradable: (Incropol L-2)

POE (3) lauryl ether

SYNONYMS:

Laureth-3 (CTFA)

PEG-3 lauryl ether

PEG (3) lauryl ether

EMPIRICAL FORMULA:

$C_{18}H_{38}O_4$

STRUCTURE:

$CH_3(CH_2)_{10}CH_2(OCH_2CH_2)_nOH$

where avg. $n = 3$

CAS No.:

3055-94-5; 9002-92-0 (generic)

RD No.: 977064-47-3

TRADENAME EQUIVALENTS:

AE-1214/3 [Procter & Gamble]

Dehydol LS3 [Henkel KGaA]

Empilan KB3, KC3 [Albright & Wilson/Marchon]

Hetoxol L3N [Heterene]

LA-55-3 [Hefti Ltd.]

CATEGORY:

Solubilizer, detergent, emulsifier, foam booster, superfatting agent, leveling agent, intermediate

APPLICATIONS:

Bath products: bath oils (Hetoxol L3N; LA-55-3)

Cosmetic industry preparations: (LA-55-3); lotions (Hetoxol L3N); shampoos (Hetoxol L3N)

Household detergents: (Dehydol LS3; Hetoxol L3N); cold cleaners (Dehydol LS3); detergent base/intermediate (Dehydol LS3; Empilan KB3); dishwashing (Dehydol LS3)

Industrial applications: dyes and pigments (Hetoxol L3N); industrial processing (Empilan KB3; Hetoxol L3N; LA-55-3); silicone products (Hetoxol L3N); textile/leather processing (Hetoxol L3N)

PROPERTIES:

Form:

Liquid (AE-1214/3; Dehydol LS3; Empilan KB3, KC3; Hetoxol L3N; LA-55-3)

Color:

Almost colorless (Empilan KB3)

Composition:

100% active (AE-1214/3; Empilan KB3, KC3; LA-55-3)

Solubility:

Sol. in isopropanol (Hetoxol L3N)

Sol. in min. oil (Hetoxol L3N)

Insol. in cold water (Empilan KB3)

Ionic Nature:

Nonionic (Empilan KB3, KC3; Hetoxol L3N; LA-55-3)

Sp.gr.:

0.9 (Empilan KB3)

POE (3) lauryl ether (cont'd.)

Visc.:
25 cs (Empilan KB3)
HLB:
7.5 (LA-55-3)
7.9 (Hetoxol L3N)
Hydroxyl No.:
170–176 (Hetoxol L3N)
Stability:
Stable to acids and alkalies (Empilan KB3)
pH:
6.0–7.0 (5% sol'n.) (Empilan KB3)

POE (7) lauryl ether

SYNONYMS:
Laureth-7 (CTFA)
PEG-7 lauryl ether
PEG (7) lauryl ether
POE (7) lauryl alcohol
EMPIRICAL FORMULA:
$C_{26}H_{54}O_8$
STRUCTURE:
$CH_3(CH_2)_{10}CH_2(OCH_2CH_2)_nOH$
where avg. $n = 7$
CAS No.:
3055-97-8; 9002-92-0 (generic)
RD No.: 977062-41-1
TRADENAME EQUIVALENTS:
Incropol L-7, L-7-90 [Croda Surfactants Inc.]
Macol LA-790 [Mazer]
Marlipal MG [Chemische Werke Huls AG]
Siponic L-7-90 [Alcolac]
CATEGORY:
Solubilizer, coupling agent, dispersant, emulsifier, thickener, surfactant, detergent, wetting agent, lubricant, antistat
APPLICATIONS:
Bath products: bubble bath (Siponic L-7-90)
Cosmetic industry preparations: (Incropol L-7, L-7-90; Macol LA-790); creams and lotions (Siponic L-7-90); perfumery (Siponic L-7-90); shampoos (Siponic L-7-90)
Household applications: (Incropol L-7, L-7-90; Macol LA-790); detergent base (Marlipal MG); dishwashing (Marlipal MG); soaps (Marlipal MG)

Industrial applications: (Incropol L-7, L-7-90; Macol LA-790); emulsion polymerization (Siponic L-7-90); metalworking lubricants (Macol LA-790); textile/leather processing (Incropol L-7, L-7-90; Macol LA-790; Siponic L-7-90)

PROPERTIES:

Form:

Liquid (Incropol L-7, L-7-90; Macol LA-790; Siponic L-7-90)

Color:

Colorless (Macol LA-790)

Gardner 1 max. (Incropol L-7, L-7-90)

Composition:

90% active (Incropol L-7-90; Siponic L-7-90)

100% active (Incropol L-7; Marlipal MG)

Solubility:

Sol. in isopropanol (@ 1%) (Macol LA-790)

Insol. in min. oil (@ 1%) (Macol LA-790)

Sol. in water (@ 1%) (Macol LA-790)

Ionic Nature:

Nonionic (Incropol L-7, L-7-90; Macol LA-790; Marlipal MG; Siponic L-7-90)

Pour Pt.:

5 C (Macol LA-790)

HLB:

11.0 (Macol LA-790)

12.1 (Siponic L-7-90)

Hydroxyl No.:

100 (Macol LA-790)

107–113 (Incropol L-7)

Stability:

Stable (Incropol L-7, L-7-90)

Stable to hydrolysis by strong acids and alkalies (Macol LA-790)

pH:

5.5–7.5 (3%) (Incropol L-7, L-7-90)

Biodegradable: (Incropol L-7, L-7-90)

POE (5) monococoate

SYNONYMS:

PEG-5 cocoate (CTFA)

PEG (5) monococoate

POE (5) monococoate (cont'd.)

STRUCTURE:

$$RC(=O)—(OCH_2CH_2)_nOH$$

where RCO⁻ represents the coconut acid radical and avg. $n = 5$

CAS No.:

61791-29-5 (generic)
RD No.: 977065-64-7

TRADENAME EQUIVALENTS:

Ethofat C/15 [Armak]

CATEGORY:

Dispersant, suspending agent, emulsifier, detergent, wetting agent

APPLICATIONS:

Cosmetic industry preparations: (Ethofat C/15)
Farm products: agricultural oils/sprays (Ethofat C/15)
Industrial applications: metalworking (Ethofat C/15); paint mfg. (Ethofat C/15); petroleum industry (Ethofat C/15); textile/leather processing (Ethofat C/15)

PROPERTIES:

Form:

Clear liquid (Ethofat C/15)

Color:

Gardner 9 max. (Ethofat C/15)

Composition:

97% active min. (Ethofat C/15)

Solubility:

Sol. in acetone (Ethofat C/15)
Sol. in benzene (Ethofat C/15)
Sol. in carbon tetrachloride (Ethofat C/15)
Sol. in dioxane (Ethofat C/15)
Sol. in isopropanol (Ethofat C/15)
Disp. in water (< 70 C) (Ethofat C/15)

Ionic Nature:

Nonionic (Ethofat C/15)

Sp.gr.:

0.997 (Ethofat C/15)

F.P.:

< 10 C (Ethofat C/15)

HLB:

10.6 (Ethofat C/15)

Acid No.:

1.0 max. (Ethofat C/15)

Saponification No.:
120–130 (Ethofat C/15)
Stability:
Fairly stable in mild acid or alkaline sol'ns. (Ethofat C/15)
pH:
6.0–7.5 (10% aq.) (Ethofat C/15)
Surface Tension:
33 dynes/cm (0.1%) (Ethofat C/15)

POE (15) monococoate

SYNONYMS:
PEG-15 cocoate (CTFA)
PEG (15) monococoate
STRUCTURE:

$$RC(=O)—(OCH_2CH_2)_nOH$$

where RCO⁻ represents the coconut acid radical and avg. $n = 15$
CAS No.:
61791-29-5 (generic)
RD No.: 977065-34-1
TRADENAME EQUIVALENTS:
Ethofat C/25 [Armak]
CATEGORY:
Dispersant, suspending agent, emulsifier, detergent, wetting agent
APPLICATIONS:
Cosmetic industry preparations: (Ethofat C/25)
Farm products: agricultural oils/sprays (Ethofat C/25)
Industrial applications: metalworking (Ethofat C/25); paint mfg. (Ethofat C/25); petroleum industry (Ethofat C/25); textile/leather processing (Ethofat C/25)
PROPERTIES:
Form:
Paste (Ethofat C/25)
Color:
Gardner 8 max. (Ethofat C/25)
Composition:
97% active min. (Ethofat C/25)
Ionic Nature:
Nonionic (Ethofat C/25)

POE (15) monococoate (*cont'd.*)

Sp.gr.:
1.059 (Ethofat C/25)
Flash Pt.:
> 400 F (PM) (Ethofat C/25)
Acid No.:
1 max. (Ethofat C/25)
Saponification No.:
60–70 (Ethofat C/25)
Stability:
Fairly stable in mild acid or alkaline sol'ns. (Ethofat C/25)
pH:
7.0–7.5 (10% aq.) (Ethofat C/25)

POE (5) monooleate

SYNONYMS:
PEG-5 oleate (CTFA)
PEG (5) monooleate
STRUCTURE:

$$CH_3(CH_2)_7CH{=}CH(CH_2)_7\overset{O}{\overset{\|}{C}}{-}(OCH_2CH_2)_nOH$$

where avg. $n = 5$
CAS No.:
9004-96-0 (generic)
RD No.: 977065-69-2
TRADENAME EQUIVALENTS:
Emulphor VN-430 [GAF]
Ethofat O/15 [Armak]
Hetoxamate MO5 [Heterene]
CATEGORY:
Emulsifier, dispersant, suspending agent, lubricant, detergent, wetting agent, softener
APPLICATIONS:
Cleansers: hand cleanser (Hetoxamate MO5)
Cosmetic industry preparations: (Emulphor VN-430; Ethofat O/15; Hetoxamate MO5); creams and lotions (Emulphor VN-430; Hetoxamate MO5); hair preparations (Emulphor VN-430)
Degreasers: (Emulphor VN-430)
Farm products: agricultural oils/sprays (Ethofat O/15)
Industrial applications: dyes and pigments (Hetoxamate MO5); lubricating/cutting oils (Emulphor VN-430); metalworking (Ethofat O/15); paint mfg. (Ethofat O/15);

petroleum industry (Ethofat O/15); textile/leather processing (Emulphor VN-430; Ethofat O/15; Hetoxamate MO5)
Industrial cleaners: (Emulphor VN-430); metal processing surfactants (Hetoxamate MO5)

PROPERTIES:

Form:
Liquid (Hetoxamate MO5)
Clear liquid (Emulphor VN-430; Ethofat O/15)

Color:
Dark amber (Emulphor VN-430)
Gardner 9 max. (Ethofat O/15)

Composition:
99% active min. (Ethofat O/15)
100% active (Emulphor VN-430)

Solubility:
Sol. in acetone (< 25 C) (Ethofat O/15)
Sol. in benzene (Ethofat O/15)
Sol. in butyl Cellosolve (Emulphor VN-430)
Sol. cloudy in carbon tetrachloride (Ethofat O/15)
Sol. in dioxane (Ethofat O/15)
Sol. in ethanol (Emulphor VN-430)
Sol. in isopropanol (Ethofat O/15; Hetoxamate MO5)
Sol. in white min. oil (Emulphor VN-430)
Disp. in water (Emulphor VN-430; Ethofat O/15; Hetoxamate MO5)
Sol. in xylene (Emulphor VN-430)

Ionic Nature:
Nonionic (Emulphor VN-430; Ethofat O/15; Hetoxamate MO5)

Sp.gr.:
0.992 (Ethofat O/15)
1.0 (Emulphor VN-430)

F.P.:
< 10 C (Ethofat O/15)

Pour Pt.:
13 C (Emulphor VN-430)

Solidification Pt.:
10 C (Emulphor VN-430)

Flash Pt.:
> 93 C (PMCC) (Emulphor VN-430)
> 400 F (PM) (Ethofat O/15)

Cloud Pt.:
< 10 C (1% sol'n.) (Emulphor VN-430)

HLB:
8.0 (Hetoxamate MO5)

POE (5) monooleate (cont'd.)

8.6 (Ethofat O/15)

Acid No.:

1.0 max. (Ethofat O/15)

2.0 (Hetoxamate MO5)

Saponification No.:

110–120 (Ethofat O/15)

115–125 (Hetoxamate MO5)

Stability:

Fairly stable in mild acid or alkaline sol'ns. (Ethofat O/15)

pH:

6.0–7.5 (10% aq.) (Ethofat O/15)

7.0–7.5 (Emulphor VN-430)

Surface Tension:

32 dynes/cm (0.1% conc.) (Emulphor VN-430)

35 dynes/cm (0.1%) (Ethofat O/15)

POE (10) monooleate

SYNONYMS:

PEG-10 oleate (CTFA)

PEG (10) monooleate

STRUCTURE:

$$CH_3(CH_2)_7CH{=}CH(CH_2)_7C(=O)\text{—}(OCH_2CH_2)_nOH$$

where avg. $n = 10$

CAS No.:

9004-96-0 (generic)

RD No.: 977063-22-1

TRADENAME EQUIVALENTS:

Ethofat O/20 [Armak]

CATEGORY:

Dispersant, suspending agent, emulsifier, detergent, wetting agent

APPLICATIONS:

Cosmetic industry preparations: (Ethofat O/20)

Farm products: agricultural oils/sprays (Ethofat O/20)

Industrial applications: metalworking (Ethofat O/20); paint mfg. (Ethofat O/20); petroleum industry (Ethofat O/20); textile/leather processing (Ethofat O/20)

PROPERTIES:

Form:

Clear liquid (Ethofat O/20)

Color:
Gardner 8 max. (Ethofat O/20)
Composition:
97% active min. (Ethofat O/20)
Solubility:
Sol. in acetone (< 50 C) (Ethofat O/20)
Sol. in benzene (Ethofat O/20)
Sol. in carbon tetrachloride (Ethofat O/20)
Sol. in dioxane (Ethofat O/20)
Sol. in isopropanol (Ethofat O/20)
Sol. in Stoddard solvent (> 25 C) (Ethofat O/20)
Sol. in water (> 75 C) (Ethofat O/20)
Ionic Nature:
Nonionic (Ethofat O/20)
Sp.gr.:
1.03 (Ethofat O/20)
F.P.:
< 10 C (Ethofat O/20)
Flash Pt.:
485 F (PM) (Ethofat O/20)
Acid No.:
1.0 max. (Ethofat O/20)
Saponification No.:
75–85 (Ethofat O/20)
Stability:
Fairly stable in mild acid or alkaline sol'ns. (Ethofat O/20)
pH:
6.0–7.5 (10% aq.) (Ethofat O/20)
Surface Tension:
41 dynes/cm (0.1%) (Ethofat O/20)

POE (5) monostearate

SYNONYMS:
PEG-5 stearate (CTFA)
PEG (5) monostearate
STRUCTURE:

$$CH_3(CH_2)_{16}\overset{\overset{\displaystyle O}{\|}}{C}—(OCH_2CH_2)_nOH$$

where avg. $n = 5$

POE (5) monostearate (cont'd.)

CAS No.:
9004-99-3 (generic)
RD No.: 977065-72-7

TRADENAME EQUIVALENTS:
Ethofat 60/15 [Armak]
Hetoxamate SA-5 [Heterene]
Industrol MS-5 [BASF Wyandotte]

CATEGORY:
Emulsifier, dispersant, detergent, lubricant, softener, wetting agent, suspending agent, surfactant

APPLICATIONS:
Cleansers: hand cleanser (Hetoxamate SA-5)
Cosmetic industry preparations: (Ethofat 60/15; Hetoxamate SA-5); creams and lotions (Hetoxamate SA-5)
Farm products: agricultural oils/sprays (Ethofat 60/15)
Industrial applications: dyes and pigments (Hetoxamate SA-5); metalworking (Ethofat 60/15); paint mfg. (Ethofat 60/15); petroleum industry (Ethofat 60/15); textile/leather processing (Ethofat 60/15; Hetoxamate SA-5); waxes and oils (Hetoxamate SA-5)
Industrial cleaners: metal processing surfactants (Hetoxamate SA-5)

PROPERTIES:

Form:
Soft paste (Ethofat 60/15)
Solid (Hetoxamate SA-5)
Wax (Industrol MS-5)

Color:
Gardner 3 max. (Industrol MS-5)
Gardner 8 max. (Ethofat 60/15)

Composition:
97% active min. (Ethofat 60/15)
100% active (Industrol MS-5)

Solubility:
Sol. in acetone (Ethofat 60/15)
Sol. in benzene (Ethofat 60/15)
Sol. in carbon tetrachloride (Ethofat 60/15)
Sol. in dioxane (Ethofat 60/15)
Sol. in isopropanol (Ethofat 60/15; Hetoxamate SA-5)
Disp. in water (Ethofat 60/15); disp. hot (Hetoxamate SA-5); insol. (Industrol MS-5)

Ionic Nature:
Nonionic (Ethofat 60/15; Industrol MS-5)

M.W.:
500 (Industrol MS-5)

Sp.gr.:
1.013 (Ethofat 60/15)
M.P.:
28 C (Industrol MS-5)
Flash Pt.:
475 F (PM) (Ethofat 60/15)
HLB:
8.0 (Hetoxamate SA-5)
9.1 (Industrol MS-5)
Acid No.:
1.0 max. (Ethofat 60/15)
2.0 (Hetoxamate SA-5)
2.0 max. (Industrol MS-5)
Saponification No.:
110–120 (Ethofat 60/15)
120–130 (Hetoxamate SA-5)
Stability:
Fairly stable in mild acid or alkaline sol'ns. (Ethofat 60/15)
pH:
5.5–7.0 (5% aq.) (Industrol MS-5)
6.0–7.5 (10% aq.) (Ethofat 60/15)
Surface Tension:
39 dynes/cm (0.1%) (Ethofat 60/15)
STD. PKGS.:
55-gal (420 lb net) steel drums (Industrol MS-5)

POE (10) monostearate

SYNONYMS:
PEG-10 stearate (CTFA)
PEG (10) monostearate
STRUCTURE:

$$CH_3(CH_2)_{16}C(=O)—(OCH_2CH_2)_nOH$$

where avg. $n = 10$
CAS No.:
9004-99-3 (generic)
RD No.: 977065-33-0
TRADENAME EQUIVALENTS:
Ethofat 60/20 [Armak]
Nikkol MYS-10 [Nikko]

POE (10) monostearate (cont'd.)

CATEGORY:

Emulsifier, dispersant, solubilizer, detergent, wetting agent, suspending agent

APPLICATIONS:

Cosmetic industry preparations: (Ethofat 60/20)
Farm products: agricultural oils/sprays (Ethofat 60/20)
Industrial applications: metalworking (Ethofat 60/20); paint mfg. (Ethofat 60/20); petroleum industry (Ethofat 60/20); textile/leather processing (Ethofat 60/20)

PROPERTIES:

Form:

Soft paste (Ethofat 60/20)
Solid (Nikkol MYS-10)

Color:

Gardner 11 max. (Ethofat 60/20)

Composition:

97% active min. (Ethofat 60/20)
100% conc. (Nikkol MYS-10)

Solubility:

Sol. in acetone (< 80 C) (Ethofat 60/20)
Sol. in benzene (Ethofat 60/20)
Sol. in carbon tetrachloride (< 75 C) (Ethofat 60/20)
Sol. in dioxane (< 75 C) (Ethofat 60/20)
Sol. in isopropanol (< 50 C) (Ethofat 60/20)
Disp. in water (Ethofat 60/20)

Ionic Nature:

Nonionic (Ethofat 60/20; Nikkol MYS-10)

Sp.gr.:

1.024 (Ethofat 60/20)

F.P.:

35 C (Ethofat 60/20)

HLB:

11.0 (Nikkol MYS-10)

Acid No.:

1.0 max. (Ethofat 60/20)

Saponification No.:

70–80 (Ethofat 60/20)

Stability:

Fairly stable in mild acid or alkaline sol'ns. (Ethofat 60/20)

pH:

6.0–7.5 (10% aq.) (Ethofat 60/20)

Surface Tension:

36 dynes/cm (0.1%) (Ethofat 60/20)

STD. PKGS.:

55-gal unlined steel drums (Ethofat 60/20)

SYNONYMS:
PEG-100 stearate (CTFA)
PEG (100) monostearate

STRUCTURE:

$$CH_3(CH_2)_{16}C(=O)—(OCH_2CH_2)_nOH$$

where avg. $n = 100$

CAS No.:
9004-99-3 (generic); RD No.: 977055-47-2

TRADENAME EQUIVALENTS:
Emerest 2717 [Emery]
Lipopeg 100-S [Lipo]
Mapeg S-100 [Mazer]
Myrj 59 [ICI]
RS-55-100 [Hefti Ltd.]
Simulsol M59 [Seppic]

CATEGORY:
Dispersant, emulsifier, spreading agent, lubricant, binder

APPLICATIONS:
Bath products: bath oils (Lipopeg 100-S)
Cosmetic industry preparations: (Myrj 59; RS-55-100); creams and lotions (Emerest 2717; Lipopeg 100-S)
Pharmaceutical applications: (Myrj 59; RS-55-100)

PROPERTIES:

Form:
Solid (Emerest 2717; Myrj 59; RS-55-100; Simulsol M59)
Beads (Lipopeg 100-S)
Flake (Lipopeg 100-S; Mapeg S-100)

Color:
Tan (Lipopeg 100-S)

Composition:
100% conc. (Emerest 2717; Myrj 59; RS-55-100; Simulsol M59)

Ionic Nature:
Nonionic (Emerest 2717; Myrj 59; RS-55-100; Simulsol M59)

HLB:
18.5 (RS-55-100)
18.8 (Emerest 2717; Myrj 59; Simulsol M59)
18.8 ± 1 (Lipopeg 100-S)

Acid No.:
1.0 max. (Lipopeg 100-S)

Saponification No.:
9–20 (Lipopeg 100-S)

POE (150) monostearate

SYNONYMS:

PEG-150 stearate (CTFA)
PEG 6000 monostearate

STRUCTURE:

$$CH_3(CH_2)_{16}C(=O)—(OCH_2CH_2)_nOH$$

where avg. $n = 150$

CAS No.:

9004-99-3 (generic)
RD No.: 977055-48-3

TRADENAME EQUIVALENTS:

Kessco PEG 6000 Monostearate [Armak]
Mapeg 6000 MS [Mazer]

CATEGORY:

Dispersant, surfactant, thickener, emulsifier, solubilizer, defoamer, antistat, intermediate

APPLICATIONS:

Bath products: bath oils (Kessco PEG 6000 Monostearate)
Cosmetic industry preparations: (Kessco PEG 6000 Monostearate; Mapeg 6000 MS); cream rinses (Kessco PEG 6000 Monostearate); hair preparations (Kessco PEG 6000 Monostearate); perfumery (Kessco PEG 6000 Monostearate); shampoos (Kessco PEG 6000 Monostearate)
Farm products: (Kessco PEG 6000 Monostearate)
Food applications: (Kessco PEG 6000 Monostearate)
Industrial applications: metalworking lubricants (Mapeg 6000 MS); plastics (Kessco PEG 6000 Monostearate); textile/leather processing (Mapeg 6000 MS)
Pharmaceutical applications: (Kessco PEG 6000 Monostearate; Mapeg 6000 MS)

PROPERTIES:

Form:

Flake (Mapeg 6000 MS)
Wax (Kessco PEG 6000 Monostearate)
Waxy solid (Mapeg 6000 MS)

Color:

Cream (Kessco PEG 6000 Monostearate)
Light yellow (Mapeg 6000 MS)

Solubility:

Sol. in acetone (Kessco PEG 6000 Monostearate)
Sol. in carbon tetrachloride (Kessco PEG 6000 Monostearate)
Sol. in ethyl acetate (Kessco PEG 6000 Monostearate)
Sol. in isopropanol (Kessco PEG 6000 Monostearate; Mapeg 6000 MS)
Insol. in min. oils (Mapeg 6000 MS)
Sol. in 5% Na_2SO_4 (Kessco PEG 6000 Monostearate)

Sol. in propylene glycol (Kessco PEG 6000 Monostearate); insol. (Mapeg 6000 MS)
Insol. in soybean oils (Mapeg 6000 MS)
Sol. in toluol (Kessco PEG 6000 Monostearate; Mapeg 6000 MS)
Sol. in water (Kessco PEG 6000 Monostearate; Mapeg 6000 MS)

Ionic Nature:
Nonionic (Kessco PEG 6000 Monostearate)

Sp.gr.:
1.080 (65 C) (Kessco PEG 6000 Monostearate)

M.P.:
60 C (Mapeg 6000 MS)
61 C (Kessco PEG 6000 Monostearate)

Flash Pt.:
480 F (COC) (Kessco PEG 6000 Monostearate)

Fire Pt.:
525 F (Kessco PEG 6000 Monostearate)

HLB:
18.8 (Kessco PEG 6000 Monostearate; Mapeg 6000 MS)

Acid No.:
5.0 max. (Kessco PEG 6000 Monostearate; Mapeg 6000 MS)

Iodine No.:
0.1 max. (Kessco PEG 6000 Monostearate; Mapeg 6000 MS)

Saponification No.:
7–13 (Kessco PEG 6000 Monostearate)
7–14 (Mapeg 6000 MS)

Stability:
Stable (Kessco PEG 6000 Monostearate)

pH:
5.0 (3% disp.) (Kessco PEG 6000 Monostearate)

POE (1) nonyl phenyl ether

SYNONYMS:
Ethanol, 2-(nonylphenoxy)-
Ethylene glycol nonyl phenyl ether
Nonoxynol-1 (CTFA)
2-(Nonylphenoxy) ethanol
PEG-1 nonyl phenyl ether
POE (1) nonyl phenol

STRUCTURE:
$C_9H_{19}C_6H_4OCH_2OH$

POE (1) nonyl phenyl ether (cont'd.)

CAS No.:
9016-45-9 (generic); 26027-38-3 (generic); 27986-36-3; 37205-87-1 (generic)

TRADENAME EQUIVALENTS:
Alkasurf NP-1 [Alkaril]
Cedepal CO-210 [Domtar] (1.5 EO)
Ethylan NP1 [Lankro Chem. Ltd.]
Iconol NP-1.5 [BASF Wyandotte] (1.5 EO)
Igepal CO-210 [GAF] (1.5 EO)
Merpoxen NO15 [Kempen] (1.5 EO)
Norfox NP-1 [Norman, Fox & Co.] (1.5 EO)
Peganol NP1.5 [Borg-Warner] (1.5 EO)
Rexol 25/1 [Hart Chem. Ltd.]
Surfonic N-10 [Jefferson]
T-Det N-1.5 [Thompson-Hayward] (1.5 EO)
Triton N-17 [Rohm & Haas] (1.5 EO)
Trycol NP-1 [Emery]

CATEGORY:
Emulsifier, coemulsifier, wetting agent, detergent, dispersant, penetrant, solubilizer, retardant, defoamer, foam stabilizer, stabilizer, intermediate, surfactant

APPLICATIONS:
Cosmetic industry preparations: (Surfonic N-10); hair color preparations (Alkasurf NP-1)
Food applications: indirect food additives (Surfonic N-10)
Household detergents: (Igepal CO-210; Peganol NP1.5; Rexol 25/1)
Industrial applications: industrial processing (Peganol NP1.5); lubricating/cutting oils (Peganol NP1.5; Surfonic N-10); petroleum industry (Surfonic N-10); petroleum oils (Alkasurf NP-1; Igepal CO-210; T-Det N-1.5); rubber (Surfonic N-10); textile/leather processing (Surfonic N-10)

PROPERTIES:

Form:
Liquid (Alkasurf NP-1; Cedepal CO-210; Igepal CO-210; Norfox NP-1; Rexol 25/1; Surfonic N-10; T-Det N-1.5; Trycol NP-1)
Clear liquid (Ethylan NP1; Iconol NP-1.5; Triton N-17)
Clear viscous liquid (Peganol NP1.5)
Viscous liquid (Merpoxen NO15)

Color:
Water-white (Merpoxen NO15)
Straw (Ethylan NP1)
Pale yellow (T-Det N-1.5)
Yellow (Igepal CO-210)
APHA 250 (Triton N-17)
Gardner 2 max. (Iconol NP-1.5; Peganol NP1.5)

Odor:
Mild (Triton N-17)
Aromatic (Igepal CO-210)

Composition:
99% active min. (Alkasurf NP-1; Iconol NP-1.5)
99–100% active (Triton N-17)
99.5% active min. (Peganol NP1.5; T-Det N-1.5)
100% active (Ethylan NP1; Igepal CO-210; Merpoxen NO15; Rexol 25/1; Surfonic N-10; Trycol NP-1)
100% conc. (Cedepal CO-210; Norfox NP-1)

Solubility:
Sol. in acetone (Surfonic N-10)
Sol. in aromatic solvent (@ 10%) (Alkasurf NP-1)
Sol. in butyl Cellosolve (Igepal CO-210)
Sol. in carbon tetrachloride (Surfonic N-10)
Sol. in corn oil (Igepal CO-210)
Sol. in deodorized kerosene (Igepal CO-210)
Sol. in dibutyl phthalate (Igepal CO-210)
Sol. in ethanol (Igepal CO-210)
Sol. in ethylene dichloride (Igepal CO-210)
Sol. in methanol (Surfonic N-10)
Sol. in min. oil (Surfonic N-10); (@ 10%) (Alkasurf NP-1); (@ 5%) (Trycol NP-1); sol. in white min. oil (Peganol NP1.5); sol. in low-viscosity white min. oil (Igepal CO-210)
Sol. in min. spirits (@ 10%) (Alkasurf NP-1)
Sol. in heavy aromatic naphtha (Igepal CO-210)
Sol. in oil (Cedepal CO-210; Ethylan NP1; Iconol NP-1.5; Norfox NP-1; T-Det N-1.5)
Sol. in perchloroethylene (Igepal CO-210; Peganol NP1.5); (@ 10%) (Alkasurf NP-1); (@ 5%) (Trycol NP-1)
Sol. in Stoddard solvent (Igepal CO-210); (@ 5%) (Trycol NP-1)
Negligible sol. in water (Triton N-17); insol. in water (Iconol NP-1.5; Peganol NP1.5); (@ 10%) (Alkasurf NP-1)
Sol. in xylene (Igepal CO-210; Peganol NP1.5; Surfonic N-10)

Ionic Nature:
Nonionic (Cedepal CO-210; Ethylan NP1; Igepal CO-210; Merpoxen NO15; Norfox NP-1; Peganol NP1.5; Surfonic N-10; T-Det N-1.5; Triton N-17; Trycol NP-1)

M.W.:
264 (Surfonic N-10)
281 (Iconol NP-1.5)
286 (Triton N-17)

Sp.gr.:
0.98 (20/20 C) (Surfonic N-10)
0.984 (Peganol NP1.5; Triton N-17)

POE (1) nonyl phenyl ether (cont'd.)

0.987 (20 C) (Ethylan NP1)
0.99 (Iconol NP-1.5; Igepal CO-210); (68 F) (T-Det N-1.5)

Density:

0.99 g/ml (Alkasurf NP-1)
8.1 lb/gal (20 C) (Surfonic N-10)
8.2 lb/gal (Triton N-17); (68 F) (T-Det N-1.5)
8.3 lb/gal (Trycol NP-1)

Visc.:

8 cs (100 C) (Peganol NP1.5)
300–400 cps (Igepal CO-210)
350 cps (Iconol NP-1.5)
460 cs (Trycol NP-1)
500 cps (Brookfield, 12 rpm) (Triton N-17)
650 cs (20 C) (Ethylan NP1)
675 SUS (100 F) (Surfonic N-10)

Pour Pt.:

–19 C (Peganol NP1.5)
–11 C (Iconol NP-1.5)
–9 C (Triton N-17)
< 0 C (Ethylan NP1)
–3 ± 2 F (Igepal CO-210)
< 0 F (T-Det N-1.5)

Solidification Pt.:

–8 ± 2 F (Igepal CO-210)

Flash Pt.:

> 149 C (TOC) (Triton N-17)
> 200 C (Peganol NP1.5)
> 200 F (PMCC) (Igepal CO-210)
300 F (COC) (Ethylan NP1)
340 F (PMCC) (T-Det N-1.5)
355 F (OC) (Surfonic N-10)

Cloud Pt.:

Insol. (Peganol NP1.5)
< 25 C (Trycol NP-1)

HLB:

1.5 (T-Det N-1.5)
3.4 (Surfonic N-10)
4.0–5.0 (Norfox NP-1)
4.5 (Ethylan NP1)
4.6 (Alkasurf NP-1; Cedepal CO-210; Iconol NP-1.5; Igepal CO-210; Peganol NP1.5; Rexol 25/1; Triton N-17; Trycol NP-1)

Hydroxyl No.:

196 (Peganol NP1.5)

Stability:
Stable to acids, bases, and salts (Surfonic N-10; T-Det N-1.5)
Stable to hydrolysis by acids and alkalis (Trycol NP-1)
Stable to acids, alkalis, dilute sol'ns. of many oxidizing and reducing agents (Igepal CO-210)
Stable to electrolytes, hard water, and extremes of pH and temperature (Ethylan NP1)
Ref. Index:
1.5090 (20 C) (Surfonic N-10)
pH:
5.0–7.0 (1% aq.) (T-Det N-1.5)
5.0–8.0 (5% DW) (Alkasurf NP-1)
6.0–7.5 (5% aq.) (Iconol NP-1.5); (5% in dist. water) (Peganol NP1.5)
6.0–8.0 (5% aq.) (Triton N-17); (1% aq.) (Ethylan NP1)
TOXICITY/HANDLING:
Moderately irritating to eyes; slightly irritating to skin (Triton N-17)
Nonhazardous; however, considered to possess low acute oral and skin penetration toxicity (T-Det N-1.5)
Avoid prolonged contact with conc. form; spillages may be slippery (Ethylan NP1)
STORAGE/HANDLING:
Avoid contact with strong oxidizing and reducing agents—potentially explosive (Ethylan NP1; Igepal CO-210; T-Det N-1.5; Triton N-17)
Store at 16–52 C (Triton N-17)
STD. PKGS.:
200-kg net iron drums (Merpoxen NO15)
200-kg net mild-steel drums or bulk (Ethylan NP1)
55-gal (450 lb net) steel drums (Iconol NP-1.5)
440 lb drums (T-Det N-1.5)

POE (2) nonyl phenyl ether

SYNONYMS:
Nonoxynol-2 (CTFA)
Nonyl phenol ethoxylate (2 moles EO)
PEG-2 nonyl phenyl ether
PEG 100 nonyl phenyl ether
STRUCTURE:
$C_9H_{19}C_6H_4(OCH_2CH_2)_nOH$
where avg. $n = 2$
CAS No.:
9016-45-9 (generic); 26027-38-3 (generic); 27176-93-8; 37205-87-1 (generic)
RD No.: 977057-30-9

POE (2) nonyl phenyl ether *(cont'd.)*

TRADENAME EQUIVALENTS:
Teric N2 [ICI Australia Ltd.]
CATEGORY:
Wetting agent, dispersant, emulsifier, coemulsifier, surfactant, defoamer, intermediate, foam stabilizer
APPLICATIONS:
Farm products: herbicides (Teric N2); insecticides/pesticides (Teric N2)
Industrial applications: coatings (Teric N2)
INDUSTRIAL CLEANERS: SOLVENT CLeaners (Teric N2)
PROPERTIES:
Form:
Liquid (ITeric N2)
Color:
Hazen 100 (Teric N2)
Composition:
100% active (Teric N2)
Solubility:
Sol. in benzene (Teric N2)
Sol. in ethanol (Teric N2)
Sol. in ethyl acetate (Teric N2)
Sol. in kerosene (Teric N2)
Sol. in min. oil (Teric N2)
Sol. in olein (Teric N2)
Sol. in paraffin oil (Teric N2)
Sol. in perchloroethylene (Teric N2)
Sol. in veg. oil (Teric N2)
Disp. in water (Teric N2)
Sp.gr.:
1.001 (20 C) (Teric N2)
Visc.:
620 cps (Teric N2)
M.P.:
< 0 C (Teric N2)
HLB:
5.7 (Teric N2)
Stability:
Stable in hard or saline waters and in reasonable concs. of acids and alkalis (Teric N2)
TOXICITY/HANDLING:
May cause skin and eye irritation; spillages are slippery (Teric N2)

POE (4) nonyl phenyl ether

SYNONYMS:

Ethanol, 2- [2- [2- [2- (4-nonylphenoxy) ethoxy] ethoxy] ethoxy]-
Nonoxynol-4 (CTFA)
Nonyl phenol ethoxylate (4 moles EO)
Nonyl phenol 4 polyglycol ether
Nonylphenoxy polyethoxy ethanol (4 moles EO)
Nonylphenoxy poly (ethyleneoxy) ethanol (4 moles EO)
PEG-4 nonyl phenyl ether
PEG 200 nonyl phenyl ether
POE (4) nonyl phenol

STRUCTURE:

$C_9H_{19}C_6H_4(OCH_2CH_2)_nOH$
where avg. $n = 4$

CAS No.:

9016-45-9 (generic); 26027-38-3 (generic); 27176-93-8; 37205-87-1 (generic)
RD No.: 977057-30-9

TRADENAME EQUIVALENTS:

Alkasurf NP-4 [Alkaril]
Arkopal N-040 [Amer. Hoechst]
Carsonon N-4 [Carson]
Cedepal CO-430 [Domtar]
Chemax NP-4 [Chemax]
Chemcol NPE-40 [Chemform]
Conco NI-43 [Continental Chem.]
Hetoxide NP-4 [Heterene]
Hostapal N-040 [Amer. Hoechst]
Hyonic NP-40 [Diamond Shamrock]
Iconol NP-4 [BASF Wyandotte]
Igepal CO-430 [GAF]
Macol NP-4 [Mazer]
Makon 4 [Stepan]
Merpoxen NO 40 [Elektrochemische Fabrik Kempen]
Norfox NP-4 [Norman, Fox & Co.]
NP-55-40 [Hefti Ltd.]
Nutrol 622 [Clough Chem. Ltd.]
Peganol NP4 [Borg-Warner]
Polystep F-1 [Stepan]
Renex 647 [Atlas Chem. Ind. NV]
Rewopal HV4 [Rewo Chemische Werke GmbH]
Serdox NNP4 [Servo B.V.]
Siponic NP4 [Alcolac]
Sterox ND [Monsanto]
Surfonic N-40 [Jefferson]

POE (4) nonyl phenyl ether (cont'd.)

TRADENAME EQUIVALENTS *(cont'd.)*:

Synperonic NP4 [ICI Petrochemicals Div.]
T-Det N-4 [Thompson-Hayward]
Tergitol NP-4, NP-14 [Union Carbide]
Teric N4 [ICI Australia Ltd.]
Triton N-42 [Rohm & Haas]
Trycol NP-4 [Emery]

CATEGORY:

Emulsifier, coemulsifier, surfactant, dispersant, intermediate, detergent, wetting agent, stabilizer, foaming agent, penetrant, emollient, plasticizer, antistat, corrosion inhibitor, solubilizer, coupling agent

APPLICATIONS:

Cleansers: waterless hand cleanser (Hyonic NP-40)

Cosmetic industry preparations: (Alkasurf NP-4; Conco NI-43; Hetoxide NP-4; Makon 4; Surfonic N-40); creams and lotions (Hetoxide NP-4); perfumery (Hetoxide NP-4; Makon 4); shampoos (Makon 4)

Farm products: (Chemax NP-4; Chemcol NPE-40); agricultural oils/sprays (Hyonic NP-40; Makon 4; Sterox ND); herbicides (Serdox NNP4; T-Det N-4; Teric N4); insecticides/pesticides (Conco NI-43; NP-55-40; Serdox NNP4; T-Det N-4; Teric N4)

Food applications: indirect food applications (Hyonic NP-40; Surfonic N-40)

Household detergents: (Alkasurf NP-4; Carsonon N-4; Chemcol NPE-40; Conco NI-43; Hetoxide NP-4; T-Det N-4); light-duty cleaners (Nutrol 622)

Industrial applications: (Alkasurf NP-4; Arkopal N-040; Hostapal N-040); coatings (Teric N4); dyes and pigments (Alkasurf NP-4; Hetoxide NP-4; Makon 4; NP-55-40; Polystep F-1; Tergitol NP-14); industrial processing (Chemax NP-4); latex emulsions (Igepal CO-430; Peganol NP4; Sterox ND); lubricating/cutting oils (Alkasurf NP-4; Makon 4; Norfox NP-4; Peganol NP4; Sterox ND; Surfonic N-40); metalworking (Hetoxide NP-4); paint mfg. (T-Det N-4); paper mfg. (Makon 4); petroleum industry (Makon 4; Surfonic N-40); petroleum oils (Carsonon N-4; Chemcol NPE-40; Conco NI-43; Hyonic NP-40; Igepal CO-430; Sterox ND; T-Det N-4; Trycol NP-4); plastics (Igepal CO-430; Peganol NP4; Sterox ND); plastisols (Hyonic NP-40); polishes and waxes (Conco NI-43); polymers/polymerization (Conco NI-43); rubber (Surfonic N-40); silicone products (Hyonic NP-40); textile/leather processing (Chemcol NPE-40; Conco NI-43; Makon 4; Surfonic N-40; T-Det N-4); waxes (Alkasurf NP-4)

Industrial cleaners: (Conco NI-43; Makon 4; Surfonic N-40; Tergitol NP-4; Teric N4); bottle cleaners (Conco NI-43); drycleaning compositions (Carsonon N-4; Tergitol NP-4); metal processing surfactants (Hyonic NP-40; Makon 4; T-Det N-4); solvent cleaners (Hyonic NP-40; NP-55-40; Sterox ND; T-Det N-4; Teric N4); textile/leather cleaning (Hetoxide NP-4; Makon 4; Sterox ND; Tergitol NP-14)

PROPERTIES:

Form:

Liquid (Alkasurf NP-4; Arkopal N-040; Carsonon N-4; Cedepal CO-430; Chemax NP-4; Chemcol NPE-40; Conco NI-43; Hostapal N-040; Iconol NP-4; Igepal CO-430; Macol NP-4; Makon 4; Norfox NP-4; NP-55-40; Nutrol 622; Renex 647; Serdox NNP4; Siponic NP4; Surfonic N-40; Synperonic NP4; Teric N4; Tergitol NP-14; Trycol NP-4); (@ 30 C) (Hetoxide NP-4)
Clear liquid (Hyonic NP-40; Polystep F-1; Tergitol NP-4)
Clear viscous liquid (Peganol NP4; Sterox ND; Triton N-42)
Viscous liquid (Merpoxen NO 40)
Med. visc. liquid (Rewopal HV4)
Oily liquid (T-Det N-4)

Color:

Colorless (Tergitol NP-4)
Colorless to very slightly yellow (Triton N-42)
Water-white (Merpoxen NO 40)
Light/pale (Alkasurf NP-4; Carsonon N-4; Polystep F-1)
Light straw (Makon 4)
Pale yellow (Chemcol NPE-40; Igepal CO-430; T-Det N-4)
Yellow (Sterox ND)
APHA 100 max. (Iconol NP-4; Peganol NP4)
Gardner 1 (Trycol NP-4)
Hazen 100 (Teric N4)
Hazen 150 max. (Synperonic NP4)
Pt-Co 100 max. (Tergitol NP-14)

Odor:

Low (Alkasurf NP-4)
Aromatic (Igepal CO-430)
Mild (Sterox ND; Triton N-42)
Mild, aromatic (T-Det N-4)
Characteristic (Tergitol NP-4)

Composition:

> 99% active (Hyonic NP-40; Nutrol 622)
99% active min. (Synperonic NP4)
99.5% active (Sterox ND)
99.5% active min. (T-Det N-4)
100% active (Alkasurf NP-4; Carsonon N-4; Chemcol NPE-40; Conco NI-43; Igepal CO-430; Makon 4; Merpoxen NO 40; Polystep F-1; Surfonic N-40; Tergitol NP-4; Teric N4; Triton N-42; Trycol NP-4)
100% conc. (Arkopal N-040; Cedepal CO-430; Hostapal N-040; Macol NP-4; Norfox NP-4; NP-55-40; Renex 647; Rewopal HV4; Serdox NNP4; Siponic NP4; Tergitol NP-14)

POE (4) nonyl phenyl ether (cont'd.)

Solubility:

Sol. in acetone (Surfonic N-40)
Sol. in alcohols (Carsonon N-4; Synperonic NP4)
Sol. in aliphatic hydrocarbons (Tergitol NP-4)
Sol. in aromatic solvents (Carsonon N-4)
Sol. in benzene (Teric N4)
Sol. in butyl acetate (Tergitol NP-4)
Sol. in butyl Cellosolve (Igepal CO-430; T-Det N-4; Tergitol NP-4)
Sol. in carbon tetrachloride (Surfonic N-40; T-Det N-4)
Sol. in chlorinated hydrocarbons (Carsonon N-4)
Sol. in corn oil (Igepal CO-430; T-Det N-4; Tergitol NP-4)
Sol. in dibutyl phthalate (Igepal CO-430)
Sol. in diesel fuel (Tergitol NP-4)
Sol. in ethanol (Igepal CO-430; T-Det N-4; Teric N4)
Sol. in ethyl acetate (Teric N4)
Sol. in ethylene dichloride (Igepal CO-430)
Disp. in ethylene glycol (Tergitol NP-4)
Sol. in glycol ethers (Synperonic NP4)
Sol. in isopropanol (Hetoxide NP-4); anhyd. isopropanol (Tergitol NP-4)
Sol. in kerosene (Carsonon N-4; Synperonic NP4; T-Det N-4; Tergitol NP-4; Teric N4); sol. in deodorized kerosene (Igepal CO-430)
Sol. in methanol (Surfonic N-40)
Sol. in min. oil (Hetoxide NP-4; Synperonic NP4; T-Det N-4; Teric N4); sol./disp. (@ 5%) (Trycol NP-4)
Sol. in min. spirits (T-Det N-4)
Sol. in heavy aromatic naphtha (Igepal CO-430)
Sol. in nonpolar solvents (Sterox ND)
Sol. in oil (Alkasurf NP-4; Carsonon N-4; Cedepal CO-430; Chemcol NPE-40; Iconol NP-4; Makon 4; Norfox NP-4; Polystep F-1; Siponic NP4)
Sol. in olein (Teric N4)
Sol. in perchloroethylene (Igepal CO-430; Peganol NP4; T-Det N-4; Teric N4); (@ 5%) (Trycol NP-4)
Sol. in Stoddard solvent (Carsonon N-4; Igepal CO-430; Surfonic N-40; T-Det N-4; Tergitol NP-4); (@ 5%) (Trycol NP-4)
Sol. in toluene (T-Det N-4; Tergitol NP-4)
Sol. in veg. oil (Teric N4)
Disp. in water (Teric N4; Triton N-42); misc. with water (Sterox ND); insol. (Alkasurf NP-4; Hyonic NP-40; Iconol NP-4; Peganol NP4)
Sol. in white min. oil (Tergitol NP-4); insol. (Peganol NP4)
Sol. in xylene (Igepal CO-430; Peganol NP4; Surfonic N-40)

Ionic Nature:

Nonionic (Alkasurf NP-4; Arkopal N-040; Carsonon N-4; Cedepal CO-430; Chemcol NPE-40; Conco NI-43; Hetoxide NP-4; Hostapal N-040; Hyonic NP-40; Iconol

NP-4; Igepal CO-430; Macol NP-4; Makon 4; Merpoxen NO 40; Norfox NP-4; NP-55-40; Nutrol 622; Peganol NP4; Renex 647; Rewopal HV4; Serdox NNP4; Siponic NP4; Sterox ND; Surfonic N-40; Synperonic NP4; T-Det N-4; Tergitol NP-4, NP-14; Teric N4; Triton N-42; Trycol NP-4)

M.W.:

391 (Iconol NP-4)
396 (Sterox ND; Surfonic N-40; Tergitol NP-4, NP-14)
405 (Triton N-42)

Sp.gr.:

1.02 (Alkasurf NP-4; Carsonon N-4; Iconol NP-4; Igepal CO-430; T-Det N-4)
1.02–1.04 (Conco NI-43)
1.0216 (Sterox ND)
1.023 (Peganol NP4); (20 C) (Teric N4)
1.026 (20/20 C) (Surfonic N-40)
1.031 (20/20 C) (Tergitol NP-4, NP-14)
1.068 (Triton N-42)

Density:

1.02 g/ml (Hyonic NP-40)
1.022 g/ml (20 C) (Synperonic NP4)
8.5 lb/gal (Chemcol NPE-40; Makon 4; T-Det N-4; Triton N-42; Trycol NP-4); (20 C) (Surfonic N-40)
8.57 lb/gal (20 C) (Tergitol NP-4, NP-14)
8.6 lb/gal (Carsonon N-4)

Visc.:

175–250 cps (Carsonon N-4)
250 cps (Brookfield, 12 rpm) (Triton N-42)
250–325 cps (Igepal CO-430)
252 cps (Sterox ND)
260 cps (Makon 4)
350 cps (Iconol NP-4)
370 cps (20 C) (Teric N4)
400 cps (20 C) (Synperonic NP4)
9 cs (100 C) (Peganol NP4)
300 cs (Trycol NP-4)
445 cs (20 C) (Tergitol NP-4)
448 cs (20 C) (Tergitol NP-14)
52 SUS (210 F) (T-Det N-4)
445 SUS (100 F) (Surfonic N-40)

F.P.:

–40 C (Tergitol NP-4)
< 0 C (Synperonic NP4)

B.P.:

> 300 C (Tergitol NP-4)

POE (4) nonyl phenyl ether (cont'd.)

M.P.:
< 0 C (Teric N4)

Pour Pt.:
–27 C (Iconol NP-4)
–26 C (Triton N-42)
–25 C (Peganol NP4)
–20 C (Makon 4)
< 0 C (Chemcol NPE-40; Synperonic NP4)
26 C (Hyonic NP-40)
–24 F (Sterox ND)
–15 F (Chemax NP-4)
–15 ± 2 F (Igepal CO-430)
–5 ± 2 F (Carsonon N-4)
< 0 F (T-Det N-4)

Solidification Pt.:
–40 C (Tergitol NP-14)
–28.9 C (Makon 4)
–20 ± 2 F (Carsonon N-4; Igepal CO-430)

Flash Pt.:
149 C (TOC) (Triton N-42)
> 200 C (Peganol NP4)
> 200 F (PMCC) (Igepal CO-430)
> 400 F (TOC) (Chemcol NPE-40); (OC) (T-Det N-4)
435 F (Sterox ND); (OC) (Surfonic N-40)
480 F (COC) (Tergitol NP-4, NP-14)
500–600 F (Carsonon N-4)

Fire Pt.:
500–600 F (Carsonon N-4)

Cloud Pt.:
0 C (0.5% aq.) (Tergitol NP-4); (1% aq.) (Tergitol NP-14)
< 25 C (Trycol NP-4)

HLB:
7.6 (Hetoxide NP-4)
8.1 (Siponic NP4)
8.8 (Cedepal CO-430; Chemcol NPE-40; Igepal CO-430; Macol NP-4; Nutrol 622; Peganol NP4)
8.9 (Iconol NP-4; Sterox ND; Surfonic N-40; Synperonic NP4; T-Det N-4; Tergitol NP-4, NP-14; Teric N4; Trycol NP-4)
9.0 (Alkasurf NP-4; Hyonic NP-40; Norfox NP-4; NP-55-40; Serdox NNP4)
9.1 (Triton N-42)

Hydroxyl No.:
133–150 (Sterox ND)
135–140 (Hetoxide NP-4)

141 (Synperonic NP4)
142 (Peganol NP4; Tergitol NP-14)

Stability:

Completely stable in acid or alkaline media (Alkasurf NP-4)
Stable to hydrolysis by acids and alkalis (Trycol NP-4)
Stable to acids, alkalis, salts (Surfonic N-40; T-Det N-4)
Stable to acids, alkalis, hard water, and foam (Conco NI-43)
Stable to acids, alkalis, dilute sol'ns. of many oxidizing and reducing agents (Igepal CO-430)
Good in hard or saline waters and in reasonable concs. of acids and alkalis (Teric N4)
Chemically stable in the presence of ionic materials (Tergitol NP-4)
Good (Sterox ND)

pH:

5.0–8.0 (10%) (Tergitol NP-14); (10% aq. mix.) (Tergitol NP-4)
6.0–7.5 (1%) (Sterox ND); (5% aq.) (Iconol NP-4; Peganol NP4)
6.0–8.0 (1% aq.) (Synperonic NP4; Teric N4); (5% in 25% IPA) (Chemcol NPE-40)
7.0 (T-Det N-4); (1% aq. sol'n.) (Hyonic NP-40)
7.5 (5% in 50% alcohol) (Triton N-42)
7.7 (1% sol'n.) (Makon 4)

Surface Tension:

27.5 dynes/cm (0.1%) (Surfonic N-40)
29.0 dynes/cm (0.01% aq.) (Triton N-42)
29.7 dynes/cm (Sterox ND)

Biodegradable: (Serdox NNP4); fair biodegradability (Tergitol NP-14)

TOXICITY/HANDLING:

Causes eye burn, skin irritation (Sterox ND)
Eye irritant (Tergitol NP-14)
Eye irritant; highly toxic to aquatic life; burning can produce carbon monoxide and/or carbon dioxide (Tergitol NP-4)
May cause skin and eye irritation; spillages are slippery (Teric N4)
Substantially irritating to eyes; severely irritating to skin (Triton N-42)
Eye irritant; skin irritant on prolonged contact with concentrates (Synperonic NP4)

STORAGE/HANDLING:

Contact with conc. oxidizing or reducing agents may be explosive (Igepal CO-430; T-Det N-4)
Avoid strong oxidizing and reducing agents, excessive heat (Triton N-42)
Store at temps. below 50 C (Tergitol NP-4)

STD. PKGS.:

Steel drums and bulk (Sterox ND)
200-kg net iron drums (Merpoxen NO 40)
55-gal drums, 1- and 5-gal containers (Tergitol NP-14)
55-gal (200 l) steel drums, bulk (Hyonic NP-40)
55-gal (450 lb net) steel drums (Iconol NP-4)

POE (4) nonyl phenyl ether (cont'd.)

55-gal (460 lb net) drums, 5-gal (43 lb net) pails, 1-gal (8.5 lb net) jugs (Tergitol NP-4)
460 lb net closed-head steel drums (T-Det N-4)

POE (5) nonyl phenyl ether

SYNONYMS:
Nonoxynol-5 (CTFA)
Nonyl phenol ethoxylate (5 moles EO)
Nonyl phenol 5 polyglycol ether
Nonylphenoxy polyethoxy ethanol (5 moles EO)
Nonylphenoxy poly(ethyleneoxy) ethanol (5 moles EO)
PEG-5 nonyl phenyl ether
PEG (5) nonyl phenyl ether
POE (5) nonyl phenol

STRUCTURE:
$C_9H_{19}C_6H_4(OCH_2CH_2)_nOH$
where avg. $n = 5$

CAS No.:
9016-45-9 (generic); 26027-38-3 (generic); 37205-87-1 (generic)
RD No.: 977057-31-0

TRADENAME EQUIVALENTS:
Alkasurf NP-5 [Alkaril]
Iconol NP-5 [BASF Wyandotte]
Igepal CO-520 [GAF]
Lorapal HV5 [Dutton & Reinisch]
NP-55-50 [Hefti Ltd.]
Peganol NP5 [Borg-Warner]
Renex 648 [ICI United States]
Rewopal HV5 [Dutton & Reinisch]
Serdox NNP5 [Servo B.V.]
Steinapal HV5 [Dutton & Reinisch]
Sterox NE [Monsanto]
Synperonic NP5 [ICI Petrochem. Div.]
Tergitol NP-5 [Union Carbide]
Teric GN5 [ICI Australia Ltd.] (tech. grade)
Teric N5 [ICI Australia Ltd.]
Triton N-57 [Rohm & Haas]

CATEGORY:
Emulsifier, coemulsifier, dispersant, intermediate, raw material, coupling agent, rust inhibitor, detergent, deicing fluid, wetting agent, surfactant, defoamer, degreaser

APPLICATIONS:

Cosmetic industry preparations: (Alkasurf NP-5)

Farm products: herbicides (Serdox NNP5; Teric N5); insecticides/pesticides (Lorapal HV5; NP-55-50; Rewopal HV5; Serdox NNP5; Steinapal HV5; Teric N5)

Household detergents: (Alkasurf NP-5; Serdox NNP5; Synperonic NP5; Tergitol NP-5); detergent base (Lorapal HV5; Rewopal HV5; Steinapal HV5); liquid detergents (Tergitol NP-5); powdered detergents (Tergitol NP-5)

Industrial applications: (Alkasurf NP-5); aircraft and automotive fuels (Igepal CO-520; Peganol NP5); coatings (Teric N5); dyes and pigments (Alkasurf NP-5); lubricating/cutting oils (Alkasurf NP-5; Peganol NP5; Sterox NE); paint mfg. (Alkasurf NP-5); paper mfg. (Lorapal HV5; Rewopal HV5; Steinapal HV5); petroleum industry (Igepal CO-520; Peganol NP5; Sterox NE); printing inks (Alkasurf NP-5); textile/leather processing (Lorapal HV5; Rewopal HV5; Steinapal HV5); waxes (Alkasurf NP-5)

Industrial cleaners: (Lorapal HV5; Renex 648; Rewopal HV5; Steinapal HV5; Tergitol NP-5; Teric N5); metal processing surfactants (Renex 648); solvent cleaners (NP-55-50; Renex 648; Sterox NE; Teric N5)

PROPERTIES:

Form:

Liquid (Alkasurf NP-5; Iconol NP-5; Igepal CO-520; Lorapal HV5; NP-55-50; Rewopal HV5; Serdox NNP5; Steinapal HV5; Synperonic NP5; Tergitol NP-5; Teric N5, GN5; Triton N-57)

Clear liquid (Renex 648; Sterox NE)

Clear viscous liquid (Peganol NP5)

Color:

Light (Alkasurf NP-5)

Pale yellow (Igepal CO-520)

Yellow (Sterox NE)

APHA 100 (Triton N-57)

APHA 100 max. (Iconol NP-5; Peganol NP5)

Gardner 2 max. (Lorapal HV5; Rewopal HV5; Steinapal HV5)

Hazen 100 (Teric N5)

Hazen 150 max. (Synperonic NP5)

Hazen 250 (Teric GN5)

Odor:

Low (Alkasurf NP-5)

Aromatic (Igepal CO-520)

Mild (Sterox NE)

Composition:

99% active min. (Iconol NP-5)

99.5% active (Sterox NE)

100% active (Alkasurf NP-5; Igepal CO-520; Lorapal HV5; Renex 648; Rewopal HV5; Steinapal HV5; Teric N5; Triton N-57)

POE (5) nonyl phenyl ether (cont'd.)

100% conc. (NP-55-50; Serdox NNP5; Tergitol NP-5)

Solubility:

Sol. in alcohols (Lorapal HV5; Rewopal HV5; Steinapal HV5; Synperonic NP5)
Sol. in benzene (Teric N5)
Sol. in butyl Cellosolve (Igepal CO-520)
Sol. in chlorinated hydrocarbons (Lorapal HV5; Rewopal HV5; Steinapal HV5)
Sol. in corn oil (Igepal CO-520)
Sol. in cottonseed oil (@ 10%) (Renex 648)
Sol. in dibutyl phthalate (Igepal CO-520)
Sol. in ethanol (Igepal CO-520; Teric N5)
Sol. in ethyl acetate (Teric N5)
Sol. in ethylene dichloride (Igepal CO-520)
Sol. in glycol ethers (Synperonic NP5)
Sol. in isopropanol (@ 10%) (Renex 648)
Sol. in kerosene (Synperonic NP5; Teric N5)
Sol. in ketone (Lorapal HV5; Rewopal HV5; Steinapal HV5)
Sol. in min. oil (Lorapal HV5; Rewopal HV5; Steinapal HV5; Synperonic NP5)
Sol. in heavy aromatic naphtha (Igepal CO-520)
Sol. in nonpolar solvents (Sterox NE)
Sol. in oil (Triton N-57)
Sol. in olein (Teric N5)
Sol. in perchloroethylene (Igepal CO-520; Peganol NP5; Teric N5); (@ 10%) (Renex 648)
Sol. in propylene glycol (@ 10%) (Renex 648)
Sol. in Stoddard solvent (Igepal CO-520)
Sol. in toluene (Lorapal HV5; Rewopal HV5; Steinapal HV5)
Sol. in veg. oil (Teric N5)
Disp. in water (Iconol NP-5; Teric N5; Triton N-57); misc. with water (Sterox NE); insol./disp. (Synperonic NP5); insol. (Alkasurf NP-5; Peganol NP5); self-emulsifying in water (Renex 648)
Insol. in white min. oil (Peganol NP5)
Sol. in xylene (Igepal CO-520; Lorapal HV5; Peganol NP5; Rewopal HV5; Steinapal HV5)

Ionic Nature:

Nonionic (Alkasurf NP-5; Iconol NP-5; Igepal CO-520; Lorapal HV5; NP-55-50; Peganol NP5; Renex 648; Rewopal HV5; Serdox NNP5; Steinapal HV5; Sterox NE; Synperonic NP5; Teric N5; Triton N-57)

M.W.:

435 (Iconol NP-5)
440 avg. (Triton N-57)

Sp.gr.:

1.03 (Alkasurf NP-5; Iconol NP-5; Igepal CO-520; Renex 648)
1.031 (Peganol NP5)

1.0322 (Sterox NE)
1.035 (20 C) (Teric N5)

Density:
1.035 g/ml (20 C) (Synperonic NP5)
8.5 lb/gal (Triton N-57)

Visc.:
10 cs (100 C) (Peganol NP5)
223 cps (Renex 648)
236 cps (Sterox NE)
240 cps (Triton N-57)
240–300 cps (Igepal CO-520)
300 cps (Iconol NP-5)
350 cps (20 C) (Synperonic NP5)
355 cps (20 C) (Teric N5)

F.P.:
< 0 C (Synperonic NP5)

M.P.:
< 0 C (Teric N5)

Pour Pt.:
–30 C (Peganol NP5)
–27 C (Iconol NP-5)
< 0 C (Synperonic NP5)
–30 F (Sterox NE)
–25 F (Triton N-57)
–24 ± 2 F (Igepal CO-520)
–20 F (Renex 648)

Solidification Pt.:
–29 ± 2 F (Igepal CO-520)

Flash Pt.:
> 200 C (Peganol NP5)
> 200 F (Renex 648); (PMCC) (Igepal CO-520)
445 F (Sterox NE)
> 515 F (COC) (Triton N-57)

Cloud Pt.:
< 0 C (1%) (Triton N-57)
52–62 C (10% in 25% aq. butyl diglycol) (Lorapal HV5; Rewopal HV5; Steinapal HV5)

HLB:
10.0 (Alkasurf NP-5; Iconol NP-5; Igepal CO-520; NP-55-50; Peganol NP5; Renex 648; Serdox NNP5; Sterox NE; Tergitol NP-5; Triton N-57)
10.5 (Synperonic NP5; Teric N5)

Hydroxyl No.:
128 (Peganol NP5)

POE (5) nonyl phenyl ether (cont'd.)

130 (Synperonic NP5)

Stability:

Completely stable in acid or alkaline media (Alkasurf NP-5)
Very good stability in acids, alkalis, and in hard water (Renex 648)
Stable to acids, alkalis, dilute sol'ns. of many oxidizing and reducing agents (Igepal CO-520)
Good in hard or saline waters and in reasonable concs. of acids and alkalis (Teric N5)
Stable in presence of dilute acids and alkalis, in hard water, and oxidizing agents (Lorapal HV5; Rewopal HV5; Steinapal HV5)
Good (Sterox NE)

Ref. Index:

1.4955 (Sterox NE)

pH:

6.0–7.5 (5% aq.) (Iconol NP-5; Peganol NP5)
6.0–8.0 (1% aq.) (Synperonic NP5; Teric N5)

Surface Tension:

28.7 dynes/cm (Sterox NE)
29 dynes/cm (0.1%) (Iconol NP-5); (1%) (Triton N-57)
30 dynes/cm (0.01% sol'n.) (Igepal CO-520)

Biodegradable: (Serdox NNP5)

TOXICITY/HANDLING:

Skin and eye irritant (Sterox NE)
May cause skin and eye irritation; spillages are slippery (Teric N5)
Eye irritant; skin irritant on prolonged contact with conc. form (Synperonic NP5)

STORAGE/HANDLING:

Contact with conc. oxidizing or reducing agents may be explosive (Igepal CO-520)

STD. PKGS.:

Steel drums and bulk (Sterox NE)
55-gal (450 lb net) steel drums (Iconol NP-5)

POE (6) nonyl phenyl ether

SYNONYMS:

Nonoxynol-6 (CTFA)
Nonyl phenol ethoxylate (6 moles EO)
Nonyl phenol 6 polyglycol ether
Nonylphenoxy polyethoxy ethanol (6 moles EO)
Nonylphenoxy poly(ethyleneoxy) ethanol (6 moles EO)
PEG-6 nonyl phenyl ether
PEG 300 nonyl phenyl ether
POE (6) nonyl phenol

STRUCTURE:

$C_9H_{19}C_6H_4(OCH_2CH_2)_nOH$
where avg. $n = 6$

CAS No.:

9016-45-9 (generic); 26027-38-3 (generic); 37205-87-1 (generic)
RD No.: 977057-32-1

TRADENAME EQUIVALENTS:

Ablunol NP6 [Taiwan Surfactant]
Alkasurf NP-6 [Alkaril]
Arkopal N-060 [Amer. Hoechst]
Carsonon N-6 [Carson]
Cedepal CO-530 [Domtar]
Chemax NP-6 [Chemax]
Chemcol NPE-60 [Chemform]
Conco NI-60 [Continental]
Hostapal N-060 [Amer. Hoechst]
Hyonic NP-60 [Diamond Shamrock]
Iconol NP-6 [BASF Wyandotte]
Igepal CO-530 [GAF]
Macol NP-6 [Mazer]
Makon 6 [Stepan]
Merpoxen NO60 [Elektrochemische Fabrik Kempen]
Nissan Nonion NS-206 [Nippon Oil & Fats]
Norfox NP-6 [Norman, Fox & Co.]
NP-55-60 [Hefti Ltd.]
Peganol NP6 [Borg-Warner]
Polystep F-2 [Stepan]
Renex 697 [ICI United States]
Rewopal HV6 [Rewo Chemische Werke GmbH]
Siponic NP6 [Alcolac]
Serdox NNP6 [Servo B.V.]
Sterox NF [Monsanto]
Sterox NG [Monsanto] (6.5 EO)
Surfonic N-60 [Jefferson]
Synperonic NP6 [ICI Petrochem. Div.]
T-Det N-6 [Thompson-Hayward]
Tergitol NP-6 [Union Carbide]
Triton N-60 [Rohm & Haas]
Trycol NP-6 [Emery]

CATEGORY:

Emulsifier, coemulsifier, dispersant, detergent, wetting agent, stabilizer, penetrant, solubilizer, coupling agent, plasticizer, antistat, foaming agent, intermediate

POE (6) nonyl phenyl ether (cont'd.)

APPLICATIONS:

Cosmetic industry preparations: (Conco NI-60; Makon 6; Surfonic N-60); perfumery (Makon 6); shampoos (Makon 6)

Degreasers: (Rewopal HV6)

Farm products: (Chemax NP-6; Chemcol NPE-60; Igepal CO-530; Makon 6; Nissan Nonion NS-206; NP-55-60; Peganol NP6); agricultural oils/sprays (Hyonic NP-60); herbicides (Alkasurf NP-6; T-Det N-6); insecticides/pesticides (Alkasurf NP-6; Conco NI-60; Serdox NNP6; T-Det N-6)

Food applications: indirect food additives (Hyonic NP-60; Surfonic N-60)

Household detergents: (Alkasurf NP-6; Carsonon N-6; Chemcol NPE-60; Conco NI-60; Nissan Nonion NS-206; Serdox NNP6); detergent intermediate (Makon 6); dishwashing (Rewopal HV6); powdered detergents (Merpoxen NO60)

Industrial applications: (Arkopal N-060; Hostapal N-060); coatings (Hyonic NP-60); dyes and pigments (Alkasurf NP-6; Makon 6; NP-55-60; Polystep F-2); industrial processing (Chemax NP-6); lubricating/cutting oils (Alkasurf NP-6; Makon 6; Peganol NP6; Sterox NF, NG; Surfonic N-60); metalworking (T-Det N-6); paint mfg. (Alkasurf NP-6; Hyonic NP-60; Nissan Nonion NS-206; Sterox NF, NG); paper mfg. (Igepal CO-530; Makon 6; T-Det N-6); petroleum industry (Makon 6; Surfonic N-60); petroleum oils/products (Chemax NP-6; Chemcol NPE-60; Conco NI-60; Igepal CO-530; Sterox NF, NG; T-Det N-6); plastics (Alkasurf NP-6; Surfonic N-60); polishes and waxes (T-Det N-6); polymers/polymerization (Conco NI-60; Hyonic NP-60; Nissan Nonion NS-206); printing inks (Alkasurf NP-6; Nissan Nonion NS-206); rubber (Surfonic N-60); silicone products (Chemax NP-6; Igepal CO-530; Peganol NP6; T-Det N-6); textile/leather processing (Conco NI-60; Hyonic NP-60; Makon 6; Merpoxen NO60; Nissan Nonion NS-206; Surfonic N-60; T-Det N-6)

Industrial cleaners: (Alkasurf NP-6; Carsonon N-6; Conco NI-60; Makon 6; Nissan Nonion NS-206; Renex 697; Surfonic N-60; Tergitol NP-6); acid cleaners (Trycol NP-6); bottle washing (Conco NI-60); drycleaning compositions (Surfonic N-60; T-Det N-6); metal processing surfactants (Makon 6); solvent cleaners (NP-55-60; Renex 697; Sterox NF, NG; T-Det N-6; Trycol NP-6); textile cleaning (Chemcol NPE-60; Makon 6)

PROPERTIES:

Form:

Liquid (Alkasurf NP-6; Arkopal N-060; Carsonon N-6; Cedepal CO-530; Chemax NP-6; Chemcol NPE-60; Conco NI-60; Hostapal N-060; Iconol NP-6; Igepal CO-530; Macol NP-6; Makon 6; Nissan Nonion NS-206; Norfox NP-6; NP-55-60; Serdox NNP6; Siponic NP6; Surfonic N-60; Synperonic NP6; T-Det N-6; Triton N-60; Trycol NP-6)

Clear liquid (Hyonic NP-60; Polystep F-2; Renex 697; Sterox NF, NG; Tergitol NP-6)

Clear, viscous liquid (Peganol NP6)

Viscous liquid (Merpoxen NO60)

Med. viscosity liquid (Rewopal HV6)

Color:

Colorless (Tergitol NP-6)
Light/pale (Alkasurf NP-6; Polystep F-2)
Water-white (Merpoxen NO60)
Light straw (Makon 6)
Pale yellow (Chemcol NPE-60; Igepal CO-530; T-Det N-6)
Yellow (Sterox NF, NG)
APHA 100 max. (Iconol NP-6; Peganol NP6)
APHA 120 max. (Nissan Nonion NS-206)
APHA 125 (Triton N-60)
Gardner 2 (Trycol NP-6)
Hazen 150 max. (Synperonic NP6)

Odor:

Characteristic (Alkasurf NP-6; Tergitol NP-6)
Mild, pleasant (Carsonon N-6)
Aromatic (Igepal CO-530)
Mild (Sterox NF)
Mild, aromatic (T-Det N-6)

Composition:

99% active min. (Hyonic NP-60; Synperonic NP6)
99.5% active (Sterox NG)
99.5% active min. (T-Det N-6)
100% active (Alkasurf NP-6; Carsonon N-6; Chemcol NPE-60; Conco NI-60; Igepal CO-530; Makon 6; Merpoxen NO60; Polystep F-2; Renex 697; Surfonic N-60; Tergitol NP-6; Triton N-60)
100% conc. (Ablunol NP6; Arkopal N-060; Cedepal CO-530; Hostapal N-060; Macol NP-6; Norfox NP-6; NP-55-60; Rewopal HV6; Serdox NNP6; Siponic NP6)

Solubility:

Sol. in acetone (Surfonic N-60)
Sol. in alcohols (Carsonon N-6; Synperonic NP6)
Sol. in aliphatic solvents (Carsonon N-6)
Sol. in aromatic solvents (Carsonon N-6)
Sol. in butyl acetate (Tergitol NP-6)
Sol. in butyl Cellosolve (Igepal CO-530; T-Det N-6; Tergitol NP-6)
Sol. in carbon tetrachloride (Surfonic N-60; T-Det N-6)
Sol. in chlorinated hydrocarbons (Carsonon N-6)
Sol. in corn oil (Igepal CO-530; T-Det N-6; Tergitol NP-6)
Sol. in cottonseed oil (Renex 697)
Sol. in dibutyl phthalate (Igepal CO-530)
Sol. in diesel fuel (T-Det N-6); disp. (Tergitol NP-6)
Sol. in diethylene glycol (Nissan Nonion NS-206)
Sol. in ethanol (Igepal CO-530; T-Det N-6)

Sol. in ether (Nissan Nonion NS-206)
Sol. in ethylene dichloride (Igepal CO-530)
Disp. in ethylene glycol (Tergitol NP-6)
Sol. in fatty oil (Carsonon N-6)
Sol. in glycol ethers (Synperonic NP6)
Sol. in isopropanol (@ 10%) (Renex 697); sol. in anhyd. isopropanol (Tergitol NP-6)
Sol. in kerosene (T-Det N-6); sol./disp. (Synperonic NP6)
Sol. in methanol (Nissan Nonion NS-206; Surfonic N-60)
Sol. in min. oil (Synperonic NP6); disp. (Trycol NP-6)
Sol. in heavy aromatic naphtha (Igepal CO-530)
Sol. in nonpolar solvents (Sterox NF, NG)
Sol. in oil (Cedepal CO-530; Norfox NP-6; Polystep F-2; Serdox NNP6; Siponic NP6)
Sol. in perchloroethylene (Igepal CO-530; Peganol NP6); (@ 10%) (Renex 697)
Sol. in propylene glycol (@ 10%) (Renex 697)
Disp. in soybean oil (Nissan Nonion NS-206)
Sol. in Stoddard solvent (Igepal CO-530; Surfonic N-60; T-Det N-6; Tergitol NP-6); (@ 5%) (Trycol NP-6)
Sol. in tetrachloromethan (Nissan Nonion NS-206)
Sol. in toluene (Tergitol NP-6)
Sol. in water (Chemcol NPE-60); disp. (Carsonon N-6; Hyonic NP-60; Iconol NP-6; Makon 6; Nissan Nonion NS-206; Surfonic N-60; Synperonic NP6; Trycol NP-6); misc. (Sterox NF, NG); insol. (Alkasurf NP-6; Peganol NP6); self-emulsifying in water (Renex 697)
Disp. in white min. oil (Tergitol NP-6); insol. (Peganol NP6)
Sol. in xylene (Igepal CO-530; Nissan Nonion NS-206; Peganol NP6; Surfonic N-60; T-Det N-6); (@ 5%) (Trycol NP-6); (@ 10%) (Renex 697)

Ionic Nature:

Nonionic (Ablunol NP6; Alkasurf NP-6; Arkopal N-060; Carsonon N-6; Cedepal CO-530; Chemcol NPE-60; Conco NI-60; Hostapal N-060; Hyonic NP-60; Igepal CO-530; Macol NP-6; Makon 6; Merpoxen NO60; Nissan Nonion NS-206; Norfox NP-6; NP-55-60; Serdox NNP6; Siponic NP6; Sterox NF, NG; Surfonic N-60; Synperonic NP6; T-Det N-6; Tergitol NP-6; Triton N-60)

M.W.:

479 (Iconol NP-6)
484 (Surfonic N-60; Tergitol NP-6); avg. (Triton N-60)

Sp.gr.:

1.03–1.05 (Conco NI-60)
1.038 (Peganol NP6)
1.0385 (Sterox NF)
1.04 (Alkasurf NP-6; Iconol NP-6; Igepal CO-530; Renex 697; T-Det N-6)
1.041 (20/20 C) (Surfonic N-60)
1.0426 (Sterox NG)
1.055 (20/20 C) (Tergitol NP-6)

Density:
1.04 g/ml (Hyonic NP-60)
1.041 g/ml (20 C) (Synperonic NP6)
8.5 lb/gal (Trycol NP-6)
8.67 lb/gal (Makon 6); (20 C) (Tergitol NP-6)
8.68 lb/gal (Carsonon N-6)
8.7 lb/gal (Chemcol NPE-60; T-Det N-6; Triton N-60); (20 C) (Surfonic N-60)
Visc.:
180–240 cps (Carsonon N-6)
207 cps (Sterox NF)
229 cps (Sterox NG)
230–300 cps (Igepal CO-530)
250 cps (Renex 697)
300 cps (Iconol NP-6; Triton N-60)
355 cps (20 C) (Synperonic NP6)
10.5 cs (100 C) (Peganol NP6)
373 cs (20 C) (Tergitol NP-6)
100 cSt (100 F) (Trycol NP-6)
58 SUS (210 F) (T-Det N-6)
440 SUS (100 F) (Surfonic N-60)
F.P.:
–10 C (Tergitol NP-6)
< 0 C (Surfonic N-60; Synperonic NP6)
B.P.:
> 300 C (dec.) (Tergitol NP-6)
Pour Pt.:
–32 C (Hyonic NP-60; Peganol NP6)
–29 C (Makon 6)
–28 C (Iconol NP-6)
–10 C (Trycol NP-6)
< 10 C (Chemcol NPE-60)
–26 F (Chemax NP-6)
–26 ± 2 F (Igepal CO-530)
–25 F (Triton N-60)
–20 F (Sterox NF)
< 0 F (T-Det N-6)
0 F (Sterox NG)
Solidification Pt.:
–34.4 C (Makon 6)
–10 C max. (Nissan Nonion NS-206)
< 0 C (Synperonic NP6)
–33 to –29 F (Carsonon N-6)
–31 ± 2 F (Igepal CO-530)

POE (6) nonyl phenyl ether (cont'd.)

–21 F (Renex 697)

Flash Pt.:

> 200 C (Peganol NP6)
> 200 F (PMCC) (Igepal CO-530)
> 300 F (Renex 697)
360 F (PMCC) (Tergitol NP-6)
> 400 F (OC) (T-Det N-6); (TOC) (Chemcol NPE-60)
475 F (OC) (Surfonic N-60)
480 F (Sterox NF)
485 F (Sterox NG)
500 F (Carsonon N-6)
515 F (Trycol NP-6)

Cloud Pt.:

0 C (0.5% aq.) (Tergitol NP-6)
0 C max. (1% aq.) (Hyonic NP-60; Nissan Nonion NS-206)
< 25 C (Trycol NP-6)
53 C (1% in 12.5% aq. isopropanol) (Triton N-60)
59–61 C (5 g in 20 ml 25% butyl glycol sol'n.) (Merpoxen NO60)
< 32 F (1% sol'n.) (Renex 697)
32 F (Conco NI-60)
90–102 F (Carsonon N-6)

HLB:

10.7 (Siponic NP6)
10.8 (Ablunol NP6; Cedepal CO-530; Chemcol NPE-60; Iconol NP-6; Igepal CO-530; Macol NP-6; Peganol NP6)
10.9 (Nissan Nonion NS-206; Renex 697; Surfonic N-60; Synperonic NP6; T-Det N-6; Trycol NP-6; Tergitol NP-6; Triton N-60)
11.0 (Alkasurf NP-6; Norfox NP-6; NP-55-60; Serdox NNP6)
11.3 (Sterox NG)
12.0 (Hyonic NP-60)

Hydroxyl No.:

113 (Synperonic NP6)
116 (Peganol NP6)

Stability:

Good (Alkasurf NP-6; Sterox NF, NG)
Very stable against hyrolysis by acids and alkalis (Trycol NP-6)
Stable to acids, alkalis, hard water, and foam (Conco NI-60)
Stable to acids, alkalis, dilute sol'ns. of many oxidizing and reducing agents (Igepal CO-530)
Stable to acids, alkalis, salts (Surfonic N-60; T-Det N-6)
Very good stability in acids, alkalis, and hard water (Renex 697)
Chemically stable in the presence of ionic materials (Tergitol NP-6)

Ref. Index:
1.4933 (Sterox NG)
1.4938 (20 C) (Surfonic N-60)
1.4940 (Sterox NF)
pH:
5.0–7.5 (1% aq. mix.) (Tergitol NP-6)
6.0–7.5 (5% aq.) (Iconol NP-6; Peganol NP6)
6.0–8.0 (1% aq.) (Synperonic NP6); (5% in 25% IPA) (Chemcol NPE-60)
7.0 (T-Det N-6); (1% aq.) (Hyonic NP-60)
7.9 (1% sol'n.) (Makon 6)
Foam (Ross Miles):
10 mm initial, 7 mm after 5 min (0.05% sol'n. in dist. water) (Hyonic NP-60)
Surface Tension:
28 dynes/cm (0.01% sol'n.) (Igepal CO-530); (1%) (Triton N-60)
28.7 dynes/cm (0.1%) (Surfonic N-60)
28.8 dynes/cm (Sterox NF)
29 dynes/cm (0.1% aq.) (Iconol NP-6)
29.1 dynes/cm (Sterox NG)
30 dynes/cm (0.01% sol'n.) (Renex 697)
Wetting (Draves):
0.21% for a 25-s test (3-g hook) (Hyonic NP-60)
Biodegradable: (Serdox NNP6); > 90% (Surfonic N-60)
TOXICITY/HANDLING:
Skin and eye irritant (Sterox NF, NG)
Eye irritant; skin irritant on prolonged contact with conc. form (Synperonic NP6)
Severe eye irritant; highly toxic to aquatic life; burning can produce carbon monoxide and/or carbon dioxide (Tergitol NP-6)
STORAGE/HANDLING:
Contact with conc. oxidizing or reducing agents may be explosive (Igepal CO-530; T-Det N-6)
Store at temps. below 50 C (Tergitol NP-6)
STD. PKGS.:
Steel drums and bulk (Sterox NF, NG)
200-kg net iron drums (Merpoxen NO60)
55-gal (470 lb net) steel drums (Iconol NP-6)
55-gal (470 lb net) drums, 5-gal (43 lb net) pails, 1-gal (8.5 lb net) jugs (Tergitol NP-6)
55-gal (200 l) steel drums, bulk (Hyonic NP-60)
460 lb net closed-head steel drums (T-Det N-6)

POE (8) nonyl phenyl ether

SYNONYMS:

Nonoxynol-8 (CTFA)
Nonyl phenol ethoxylate (8 moles EO)
Nonyl phenol 8 polyglycol ether
Nonylphenoxy polyethoxy ethanol (8 moles EO)
Nonylphenoxy poly(ethyleneoxy) ethanol (8 moles EO)
PEG-8 nonyl phenyl ether
PEG 400 nonyl phenyl ether
POE (8) nonyl phenol

STRUCTURE:

$C_9H_{19}C_6H_4(OCH_2CH_2)_nOH$
where avg. $n = 8$

CAS No.:

9016-45-9 (generic); 26027-38-3 (generic); 37205-87-1 (generic)
RD No.: 977057-34-3

TRADENAME EQUIVALENTS:

Ablunol NP8 [Taiwan Surfactant]
Alkasurf NP-8 [Alkaril]
Arkopal N-080 [Amer. Hoechst]
Carsonon N-8 [Carson]
Cedepal CO-610 [Domtar]
Igepal CO-610 [GAF]
Igepal CO-620 [GAF] (8.5 EO)
Lorapal HV8 [Dutton & Reinisch]
Makon 8 [Stepan]
Merpoxen NO80 [Elektrochemische Fabrik Kempen]
Nikkol NP-7.5 [Nikko] (7.5 EO)
NP-55-80 [Hefti Ltd.]
Nutrol 611 [Clough Chem. Ltd.]
Polystep F-3 [Stepan]
Quimipol ENF80 [Quimigal-Quimica de Portugal]
Renex 688 [ICI United States]
Rewopal HV8 [Dutton & Reinisch]
Serdox NNP8.5 [Servo B.V.] (8.5 EO)
Steinapal HV8 [Dutton & Reinisch]
Surfonic N-85 [Texaco]
Synperonic NP8 [ICI Petrochem. Div.]
T-Det N-8 [Thompson-Hayward]
Tergitol NP-8 [Union Carbide]
Teric GN8 [ICI Australia Ltd.] (tech.)
Teric N8 [ICI Australia Ltd.] (8.5 EO)
Triton N-87 [Rohm & Haas] (8.5 EO)

CATEGORY:

Detergent, emulsifier, wetting agent, dispersant, solubilizer, raw material, degreasing agent, plasticizer

APPLICATIONS:

Cosmetic industry preparations: (Alkasurf NP-8; Makon 8; Nikkol NP-7.5); perfumery (Makon 8); shampoos (Makon 8)

Farm products: (Makon 8); agricultural oils/sprays (Teric N8); herbicides (Serdox NNP8.5; T-Det N-8); insecticides/pesticides (Lorapal HV8; Nikkol NP-7.5; Rewopal HV8; Serdox NNP8.5; Steinapal HV8; T-Det N-8)

Household detergents: (Ablunol NP8; Alkasurf NP-8; Carsonon N-8; Cedepal CO-610; Lorapal HV8; Merpoxen NO80; NP-55-80; Rewopal HV8; Serdox NNP8.5; Steinapal HV8; Synperonic NP8; T-Det N-8); detergent intermediate (Makon 8); dishwashing (Renex 688); hard surface cleaner (Triton N-87); heavy-duty cleaner (Igepal CO-620; Surfonic N-85; Teric N8); laundry detergent (Alkasurf NP-8; Surfonic N-85; Triton N-87); liquid detergents (Igepal CO-620; Surfonic N-85); powdered detergents (Merpoxen NO80)

Industrial applications: (Arkopal N-080; Nikkol NP-7.5); construction (Serdox NNP8.5; Teric N8); dyes and pigments (Alkasurf NP-8; Makon 8; NP-55-80; T-Det N-8); electroplating (Alkasurf NP-8); lubricating/cutting oils (Alkasurf NP-8; Makon 8); paint mfg. (Alkasurf NP-8; T-Det N-8; Teric N8); paper mfg. (Alkasurf NP-8; Carsonon N-8; Lorapal HV8; Makon 8; Rewopal HV8; Steinapal HV8; T-Det N-8); petroleum industry (Makon 8; T-Det N-8); printing inks (Alkasurf NP-8); textile/leather processing (Alkasurf NP-8; Carsonon N-8; Lorapal HV8; Makon 8; Merpoxen NO80; NP-55-80; Rewopal HV8; Serdox NNP8.5; Steinapal HV8; T-Det N-8)

Industrial cleaners: (Ablunol NP8; Alkasurf NP-8; Carsonon N-8; Lorapal HV8; Makon 8; NP-55-80; Rewopal HV8; Serdox NNP8.5; Steinapal HV8; T-Det N-8); dry cleaning (Quimipol ENF80); equipment cleaning (Surfonic N-85); metal processing surfactants (Makon 8); solvent cleaners (Teric N8); textile cleaning (Alkasurf NP-8; Makon 8; Quimipol ENF80; Renex 688; Serdox NNP8.5; Synperonic NP8)

PROPERTIES:

Form:

Liquid (Ablunol NP8; Alkasurf NP-8; Arkopal N-080; Carsonon N-8; Cedepal CO-610; Igepal CO-610, CO-620; Lorapal HV8; Makon 8; Nikkol NP-7.5; NP-55-80; Nutrol 611; Quimipol ENF80; Renex 688; Rewopal HV8; Serdox NNP8.5; Steinapal HV8; Surfonic N-85; Synperonic NP8; Teric N8)

Clear liquid (Polystep F-3; Tergitol NP-8; Triton N-87)

Essentially clear, viscous liquid (T-Det N-8)

Viscous liquid (Merpoxen NO80)

Color:

Light/pale (Alkasurf NP-8; Carsonon N-8; Polystep F-3)

Water-white (Merpoxen NO80)

Light straw (Makon 8; Triton N-87)
Pale yellow (Igepal CO-610)
Gardner 2 max. (Lorapal HV8; Rewopal HV8; Steinapal HV8)
Hazen 100 (Teric N8)
Hazen 150 max. (Synperonic NP8)
Hazen 250 (Teric GN8)
Pt-Co 50 max. (Tergitol NP-8)

Odor:
Low (Alkasurf NP-8)
Aromatic (Igepal CO-610)
Mild (Triton N-87)
Mild, characteristic (Tergitol NP-8)
Mild, pleasant (Carsonon N-8)
Mild, aromatic (T-Det N-8)

Composition:
99% active min. (Nutrol 611; Synperonic NP8)
99.5% active min. (T-Det N-8)
100% active (Alkasurf NP-8; Carsonon N-8; Igepal CO-610; Lorapal HV8; Makon 8; Merpoxen NO80; Polystep F-3; Renex 688; Rewopal HV8; Steinapal HV8; Tergitol NP-8; Teric N8; Triton N-87)
100% conc. (Ablunol NP8; Arkopal N-080; Cedepal CO-610; Igepal CO-620; Nikkol NP-7.5; NP-55-80; Quimipol ENF80; Serdox NNP8.5; Surfonic N-85)

Solubility:
Sol. in alcohols (Carsonon N-8; Lorapal HV8; Rewopal HV8; Steinapal HV8; Synperonic NP8)
Sol. in aromatics (Carsonon N-8; Triton N-87)
Sol. in benzene (Teric N8)
Sol. in butyl acetate (Tergitol NP-8)
Sol. in butyl Cellosolve (Igepal CO-610; Tergitol NP-8)
Sol. in chlorinated hydrocarbons (Carsonon N-8; Lorapal HV8; Rewopal HV8; Steinapal HV8; Triton N-87)
Sol. in corn oil (Igepal CO-610); disp. (Tergitol NP-8)
Sol. hazy in cottonseed oil (Renex 688)
Sol. in dibutyl phthalate (Igepal CO-610)
Slightly sol. in diesel fuel (Tergitol NP-8)
Sol. in ethanol (Igepal CO-610; Teric N8)
Sol. in ethyl acetate (Teric N8)
Sol. in ethylene dichloride (Igepal CO-610)
Sol. in ethylene glycol (Igepal CO-610; Tergitol NP-8)
Sol. in glycol ethers (Synperonic NP8)
Sol. in isopropanol (@ 10%) (Renex 688); anhyd. isopropanol (Tergitol NP-8)
Sol./disp. in kerosene (Synperonic NP8)
Sol. in ketones (Lorapal HV8; Rewopal HV8; Steinapal HV8)

Sol. in min. oil (Lorapal HV8; Rewopal HV8; Steinapal HV8)
Sol. in heavy aromatic naphtha (Igepal CO-610)
Sol. in olein (Teric N8)
Sol. in perchloroethylene (Igepal CO-610; Teric N8); (@ 10%) (Renex 688)
Sol. in polar organic solvents (Triton N-87)
Sol. in propylene glycol (@ 10%) (Renex 688)
Sol. in toluene (Lorapal HV8; Rewopal HV8; Steinapal HV8; Tergitol NP-8)
Sol. in veg. oil (Teric N8)
Sol. in water (Ablunol NP8; Alkasurf NP-8; Carsonon N-8; Cedepal CO-610; Igepal CO-610, CO-620; Synperonic NP8; Tergitol NP-8; Teric N8; Triton N-87); sol. hazy (Renex 688); borderline water-sol. (Polystep F-3); sol. in water at room temp.; disp. at higher temps. (Makon 8)
Sol. in xylene (Igepal CO-610; Lorapal HV8; Rewopal HV8; Steinapal HV8); (@ 10%) (Renex 688)

Ionic Nature:
Nonionic (Ablunol NP8; Arkopal N080; Carsonon N-8; Cedepal CO-610; Igepal CO-610; Lorapal HV8; Makon 8; Merpoxen NO80; Nikkol NP-7.5; NP-55-80; Nutrol 611; Polystep F-3; Quimipol ENF80; Renex 688; Rewopal HV8; Serdox NNP8.5; Steinapal HV8; Surfonic N-85; T-Det N-8; Tergitol NP-8; Teric N8; Triton N-87)

M.W.:
572 (Tergitol NP-8)
594 (Triton N-87)

Sp.gr.:
1.05 (Alkasurf NP-8; Igepal CO-610; Renex 688; T-Det N-8)
1.055 (Triton N-87)
1.056 (20/20 C) (Tergitol NP-8); (20 C) (Teric N8)

Density:
1.053 g/ml (20 C) (Synperonic NP8)
8.75 lb/gal (Carsonon N-8; T-Det N-8)
8.76 lb/gal (Makon 8)
8.8 lb/gal (Triton N-87)

Visc.:
200–300 cps (Carsonon N-8)
205 cps (Makon 8)
230–290 cps (Igepal CO-610)
240 cps (Renex 688)
260 cps (Brookfield, 12 rpm) (Triton N-87)
350 cps (20 C) (Teric N8)
355 cps (20 C) (Synperonic NP8)
325 cs (20 C) (Tergitol NP-8)
63 SUS (210 F) (T-Det N-8)

F.P.:
< 0 C (Synperonic NP8)

POE (8) nonyl phenyl ether (cont'd.)

30 F (T-Det N-8)

B.P.:

249 C (Triton N-87)
250 C (dec.) (Tergitol NP-8)

M.P.:

0 ± 2 C (Teric N8)

Pour Pt.:

–5 C (Makon 8)
< 0 C (Synperonic NP8)
1.7 C (Triton N-87)
12 F (Renex 688)
37 ± 2 F (Igepal CO-610)
41–45 F (Carsonon N-8)

Solidification Pt.:

–3 C (Tergitol NP-8)
0 C (Makon 8)
21 F (Triton N-87)
30–34 F (Carsonon N-8)
32 ± 2 F (Igepal CO-610)

Flash Pt.:

> 218 C (PMCC) (Triton N-87)
> 200 F (PMCC) (Igepal CO-610)
> 300 F (Renex 688)
400 F (PMCC) (Tergitol NP-8)
435 F (Carsonon N-8)
500 F (OC) (T-Det N-8)

Fire Pt.:

465 F (Carsonon N-8)

Cloud Pt.:

24 C (1% sol'n.) (Makon 8)
30–34 C (1% aq.) (Synperonic NP8)
32 ± 2 C (1% in hard water) (Teric N8)
40–46 C (1% aq.) (Igepal CO-620)
41–45 C (10% in 25% aq. butyl diglycol) (Lorapal HV8; Rewopal HV8; Steinapal HV8)
42 C (1% aq.) (Triton N-87)
43 C (0.5% aq.) (Tergitol NP-8)
44–46 C (1% aq.) (Merpoxen NO80)
72–82 F (1% sol'n.) (Igepal CO-610)
78 F (1%) (T-Det N-8)
82–86 F (Carsonon N-8)
87 F (1% sol'n.) (Renex 688)

HLB:
12.0 (Alkasurf NP-8; NP-55-80)
12.2 (Ablunol NP8; Cedepal CO-610; Igepal CO-610; Nutrol 611; Quimipol ENF80)
12.3 (Renex 688; Synperonic NP8; Tergitol NP-8; Teric N8)
12.4 (Surfonic N-85; T-Det N-8)
12.5 (Serdox NNP8.5)
12.6 (Igepal CO-620; Triton N-87)
14.0 (Nikkol NP-7.5)
Hydroxyl No.:
95 (Synperonic NP8)
Stability:
Stable to acids, bases, and salts (T-Det N-8)
Very good stability in acids, alkalis, and hard water (Renex 688)
Good stability in hard or saline waters and in reasonable concs. of acids and alkalis (Teric N8)
Stable to acids, alkalis, dilute sol'ns. of many oxidizing and reducing agents (Igepal CO-610)
Stable in presence of dilute acids and alkalis, in hard water, and oxidizing agents (Lorapal HV8; Rewopal HV8; Steinapal HV8)
pH:
5.0–7.0 (1% solids) (Lorapal HV8; Rewopal HV8; Steinapal HV8); 5% aq. (Triton N-87)
5.0–8.0 (10% aq.) (Tergitol NP-8)
6.0 (1%) (T-Det N-8)
6.0–8.0 (1% aq.) (Synperonic NP8; Teric N8)
7.0 (1% sol'n.) (Makon 8)
Foam (Ross Miles):
15 mm initial, 5 mm after 5 min (0.02% aq., 160 F) (Triton N-87)
121 mm initial, 77 mm after 5 min (40 C) (Tergitol NP-8)
Surface Tension:
29.4 dynes/cm (0.1%, 20 C) (Synperonic NP8; Teric N8)
30 dynes/cm (0.01% sol'n.) (Igepal CO-610; Renex 688); (0.1%) (Tergitol NP-8; Triton N-87)
Wetting (Draves):
8 s (0.1% conc.) (Tergitol NP-8)
0.06% for 24-s wetting (Triton N-87)
Biodegradable: (Triton N-87)
TOXICITY/HANDLING:
Eye irritant (T-Det N-8)
Eye irritant; skin irritant on prolonged contact with conc. form (Synperonic NP8)
Severe eye irritant; burning can produce carbon dioxide and/or carbon monoxide; toxic to fish (Tergitol NP-8)
Severely irritating to eyes (possible permanent injury); moderately irritating to skin

POE (8) nonyl phenyl ether (cont'd.)

(Triton N-87)

STORAGE/HANDLING:

Contact with conc. oxidizing or reducing agents may be explosive (Igepal CO-610; T-Det N-8)

Avoid strong oxidizing and reducing agents and excessive heat (Triton N-87)

STD. PKGS.:

200-kg net iron drums (Merpoxen NO80)

460 lb net closed-head steel drums (T-Det N-8)

POE (9) nonyl phenyl ether

SYNONYMS:

Nonoxynol-9 (CTFA)

Nonyl phenol ethoxylate (9 moles EO)

Nonyl phenol 9 polyglycol ether

Nonylphenoxy polyethoxy ethanol (9 moles EO)

Nonylphenoxy poly(ethyleneoxy) ethanol (9 moles EO)

PEG-9 nonyl phenyl ether

PEG 450 nonyl phenyl ether

POE (9) nonyl phenol

EMPIRICAL FORMULA:

$C_{33}H_{60}O_{10}$

STRUCTURE:

$C_9H_{19}C_6H_4(OCH_2CH_2)_nOH$

where avg. $n = 9$

CAS No.:

9016-45-9 (generic); 26027-38-3 (generic); 26571-11-9; 37205-87-1 (generic)

RD No.: 977004-65-1

TRADENAME EQUIVALENTS:

Ablunol NP9 [Taiwan Surfactant]

Alkasurf NP-9 [Alkaril]

Arkopal N-090 [Amer. Hoechst]

Carsonon N-9 [Carson] (9–10 EO)

Cedepal CO-630 [Domtar]

Chemax NP-9 [Chemax]

Conco NI-90 [Continental Chem.]

Conco NI-100 [Continental Chem.] (9–10 EO)

Empilan NP9 [Albright & Wilson/Marchon]

Gradonic N-95 [Graden] (9.5 EO)

Hetoxide NP-9 [Heterene]

Hostapal N-090 [Amer. Hoechst]

TRADENAME EQUIVALENTS *(cont'd.)*:

Hyonic NP-90 [Diamond Shamrock]
Iconol NP-9 [BASF Wyandotte]
Igepal CO-630 [GAF]
Lorapal HV9 [Dutton & Reinisch]
Macol NP-9.5 [Mazer] (9.5 EO)
Merpoxen NO90 [Elektrochemische Fabrik Kempen]
Merpoxen NO95 [Elektrochemische Fabrik Kempen] (9.5 EO)
Neutronyx 600 [Onyx] (9.5 EO)
Norfox NP-9 [Norman, Fox & Co.]
NP-55-90 [Hefti Ltd.]
NP-55-95 [Hefti Ltd.] (9.5 EO)
Nutrol 600 [Clough Chem. Ltd.]
Peganol NP9 [Borg-Warner]
Renex 698 [ICI United States] (9–9.5 EO)
Rewopal HV9 [Dutton & Reinisch]
Serdox NNP9 [Servo B.V.]
Siponic NP9 [Alcolac]
Steinapal HV9 [Dutton & Reinisch]
Sterox NJ [Monsanto] (9.5 EO)
Surfonic N-95 [Jefferson] (9.5 EO)
Synperonic NP9 [ICI Petrochem. Div.]
T-Det N-9.5 [Thompson-Hayward] (9.5 EO)
Tergitol NP-9, TP-9 [Union Carbide]
Teric GN9 [ICI Australia Ltd.] (tech.)
Teric N9 [ICI Australia Ltd.]
Triton N-101 [Rohm & Haas] (9–10 EO)
Trycol NP-9 [Emery]

CATEGORY:

Detergent, wetting agent, emulsifier, dispersant, surfactant, stabilizer, foaming agent, penetrant, emollient, solubilizer, viscosity control agent, intermediate, raw material, demulsifier, corrosion inhibitor, degreasing agent

APPLICATIONS:

Cosmetic industry preparations: (Alkasurf NP-9; Conco NI-90, NI-100; Hetoxide NP-9); perfumery (Hetoxide NP-9)

Degreasers: (Hyonic NP-90)

Farm products: (Chemax NP-9; Hyonic NP-90; Surfonic N-95); agricultural oils/sprays (Sterox NJ; Teric N9); herbicides (Alkasurf NP-9; T-Det N-9.5); insecticides/pesticides (Alkasurf NP-9; Conco NI-90, NI-100; Empilan NP9; Lorapal HV9; Macol NP-9.5; Rewopal HV9; Steinapal HV9; T-Det N-9.5; Triton N-101)

Food applications: indirect food additives (Hyonic NP-90; Surfonic N-95); fruit/vegetable washing (Sterox NJ)

Household detergents: (Alkasurf NP-9; Carsonon N-9; Cedepal CO-630; Conco NI-

POE (9) nonyl phenyl ether (cont'd.)

90, NI-100; Empilan NP9; Hetoxide NP-9; Hyonic NP-90; Igepal CO-630; Lorapal HV9; Neutronyx 600; Peganol NP9; Rewopal HV9; Siponic NP9; Steinapal HV9; Synperonic NP9; T-Det N-9.5; Triton N-101); detergent base (NP-55-90, NP-55-95); dishwashing (Neutronyx 600); hard surface cleaner (Renex 698; Sterox NJ); heavy-duty cleaner (Norfox NP-9; Teric N9); laundry detergent (Alkasurf NP-9; Igepal CO-630; Neutronyx 600; Renex 698; Sterox NJ); liquid detergents (Sterox NJ); powdered detergents (Merpoxen NO90, NO95; Sterox NJ)

Industrial applications: (Arkopal N-090; Hostapal N-090); construction (Empilan NP9; Surfonic N-95; Teric N9); dust control (Surfonic N-95); dyes and pigments (Alkasurf NP-9; Hetoxide NP-9; Renex 698; Tergitol TP-9); electroplating (Alkasurf NP-9); industrial processing (Chemax NP-9; NP-55-90); lubricating/cutting oils (Alkasurf NP-9; Surfonic N-95); metalworking (Empilan NP9; Hetoxide NP-9; Igepal CO-630; Peganol NP9); paint mfg. (Alkasurf NP-9; Igepal CO-630; Macol NP-9.5; Siponic NP9; Sterox NJ; T-Det N-9.5; Teric N9); paper mfg. (Ablunol NP9; Alkasurf NP-9; Carsonon N-9; Hyonic NP-90; Igepal CO-630; Lorapal HV9; NP-55-90; NP-55-95; Peganol NP9; Rewopal HV9; Siponic NP9; Steinapal HV9; Sterox NJ; Surfonic N-95; T-Det N-9.5); petroleum industry (Conco NI-90, NI-100; Empilan NP9; Surfonic N-95; T-Det N-9.5); photography (Surfonic N-95); polishes and waxes (Conco NI-90, NI-100); polymers/polymerization (Conco NI-90, NI-100; Siponic NP9); printing inks (Alkasurf NP-9); rubber (Surfonic N-95); silicone products (Chemax NP-9); textile/leather processing (Ablunol NP9; Alkasurf NP-9; Carsonon N-9; Conco NI-90, NI-100; Hyonic NP-90; Igepal CO-630; Lorapal HV9; Macol NP-9.5; Merpoxen NO90, NO95; NP-55-90; NP-55-95; Peganol NP9; Rewopal HV9; Siponic NP9; Steinapal HV9; Sterox NJ; Surfonic N-95; T-Det N-9.5; Triton N-101)

Industrial cleaners: (Alkasurf NP-9; Carsonon N-9; Conco NI-90, NI-100; Hyonic NP-90; Igepal CO-630; Lorapal HV9; Peganol NP9; Rewopal HV9; Steinapal HV9; T-Det N-9.5; Triton N-101); bottle cleaners (Conco NI-90, NI-100); dairy/food plant cleaners (Igepal CO-630; Sterox NJ); institutional cleaners (Peganol NP9); lime soap dispersion (Surfonic N-95); metal processing surfactants (Ablunol NP9; Renex 698; Sterox NJ); sanitizers/germicides (Igepal CO-630; Neutronyx 600; Norfox NP-9; Renex 698); solvent cleaners (Teric N9); textile cleaning (Hetoxide NP-9; Igepal CO-630; Renex 698; Synperonic NP9; Tergitol TP-9)

PROPERTIES:

Form:

Liquid (Ablunol NP9; Alkasurf NP-9; Arkopal N-090; Carsonon N-9; Cedepal CO-630; Chemax NP-9; Conco NI-90, NI-100; Gradonic N-95; Hostapal N-090; Iconol NP-9; Igepal CO-630; Lorapal HV9; Macol NP-9.5; Neutronyx 600; Norfox NP-9; NP-55-90; NP-55-95; Nutrol 600; Rewopal HV9; Serdox NNP9; Siponic NP9; Steinapal HV9; Surfonic N-95; Synperonic NP9; Tergitol TP-9; Teric N9; Triton N-101; Trycol NP-9); (@ 30 C) (Hetoxide NP-9)

Clear liquid (Hyonic NP-90; Renex 698; Tergitol NP-9)

Clear, viscous liquid (Peganol NP9; Sterox NJ)

Essentially clear, viscous liquid (T-Det N-9.5)
Viscous liquid (Merpoxen NO90, NO95)
Liquid to soft paste (Empilan NP9)

Color:

Almost colorless (Gradonic N-95; Igepal CO-630)
Light/pale (Alkasurf NP-9; Carsonon N-9)
Water-white (Merpoxen NO90, NO95)
Pale straw (Empilan NP9)
Slightly yellow (Tergitol NP-9)
Yellow (Sterox NJ)
APHA 70 max. (Iconol NP-9)
APHA 100 (Triton N-101)
APHA 100 max. (Peganol NP9)
Gardner 1 (Trycol NP-9)
Gardner 2 max. (Lorapal HV9; Rewopal HV9; Steinapal HV9)
Hazen 100 (Teric N9)
Hazen 150 max. (Synperonic NP9)
Hazen 250 (Teric GN9)
Pt-Co 50 max. (Tergitol TP-9)

Odor:

Low (Alkasurf NP-9)
Aromatic (Igepal CO-630)
Mild (Gradonic N-95; Sterox NJ)
Mild, aromatic (T-Det N-9.5)
Mild, pleasant (Carsonon N-9)
Mild, characteristic (Tergitol NP-9)

Composition:

> 99% active (Hyonic NP-90; Nutrol 600; Synperonic NP9)
99.5% active (Sterox NJ)
99.5% active min. (T-Det N-9.5)
100% active (Alkasurf NP-9; Carsonon N-9; Conco NI-90, NI-100; Empilan NP9; Gradonic N-95; Igepal CO-630; Lorapal HV9; Merpoxen NO90, NO95; Renex 698; Rewopal HV9; Steinapal HV9; Surfonic N-95; Tergitol NP-9; Teric N9; Triton N-101; Trycol NP-9)
100% conc. (Ablunol NP9; Arkopal N-090; Cedepal CO-630; Hostapal N-090; Macol NP-9.5; Neutronyx 600; Norfox NP-9; NP-55-90; NP-55-95; Serdox NNP9; Siponic NP9; Tergitol TP-9)

Solubility:

Sol. in acetone (Surfonic N-95)
Sol. in alcohols (Carsonon N-9; Lorapal HV9; Rewopal HV9; Steinapal HV9; Synperonic NP9)
Sol. in aromatics (Carsonon N-9)
Sol. in benzene (Teric N9)

POE (9) nonyl phenyl ether (cont'd.)

Sol. in butyl acetate (Tergitol NP-9)
Sol. in butyl Cellosolve (Igepal CO-630; T-Det N-9.5; Tergitol NP-9)
Sol. in carbon tetrachloride (Surfonic N-95; T-Det N-9.5)
Sol. in dilute caustic sol'ns. (Gradonic N-95)
Sol. in chlorinated hydrocarbons (Carsonon N-9; Lorapal HV9; Rewopal HV9; Steinapal HV9)
Sol. in corn oil (Igepal CO-630); slightly sol. (Tergitol NP-9)
Sol. hazy in cottonseed oil (@ 10%) (Renex 698)
Sol. in dibutyl phthalate (Igepal CO-630)
Sol. in ethanol (Igepal CO-630; T-Det N-9.5)
Sol. in ethyl acetate (Teric N9)
Sol. in ethylene dichloride (Igepal CO-630)
Sol. in ethylene glycol (Igepal CO-630; T-Det N-9.5; Tergitol NP-9)
Sol. in glycol ethers (Synperonic NP9)
Sol. in dilute inorganic salt sol'ns. (Gradonic N-95)
Sol. in isopropanol (Hetoxide NP-9); (@ 10%) (Renex 698); anhyd. isopropanol (Tergitol NP-9)
Disp. in kerosene (Synperonic NP9)
Sol. in ketone (Lorapal HV9; Rewopal HV9; Steinapal HV9)
Sol. in methanol (Surfonic N-95)
Sol. in min. acid sol'ns. (Gradonic N-95)
Sol. in min. oil (Lorapal HV9; Rewopal HV9; Steinapal HV9)
Sol. in heavy aromatic naphtha (Igepal CO-630)
Sol. in olein (Teric N9)
Sol. in perchloroethylene (Igepal CO-630; Peganol NP9; Teric N9); (@ 5%) (Trycol NP-9); (@ 10%) (Renex 698)
Sol. in propylene glycol (@ 10%) (Renex 698)
Sol. in Stoddard solvent (@ 5%) (Trycol NP-9)
Sol. in toluene (Lorapal HV9; Rewopal HV9; Steinapal HV9; T-Det N-9.5; Tergitol NP-9)
Sol. in veg. oil (Teric N9)
Sol. in water (Ablunol NP9; Alkasurf NP-9; Carsonon N-9; Cedepal CO-630; Gradonic N-95; Hetoxide NP-9; Hyonic NP-90; Iconol NP-9; Igepal CO-630; Peganol NP9; Sterox NJ; Surfonic N-95; Synperonic NP9; T-Det N-9.5; Tergitol NP-9); (@ 5%) (Trycol NP-9); sol. hazy (@ 10%) (Renex 698)
Sol. in xylene (Igepal CO-630; Lorapal HV9; Peganol NP9; Rewopal HV9; Steinapal HV9; Surfonic N-95; T-Det N-9.5); (@ 10%) (Renex 698)

Ionic Nature:

Nonionic (Ablunol NP9; Alkasurf NP-9; Arkopal N090; Carsonon N-9; Cedepal CO-630; Conco NI-90, NI-100; Empilan NP9; Gradonic N-95; Hetoxide NP-9; Hostapal N-090; Hyonic NP-90; Iconol NP-9; Igepal CO-630; Lorapal HV9; Macol NP-9.5; Merpoxen NO90, NO95; Neutronyx 600; Norfox NP-9; NP-55-90; NP-55-95; Nutrol 600; Peganol NP9; Renex 698; Rewopal HV9; Serdox NNP9; Siponic NP9;

Steinapal HV9; Sterox NJ; Surfonic N-95; Synperonic NP9; T-Det N-9.5; Tergitol NP-9, TP-9; Teric N9; Triton N-101; Trycol NP-9)

M.W.:

611 (Iconol NP-9)
616 (Tergitol NP-9, TP-9)
625 (Sterox NJ)
632 (Surfonic N-95)
642 avg. (Triton N-101)

Sp.gr.:

1.04–1.08 (Conco NI-90)
1.05 (Neutronyx 600)
1.05–1.07 (Conco NI-100)
1.0538 (Sterox NJ)
1.055 (Peganol NP9)
1.057 (20/20 C) (Tergitol NP-9, TP-9)
1.06 (Alkasurf NP-9; Iconol NP-9; Igepal CO-630; Renex 698; T-Det N-9.5); (20 C) (Teric N9)
1.061 (20/20 C) (Surfonic N-95)

Density:

1.0 g/cc (Empilan NP9)
1.058 g/ml (20 C) (Synperonic NP9)
1.06 g/ml (Hyonic NP-90)
8.7 lb/gal (Triton N-101)
8.80 lb/gal (T-Det N-9.5; Trycol NP-9); (20 C) (Surfonic N-95; Tergitol NP-9, TP-9)
8.83 lb/gal (Carsonon N-9)

Visc.:

180–250 cps (Carsonon N-9)
225–300 cps (Igepal CO-630)
240 cps (Triton N-101)
241 cps (Sterox NJ)
245 cps (Renex 698)
300 cps (Iconol NP-9)
330 cps (20 C) (Teric N9)
340 cps (20 C) (Synperonic NP9)
11 cs (100 C) (Peganol NP9)
300 cs (Empilan NP9)
318 cs (20 C) (Tergitol NP-9, TP-9)
325 cs (Trycol NP-9)
68 SUS (210 F) (T-Det N-9.5)
510 SUS (100 F) (Surfonic N-95)

F.P.:

< 0 C (Synperonic NP9)
5 C (Surfonic N-95)

POE (9) nonyl phenyl ether (cont'd.)

41 F (T-Det N-9.5)

B.P.:

> 250 C (dec.) (Tergitol NP-9)

M.P.:

0 ± 2 C (Teric N9)

Pour Pt.:

–1 C (Peganol NP9)
0 C (Synperonic NP9)
2 C (Hyonic NP-90)
4 C (Iconol NP-9)
26 F (Sterox NJ)
28 F (Renex 698)
31 F (Chemax NP-9)
31 ± 2 F (Igepal CO-630)
35–39 F (Carsonon N-9)
40 F (Triton N-101)

Solidification Pt.:

0 C (Tergitol NP-9, TP-9)
24–38 F (Carsonon N-9)
26 ± 2 F (Igepal CO-630)

Flash Pt.:

> 200 F (Neutronyx 600); (PMCC) (Igepal CO-630)
300 F (TOC) (Triton N-101)
> 300 F (Renex 698)
500 F (OC) (Surfonic N-95; T-Det N-9.5)
540 F (Carsonon N-9; Sterox NJ); (COC) (Tergitol NP-9, TP-9)

Fire Pt.:

600 F (Carsonon N-9)

Cloud Pt.:

51–56 C (Sterox NJ); (1% aq.) (Synperonic NP9)
52 ± 2 C (1% in hard water) (Teric N9)
52–54 C (1.0% aq.) (Merpoxen NO90)
52–56 C (1.0% aq.) (Iconol NP-9; Peganol NP9)
54 C (Neutronyx 600); (0.5% aq.) (Tergitol NP-9, TP-9); (1% aq.) (Hyonic NP-90; Triton N-101)
54.2 C (1% aq.) (Surfonic N-95)
55 C (Trycol NP-9); (1% aq.) (Empilan NP9)
56–58 C (1% aq.) (Merpoxen NO95)
59–62 C (2% in dist. water) (Lorapal HV9; Rewopal HV9; Steinapal HV9)
87 F (Conco NI-90)
125–133 F (Conco NI-100)
126–133 F (1% sol'n.) (Igepal CO-630)
129 F (1% sol'n.) (Renex 698)

130 F (Carsonon N-9); (5% sol'n.) (Gradonic N-95)
135 F (1%) (T-Det N-9.5)

HLB:

12.7 (Siponic NP9)
12.8 (Synperonic NP9; Teric N9)
12.9 (Surfonic N-95; Tergitol NP-9, TP-9; Trycol NP-9)
13.0 (Ablunol NP9; Cedepal CO-630; Hetoxide NP-9; Hyonic NP-90; Iconol NP-9; Igepal CO-630; Macol NP-9.5; Norfox NP-9; NP-55-90; NP-55-95; Nutrol 600; Peganol NP9; Renex 698; Serdox NNP9; Sterox NJ)
13.1 (T-Det N-9.5)
13.4 (Alkasurf NP-9; Triton N-101)

Hydroxyl No.:

85–95 (Hetoxide NP-9)
89 (Synperonic NP9)
90 (Empilan NP9)
91 (Peganol NP9; Tergitol TP-9)

Stability:

Good (Alkasurf NP-9; Sterox NJ)
Stable in acid and alkaline systems (Neutronyx 600)
Stable to hydrolysis by acids and alkalis (Gradonic N-95; Trycol NP-9)
Very good stability in acids, alkalis, and in hard water (Renex 698)
Stable to acids, alkalis, and salts (Surfonic N-95; T-Det N-9.5; Tergitol NP-9)
Stable to acids, alkalis, hard water, and foam (Conco NI-90, NI-100)
Good stability in hard or saline waters and in reasonable concs. of acids and alkalis (Teric N9)
Stable to acids, alkalis, dilute sol'ns. of many oxidizing and reducing agents (Igepal CO-630)
Stable in presence of dilute acids and alkalis, in hard water, and oxidizing agents (Lorapal HV9; Rewopal HV9; Steinapal HV9)

Ref. Index:

1.4884 (Sterox NJ)

pH:

Neutral (1% sol'n.) (Carsonon N-9)
5.0–8.0 (10% aq.) (Tergitol NP-9, TP-9)
6.0–7.5 (1%) (Sterox NJ); (5% aq.) (Iconol NP-9; Peganol NP9)
6.0–8.0 (1% aq.) (Synperonic NP9; Teric N9)
7.0 (1% aq.) (Hyonic NP-90)

Foam (Ross Miles):

52 mm initial, 47 mm after 5 min (0.05% sol'n.) (Hyonic NP-90)
135 mm initial, 30 mm after 5 min (50 C) (Tergitol NP-9)

Surface Tension:

29 dynes/cm (1%) (Triton N-101)
29.8 dynes/cm (Sterox NJ)

POE (9) nonyl phenyl ether (cont'd.)

30 dynes/cm (0.01% sol'n.) (Renex 698); (0.1% aq.) (Tergitol NP-9)
30.6 dynes/cm (0.1%, 20 C) (Synperonic NP9; Teric N9)
30.8 dynes/cm (0.1%) (Surfonic N-95)
31 dynes/cm (0.01% sol'n.) (Hyonic NP-90; Igepal CO-630)
31.2 dynes/cm (0.1%) (Tergitol TP-9)
32 dynes/cm (0.1% aq.) (Iconol NP-9)

Wetting (Draves):
8 s (0.1% conc.) (Tergitol NP-9)
0.04% for a 25-s test (3-g hook) (Hyonic NP-90)

Biodegradable: (Serdox NNP9); > 90% (Surfonic N-95); fair biodegradability (Tergitol TP-9)

TOXICITY/HANDLING:
Eye irritant (T-Det N-9.5; Tergitol TP-9)
Causes skin irritation, eye burns (Sterox NJ)
May cause skin and eye irritation; spillages are slippery (Teric N9)
Eye irritant; skin irritant on prolonged contact with conc. form (Synperonic NP9)
Severe eye irritant; highly toxic to aquatic life; burning can produce carbon monoxide and/or carbon dioxide (Tergitol NP-9)

STORAGE/HANDLING:
Contact with conc. oxidizing or reducing agents may be explosive (Igepal CO-630; T-Det N-9.5)
Store at temps. below 50 C (Tergitol NP-9)
May turn cloudy below 50 F—mix before using (Sterox NJ)

STD. PKGS.:
Steel drums and bulk (Sterox NJ)
200-kg net steel drums (Merpoxen NO90)
200-kg net iron drums or road tankers (Merpoxen NO95)
55-gal drums, 1- and 5-gal containers (Tergitol TP-9)
55-gal (470 lb net) steel drums (Iconol NP-9)
55-gal (580 lb net) drums; 5-gal (44 lb net) pails; 1-gal (8.5 lb net) jugs (Tergitol NP-9)
55-gal (200 l) steel drums, bulk (Hyonic NP-90)
460-lb net closed-head steel drums (T-Det N-9.5)

POE (10) nonyl phenyl ether

SYNONYMS:
Nonoxynol-10 (CTFA)
Nonyl phenol ethoxylate (10 moles EO)
Nonyl phenol 10 polyglycol ether

SYNONYMS *(cont'd.)*:

Nonylphenoxy polyethoxy ethanol (10 moles EO)
Nonylphenoxy poly(ethyleneoxy) ethanol (10 moles EO)
PEG-10 nonyl phenyl ether
PEG 500 nonyl phenyl ether
POE (10) nonyl phenol

STRUCTURE:

$C_9H_{19}C_6H_4(OCH_2CH_2)_nOH$
where avg. $n = 10$

CAS No.:

9016-45-9 (generic); 26027-38-3 (generic); 27177-08-8; 37205-87-1 (generic)
RD No.: 977057-35-4

TRADENAME EQUIVALENTS:

Alkasurf NP-10 [Alkaril]
Arkopal N-100 [Amer. Hoechst]
Carsonon N-10 [Carson] (10–11 EO)
Cedepal CO-710 [Domtar] (10–11 EO)
Chemax NP-10 [Chemax]
Chemcol NPE-100 [Chemform]
Conco NI-110 [Continental Chem.] (10–11 EO)
Hostapal N-100 [Amer. Hoechst]
Hyonic NP-100 [Diamond Shamrock]
Hyonic PE-100 [Diamond Shamrock]
Iconol NP-10 [BASF Wyandotte]
Igepal CO-660 [GAF]
Igepal CO-710 [GAF] (10–11 EO)
Lorapal HV10 [Dutton & Reinisch]
Makon 10 [Stepan]
Merpoxen NO100 [Elektrochemische Fabrik Kempen]
Nikkol NP-10 [Nikko]
Peganol NP10 [Borg-Warner]
Polystep F-4 [Stepan]
Renex 690 [ICI United States]
Rewopal HV10 [Dutton & Reinisch]
Serdox NNP10 [Servo B.V.]
Siponic NP10.5 [Alcolac]
Steinapal HV10 [Dutton & Reinisch]
Sterox NK [Monsanto] (10.5 EO)
Surfonic N-100, N-102 [Jefferson]
Synperonic NP10 [ICI Petrochem. Div.]
T-Det N-10.5 [Thompson-Hayward] (10.5 EO)
Tergitol NP10 [Union Carbide] (10.5 EO)
Tergitol NPX [Union Carbide] (10.5 EO)

POE (10) nonyl phenyl ether (cont'd.)

TRADENAME EQUIVALENTS *(cont'd.):*

Teric GN10 [ICI Australia Ltd.] (tech. grade)
Teric N10 [ICI Australia Ltd.]

CATEGORY:

Dispersant, solubilizer, emulsifier, coemulsifier, detergent, raw material, base, wetting agent, rewetting agent, degreasing agent, foaming agent, penetrant, demulsifier, stabilizer, surfactant, intermediate, corrosion inhibitor

APPLICATIONS:

Cleansers: waterless hand cleaner (T-Det N-10.5)

Cosmetic industry preparations: (Alkasurf NP-10; Conco NI-110; Makon 10; Nikkol NP-10; Surfonic N-100, N-102); perfumery (Makon 10); shampoos (Makon 10)

Farm products: (Chemax NP-10; Chemcol NPE-100; Hyonic NP-100; Makon 10; Surfonic N-100, N-102); agricultural oils/sprays (Sterox NK; Teric N10); insecticides/pesticides (Conco NI-110; Lorapal HV10; Nikkol NP-10; Rewopal HV10; Steinapal HV10)

Food applications: indirect food additives (Hyonic NP-100; Surfonic N-100, N-102); fruit/vegetable washing (Sterox NK)

Household detergents: (Alkasurf NP-10; Carsonon N-10; Chemcol NPE-100; Conco NI-110; Igepal CO-660, CO-710; Lorapal HV10; Merpoxen NO100; Peganol NP10; Rewopal HV10; Siponic NP10.5; Steinapal HV10; Sterox NK; Synperonic NP10; T-Det N-10.5; Tergitol NP10); all-purpose cleaner (T-Det N-10.5); detergent base (Hyonic NP-100, PE-100; Makon 10); dishwashing (Hyonic NP-100, PE-100); hard surface cleaner (Renex 690; T-Det N-10.5); heavy-duty cleaner (Teric N10); laundry detergent (Alkasurf NP-10; Hyonic NP-100, PE-100; Igepal CO-660, CO-710; Renex 690; Sterox NK; T-Det N-10.5); liquid detergents (Tergitol NP10); powdered detergents (Merpoxen NO100; Tergitol NP10)

Industrial applications: (Arkopal N-100; Chemax NP-10; Hostapal N-100; Nikkol NP-10; Tergitol NP10); construction (Teric N10); dyes and pigments (Alkasurf NP-10; Makon 10; Renex 690; Tergitol NPX; Teric N10); electroplating (Alkasurf NP-10); industrial processing (Chemax NP-10; Hyonic PE-100; Surfonic N-100, N-102); lubricating/cutting oils (Makon 10; Surfonic N-100, N-102); metalworking (Igepal CO-660, CO-710; Peganol NP10); paint mfg. (Alkasurf NP-10; Igepal CO-660, CO-710; Peganol NP10; Siponic NP10.5; Sterox NK; Teric N10); paper mfg. (Alkasurf NP-10; Carsonon N-10; Hyonic NP-100, PE-100; Igepal CO-660, CO-710; Lorapal HV10; Makon 10; Peganol NP10; Rewopal HV10; Serdox NNP10; Siponic NP10.5; Steinapal HV10; Sterox NK); petroleum industry (Chemcol NPE-100; Conco NI-110; Makon 10; Surfonic N-100, N-102); plastics (Hyonic NP-100); polishes and waxes (Conco NI-110); polymers/polymerization (Conco NI-110; Polystep F-4; Siponic NP10.5); rubber (Surfonic N-100, N-102); silicone products (Chemax NP-10); textile/leather processing (Alkasurf NP-10; Carsonon N-10; Chemcol NPE-100; Conco NI-110; Hyonic NP-100, PE-100; Igepal CO-660, CO-710; Lorapal HV10; Makon 10; Merpoxen NO100; Peganol NP10; Rewopal HV10; Serdox NNP10; Siponic NP10.5; Steinapal HV10; Sterox

NK; Surfonic N-100, N-102; T-Det N-10.5; Tergitol NPX)

Industrial cleaners: (Alkasurf NP-10; Carsonon N-10; Conco NI-110; Hyonic NP-100, PE-100; Lorapal HV10; Makon 10; Peganol NP10; Rewopal HV10; Steinapal HV10; Sterox NK; T-Det N-10.5); bottle cleaners (Conco NI-110); dairy/food plant cleaners (Igepal CO-660, CO-710; Sterox NK); institutional cleaners (Peganol NP10); janitorial cleaners (T-Det N-10.5); metal processing surfactants (Makon 10; Renex 690; Sterox NK; T-Det N-10.5); sanitizers/germicides (Igepal CO-660, CO-710; Renex 690); solvent cleaners (Teric N10); specialty detergents (T-Det N-10.5); textile cleaning (Igepal CO-660, CO-710; Makon 10; Peganol NP10; Renex 690; Tergitol NPX)

PROPERTIES:

Form:

Liquid (Alkasurf NP-10; Arkopal N-100; Carsonon N-10; Cedepal CO-710; Chemax NP-10; Chemcol NPE-100; Conco NI-110; Hostapal N-100; Iconol NP-10; Igepal CO-660, CO-710; Lorapal HV10; Makon 10; Nikkol NP-10; Renex 690; Rewopal HV10; Serdox NNP10; Steinapal HV10; Surfonic N-100, N-102; Synperonic NP10; Tergitol NPX; Teric N10)

Clear liquid (Hyonic NP-100, PE-100; Peganol NP10; Polystep F-4; Tergitol NP10)

Essentially clear, viscous liquid (T-Det N-10.5)

Clear viscous liquid (Sterox NK)

Viscous liquid (Merpoxen NO100)

Semisolid (Siponic NP10.5)

Color:

Water-white (Merpoxen NO100)

Light/pale (Alkasurf NP-10; Carsonon N-10; Polystep F-4)

Colorless to pale yellow (Chemcol NPE-100)

Light straw (Makon 10)

Pale yellow (Igepal CO-660, CO-710; Tergitol NP10)

Yellow (Sterox NK)

APHA 70 max. (Iconol NP-10)

APHA 100 max. (Peganol NP10)

Gardner 2 max. (Lorapal HV10; Rewopal HV10; Steinapal HV10)

Hazen 100 (Teric N10)

Hazen 150 max. (Synperonic NP10)

Hazen 250 (Teric GN10)

Pt-Co 50 max. (Tergitol NPX)

Odor:

Low (Alkasurf NP-10)

Mild (Sterox NK)

Mild, pleasant (Carsonon N-10)

Mild, aromatic (T-Det N-10.5)

Mild, characteristic (Tergitol NP10)

POE (10) nonyl phenyl ether (cont'd.)

Composition:

> 99% active (Hyonic NP-100, PE-100; Synperonic NP10)
99.5% active (Sterox NK)
99.5% active min. (T-Det N-10.5)
100% active (Alkasurf NP-10; Carsonon N-10; Chemcol NPE-100; Conco NI-110; Igepal CO-660, CO-710; Lorapal HV10; Makon 10; Merpoxen NO100; Polystep F-4; Renex 690; Rewopal HV10; Steinapal HV10; Surfonic N-100, N-102; Tergitol NP10; Teric N10)
100% conc. (Arkopal N-100; Cedepal CO-710; Hostapal N-100; Nikkol NP-10; Serdox NNP10; Siponic NP10.5; Tergitol NPX)

Solubility:

Sol. in acetone (Surfonic N-100, N-102)
Sol. in alcohols (Carsonon N-10; Lorapal HV10; Rewopal HV10; Steinapal HV10; Synperonic NP10)
Sol. in aromatics (Carsonon N-10)
Sol. in benzene (Teric N10)
Sol. in butyl acetate (Tergitol NP10)
Sol. in butyl Cellosolve (Igepal CO-660, CO-710; T-Det N-10.5; Tergitol NP10); (@ 10%) (Tergitol NPX)
Sol. in carbon tetrachloride (Surfonic N-100, N-102)
Sol. in chlorinated hydrocarbons (Carsonon N-10; Lorapal HV10; Rewopal HV10; Steinapal HV10)
Sol. in corn oil (Igepal CO-660); slightly sol. (Tergitol NP10)
Sol. hazy in cottonseed oil (@ 10%) (Renex 690)
Sol. in dibutyl phthalate (Igepal CO-660, CO-710)
Sol. in ethanol (Igepal CO-660, CO-710; T-Det N-10.5)
Sol. in ethyl acetate (Teric N10)
Sol. in ethyl Icinol (Teric N10)
Sol. in ethylene dichloride (Igepal CO-660, CO-710)
Sol. in ethylene glycol (Igepal CO-710; T-Det N-10.5; Tergitol NP10)
Sol. in glycol ethers (Synperonic NP10)
Sol. in HAN (T-Det N-10.5)
Sol. in isopropanol (@ 10%) (Renex 690; Tergitol NPX); sol. in anhyd. isopropanol (Tergitol NP10)
Sol. in ketone (Lorapal HV10; Rewopal HV10; Steinapal HV10)
Sol. in methanol (Surfonic N-100, N-102)
Sol. in min. oil (Lorapal HV10; Rewopal HV10; Steinapal HV10)
Sol. in heavy aromatic naphtha (Igepal CO-660, CO-710)
Sol. in olein (Teric N10)
Sol. in perchloroethylene (Igepal CO-660, CO-710; Peganol NP10; Teric N10); (@ 10%) (Renex 690; Tergitol NPX)
Sol. in propylene glycol (@ 10%) (Renex 690)
Sol. in toluene (Lorapal HV10; Rewopal HV10; Steinapal HV10; T-Det N-10.5;

Tergitol NP10)
Sol. in veg. oil (Teric N10)
Sol. in water (Alkasurf NP-10; Carsonon N-10; Cedepal CO-710; Hyonic NP-100; Iconol NP-10; Igepal CO-660, CO-710; Makon 10; Peganol NP10; Sterox NK; Surfonic N-100, N-102; Synperonic NP10; T-Det N-10.5; Tergitol NP10; Teric N10); (@ 10%) (Tergitol NPX); sol. hazy (@ 10%) (Renex 690)
Sol. in xylene (Igepal CO-660, CO-710; Lorapal HV10; Peganol NP10; Rewopal HV10; Steinapal HV10; Surfonic N-100, N-102; T-Det N-10.5); (@ 10%) (Renex 690; Tergitol NPX)

Ionic Nature:
Nonionic (Alkasurf NP-10; Arkopal N-100; Carsonon N-10; Cedepal CO-710; Chemcol NPE-100; Conco NI-110; Hyonic NP-100, PE-100; Iconol NP-10; Igepal CO-660, CO-710; Lorapal HV10; Makon 10; Merpoxen NO100; Nikkol NP-10; Peganol NP10; Polystep F-4; Renex 690; Rewopal HV10; Serdox NNP10; Siponic NP10.5; Steinapal HV10; Sterox NK; Surfonic N-100, N-102; Synperonic NP10; T-Det N-10.5; Tergitol NPX; Teric N10)

M.W.:
655 (Iconol NP-10)
660 (Surfonic N-100)
668 (Surfonic N-102)
673 (Sterox NK)
682 (Tergitol NP10); (avg.) (Tergitol NPX)

Sp.gr.:
1.05–1.07 (Conco NI-110)
1.057 (Peganol NP10)
1.06 (Alkasurf NP-10; Iconol NP-10; Igepal CO-660, CO-710; Renex 690; T-Det N-10.5)
1.062 (Sterox NK); (20/20 C) (Tergitol NP10, NPX)
1.063 (20 C) (Teric N10)
1.064 (20/20 C) (Surfonic N-100)
1.065 (20/20 C) (Surfonic N-102)

Density:
1.061 g/ml (20 C) (Synperonic NP10)
1.07 g/ml (Hyonic NP-100)
8.8 lb/gal (T-Det N-10.5); (20 C) (Surfonic N-100, N-102)
8.84 lb/gal (Carsonon N-10; Tergitol NP10); (20 C) (Tergitol NPX)
8.85 lb/gal (Makon 10)
8.9 lb/gal (Hyonic PE-100)

Visc.:
12 cs (100 C) (Peganol NP10)
180–250 cps (Carsonon N-10)
225–275 cps (Igepal CO-660)
235 cps (Makon 10)

POE (10) nonyl phenyl ether (cont'd.)

240–300 cps (Igepal CO-710)
260 cps (Renex 690)
300 cps (Iconol NP-10)
304 cps (Sterox NK)
318 cs (20 C) (Tergitol NPX)
327 cs (20 C) (Tergitol NP10)
360 cps (20 C) (Synperonic NP10; Teric N10)
69 SUS (210 F) (Surfonic N-100)

F.P.:
5 C (Synperonic NP10)
8 C (Surfonic N-100)
9 C (Surfonic N-102)
49 F (T-Det N-10.5)

B.P.:
> 250 C (dec.) (Tergitol NP10)

M.P.:
5 ± 2 C (Teric N10)

Pour Pt.:
2.8 C (Makon 10)
4 C (Hyonic NP-100, PE-100)
5 C (Synperonic NP10)
8 C (Peganol NP10)
9 C (Iconol NP-10)
33 F (Renex 690)
35–39 F (Carsonon N-10)
36 F (Sterox NK)
46 ± 2 F (Igepal CO-660)
49 F (Chemax NP-10)
49 ± 2 F (Igepal CO-710)

Solidification Pt.:
4 C (Makon 10)
7 C (Tergitol NP10, NPX)
24–28 F (Carsonon N-10)
41 ± 2 F (Igepal CO-660)
45 ± 2 F (Igepal CO-710)

Flash Pt.:
> 200 C (Peganol NP10)
> 200 F (PMCC) (Igepal CO-660, CO-710)
> 300 F (Renex 690)
500 F (OC) (T-Det N-10.5); (COC) (Tergitol NP10, NPX)
510 F (OC) (Surfonic N-100, N-102)
540 F (Carsonon N-10)
560 F (Sterox NK)

Fire Pt.:
600 F (Carsonon N-10)
Cloud Pt.:
54 C (1% sol'n.) (Makon 10)
58–64 C (1% aq.) (Chemcol NPE-100)
60–62 C (Merpoxen NO100)
60–65 C (1% aq.) (Iconol NP-10)
61–65 C (1% aq.) (Peganol NP10)
62–67 C (1% aq.) (Synperonic NP10)
63 C (0.5% aq.) (Tergitol NP10, NPX)
65 C (1% aq.) (Surfonic N-100)
66–70 C (Sterox NK)
67 ± 2 C (1% in hard water) (Teric N10)
68 C (1% aq.) (Hyonic NP-100, PE-100)
70–73 C (2% in dist. water) (Lorapal HV10; Rewopal HV10; Steinapal HV10)
81 C (1% aq.) (Surfonic N-102)
140–149 F (1% sol'n.) (Igepal CO-660)
143–146 F (Carsonon N-10)
150 F (1% sol'n.) (Renex 690)
155 F (T-Det N-10.5)
158–165 F (Conco NI-110); (1% sol'n.) (Igepal CO-710)
HLB:
10.4 (T-Det N-10.5)
13.2 (Hyonic NP-100; Igepal CO-660; Peganol NP10; Surfonic N-100)
13.3 (Renex 690; Synperonic NP10; Teric N10)
13.4 (Chemcol NPE-100; Surfonic N-102)
13.5 (Alkasurf NP-10; Iconol NP-10; Serdox NNP10; Sterox NK)
13.6 (Cedepal CO-710; Igepal CO-710; Tergitol NP-10, NPX)
14.9 (Siponic NP10.5)
16.5 (Nikkol NP-10)
Hydroxyl No.:
83 (Tergitol NPX)
85 (Peganol NP10; Synperonic NP10)
Stability:
Good (Alkasurf NP-10; Sterox NK)
Very good stability in acids, alkalis, and hard water (Renex 690)
Stable to acids, bases, salts (Surfonic N-100, N-102; T-Det N-10.5)
Stable to acids, alkalis, hard water, and foam (Conco NI-110)
Good stability in hard or saline waters and in reasonable concs. of acids and alkalis (Teric N10)
Stable in presence of dilute acids and alkalis, in hard water, and oxidizing agents (Lorapal HV10; Rewopal HV10; Steinapal HV10)
Stable to acids, alkalis, dilute sol'ns. of many oxidizing and reducing agents (Igepal

CO-660, CO-710)

Chemically stable in presence of ionic materials (Tergitol NP10)

Storage Stability:

Haze may form during storage (Peganol NP10)

Ref. Index:

1.4884 (20 C) (Surfonic N-102)

1.4888 (20 C) (Surfonic N-100)

pH:

Neutral (1% sol'n.) (Carsonon N-10)

5.0–7.0 (1% solids) (Lorapal HV10; Rewopal HV10; Steinapal HV10)

5.0–8.0 (10%) (Tergitol NPX); (10% aq.) (Tergitol NP10)

6.0–7.5 (1%) (Sterox NK); (5% aq.) (Iconol NP-10; Peganol NP10)

6.0–8.0 (1% aq.) (Synperonic NP10; Teric N10); (5% in 25% IPA) (Chemcol NPE-100)

7.0 (1%) (T-Det N-10.5); (1% aq.) (Hyonic NP-100, PE-100)

8.2 (Makon 10)

Foam (Ross Miles):

90 mm initial, 80 mm after 5 min (0.05% sol'n.) (Hyonic NP-100)

131 mm initial, 21 mm after 5 min (50 C) (Tergitol NP10)

Surface Tension:

30.6 dynes/cm (0.1%, 20 C) (Synperonic NP10; Teric N10)

30.9 dynes/cm (Sterox NK)

31 dynes/cm (0.01% sol'n.) (Igepal CO-660; Renex 690); (0.1%) (Surfonic N-100); (0.1% aq.) (Tergitol NP10)

31.2 dynes/cm (0.1%) (Surfonic N-102)

32 dynes/cm (0.01% sol'n.) (Hyonic NP-100, PE-100; Igepal CO-710)

34 dynes/cm (0.1% aq.) (Iconol NP-10)

Wetting (Draves):

0.05% for a 25-s test (3-g hook) (Hyonic NP-100)

9 s (0.1% conc.) (Tergitol NP10)

Biodegradable: Serdox NNP10; > 90% (Surfonic N-100, N-102); fair (Tergitol NPX)

TOXICITY/HANDLING:

Eye irritant (T-Det N-10.5; Tergitol NPX)

May cause skin and eye irritation; spillages are slippery (Teric N10)

Causes skin irritation, eye burns (Sterox NK)

Eye irritant; skin irritant on prolonged contact with conc. form (Synperonic NP10)

Severe eye irritant; highly toxic to aquatic life; burning can produce carbon monoxide and/or carbon dioxide (Tergitol NP10)

STORAGE/HANDLING:

Contact with conc. oxidizing or reducing agents may be explosive (Igepal CO-660, CO-710; T-Det N-10.5)

May turn cloudy below 55 F—mix before using (Sterox NK)

Store at temps. below 50 C (Tergitol NP10)

STD. PKGS.:

Steel drums and bulk (Sterox NK)
200-kg net steel drums (Merpoxen NO100)
55-gal drums, 1- and 5-gal containers (Tergitol NPX)
55-gal steel drums or bulk (Hyonic PE-100)
55-gal (200 l) steel drums, bulk (Hyonic NP-100)
55-gal (470 lb net) steel drums (Iconol NP-10)
55-gal (480 lb net) drums, 5-gal (44 lb net) pails, 1-gal (8.5 lb net) jugs (Tergitol NP10)
460-lb net closed-head steel drums (T-Det N-10.5)

POE (11) nonyl phenyl ether

SYNONYMS:

Nonoxynol-11 (CTFA)
Nonyl phenol ethoxylate (11 moles EO)
Nonyl phenol polyglycol ether (11 moles EO)
Nonylphenoxy polyethoxy ethanol (11 moles EO)
Nonylphenoxy poly(ethyleneoxy) ethanol (11 moles EO)
PEG-11 nonyl phenyl ether
PEG (11) nonyl phenyl ether
POE (11) nonyl phenol
Polyethoxylated nonylphenol (11 EO)

STRUCTURE:

$C_9H_{19}C_6H_4(OCH_2CH_2)_nOH$
where avg. $n = 11$

CAS No.:

9016-45-9 (generic); 26027-38-3 (generic); 37205-87-1 (generic)
RD No.: 977065-12-5

TRADENAME EQUIVALENTS:

Alkasurf NP-11 [Alkaril]
Arkopal N110 [Amer. Hoechst]
Carsonon N-11 [Lonza]
Hyonic NP-110 [Diamond Shamrock]
Macol NP-11 [Mazer]
Merpoxen NO110 [Elektrochemische Fabrik Kempen]
Neutronyx 656 [Onyx]
Norfox NP-11 [Norman, Fox & Co.]
Nutrol 656 [Clough Chem. Ltd.]
Serdox NNP11 [Servo B.V.]
Sterox NL [Monsanto]
Teric N11 [ICI Australia Ltd.]

POE (11) nonyl phenyl ether (cont'd.)

TRADENAME EQUIVALENTS *(cont'd.):*

Triton N-111 [Rohm & Haas]
Trycol NP-11 [Emery]

CATEGORY:

Dispersant, solubilizer, detergent, wetting agent, rewetting agent, emulsifier, surfactant, penetrant, stabilizer

APPLICATIONS:

Cosmetic industry preparations: (Alkasurf NP-11)

Farm products: agricultural oils/sprays (Hyonic NP-110; Teric N11)

Food applications: indirect food contact additive (Hyonic NP-110)

Household detergents: (Alkasurf NP-11; Carsonon N-11; Hyonic NP-110; Merpoxen NO110; Neutronyx 656; Sterox NL; Trycol NP-11); dishwashing (Hyonic NP-110; Neutronyx 656); heavy-duty cleaner (Norfox NP-11; Teric N11); laundry detergent (Alkasurf NP-11; Hyonic NP-110; Neutronyx 656; Sterox NL); light-duty cleaners (Nutrol 656); liquid detergents (Sterox NL); powdered detergents (Merpoxen NO110; Sterox NL)

Industrial applications: (Arkopal N110); construction (Alkasurf NP-11; Teric N11); dyes and pigments (Alkasurf NP-11); electroplating (Alkasurf NP-11); industrial processing (Trycol NP-11); latex preparations (Hyonic NP-110); paint mfg. (Alkasurf NP-11; Teric N11); paper mfg. (Hyonic NP-110; Serdox NNP11); textile/leather processing (Alkasurf NP-11; Hyonic NP-110; Merpoxen NO110; Serdox NNP11; Sterox NL; Triton N-111); wood pulping (Alkasurf NP-11)

Industrial cleaners: (Alkasurf NP-11; Carsonon N-11; Sterox NL); metal processing surfactants (Alkasurf NP-11); sanitizers/germicides (Neutronyx 656; Norfox NP-11); solvent cleaners (Teric N11)

PROPERTIES:

Form:

Liquid (Alkasurf NP-11; Arkopal N110; Carsonon N-11; Macol NP-11; Neutronyx 656; Norfox NP-11; Nutrol 656; Serdox NNP11; Teric N11; Triton N-111; Trycol NP-11)

Clear liquid (Hyonic NP-110; Sterox NL)

Viscous liquid (Merpoxen NO110)

Color:

Water-white (Merpoxen NO110)
Light (Alkasurf NP-11)
Yellow (Sterox NL)
APHA 100 (Triton N-111)
Gardner 2 (Trycol NP-11)
Hazen 100 (Teric N11)

Odor:

Low (Alkasurf NP-11)
Mild (Sterox NL)

Composition:
99% conc. min. (Nutrol 656)
> 99% active (Hyonic NP-110)
99.5% active (Sterox NL)
100% active (Alkasurf NP-11; Teric N11; Triton N-111)
100% conc. (Arkopal N110; Carsonon N-11; Macol NP-11; Neutronyx 656; Norfox NP-11; Serdox NNP11)

Solubility:
Sol. in benzene (Teric N11)
Sol. in ethanol (Teric N11)
Sol. in ethyl acetate (Teric N11)
Sol. in ethyl Icinol (Teric N11)
Sol. in olein (Teric N11)
Sol. in perchloroethylene (Teric N11)
Sol. in Stoddard solvent (@ 5%) (Trycol NP-11)
Sol. in water (Alkasurf NP-11; Carsonon N-11; Hyonic NP-110; Macol NP-11; Sterox NL; Teric N11; Triton N-111); (@ 5%) (Trycol NP-11)
Sol. in xylene (@ 5%) (Trycol NP-11)

Ionic Nature:
Nonionic (Alkasurf NP-11; Carsonon N-11; Hyonic NP-110; Macol NP-11; Merpoxen NO110; Neutronyx 656; Norfox NP-11; Nutrol 656; Serdox NNP11; Teric N11; Triton N-111)

M.W.:
704 avg. (Triton N-111)

Sp.gr.:
1.05 (Neutronyx 656)
1.06 (Alkasurf NP-11)
1.065 (Sterox NL)
1.069 (20 C) (Teric N11)

Density:
1.07 g/ml (Hyonic NP-110)
8.7 lb/gal (Trycol NP-11)
8.8 lb/gal (Triton N-111)

Visc.:
116 cSt (100 F) (Trycol NP-11)
283 cps (Sterox NL)
310 cps (Triton N-111)
410 cps (20 C) (Teric N11)

M.P.:
7 ± 2 C (Teric N11)

Pour Pt.:
5 C (Trycol NP-11)
13 C (Hyonic NP-110)

34 F (Sterox NL)
55 F (Triton N-111)

Flash Pt.:

> 200 F (Neutronyx 656)
> 300 F (TOC) (Triton N-111)
480 F (Trycol NP-11)
535 F (Sterox NL)

Cloud Pt.:

71 C (Neutronyx 656; Trycol NP-11)
72 C (1%) (Triton N-111); (1% aq.) (Hyonic NP-110)
72–74 C (Merpoxen NO110)
74 ± 2 C (1% in hard water) (Teric N11)
74–79 C (Sterox NL)

HLB:

13–14 (Norfox NP-11)
13.5 (Trycol NP-11)
13.6 (Hyonic NP-110; Nutrol 656)
13.7 (Teric N11)
13.8 (Alkasurf NP-11; Triton N-111)
14.0 (Serdox NNP11)
14.2 (Macol NP-11)

Stability:

Good (Alkasurf NP-11; Sterox NL)
Stable in acid and alkaline systems (Neutronyx 656)
Good in hard or saline waters and in reasonable concs. of acids and alkalis (Teric N11)
Very stable against hydrolysis by acids and alkalis (Trycol NP-11)

pH:

6.0–8.0 (1% aq.) (Teric N11)
7.0 (1% aq.) (Hyonic NP-110)

Foam (Ross Miles):

105 mm initial, 95 mm after 5 min (0.05% sol'n.) (Hyonic NP-110)

Surface Tension:

32.3 dynes/cm (Sterox NL)
34 dynes/cm (0.01%) (Hyonic NP-110)
34.8 dynes/cm (0.1%, 20 C) (Teric N11)

Wetting (Draves):

0.07% for a 25-s test (3-g hook) (Hyonic NP-110)

Biodegradable: (Serdox NNP11)

TOXICITY/HANDLING:

May cause skin and eye irritation; spillages are slippery (Teric N11)

STD. PKGS.:

200-kg net steel drums (Merpoxen NO110)
55-gal (200 l) steel drums, bulk (Hyonic NP-110)

POE (12) nonyl phenyl ether

SYNONYMS:

Nonoxynol-12 (CTFA)
Nonyl phenol ethoxylate (12 moles EO)
Nonyl phenol polyglycol ether (12 EO)
Nonylphenoxy polyethoxy ethanol (12 moles EO)
Nonylphenoxy poly(ethyleneoxy) ethanol (12 moles EO)
PEG-12 nonyl phenyl ether
PEG 600 nonyl phenyl ether
POE (12) nonyl phenol
Polyethoxylated nonylphenol (12 EO)

STRUCTURE:

$C_9H_{19}C_6H_4(OCH_2CH_2)_nOH$
where avg. $n = 12$

CAS No.:

9016-45-9 (generic); 26027-38-3 (generic); 37205-87-1 (generic)
RD No.: 977061-17-8

TRADENAME EQUIVALENTS:

Alkasurf NP-12 [Alkaril]
Carsonon N-12 [Lonza]
Hyonic NP-120 [Diamond Shamrock]
Igepal CO-720 [GAF]
Makon 12 [Stepan]
Merpoxen NO120 [Elektrochemische Fabrik Kempen]
Nissan Nonion NS-212 [Nippon Oil & Fats]
NP-55-120 [Hefti Ltd.]
Peganol NP12 [Borg-Warner]
Polystep F-5 [Stepan]
Renex 682 [ICI United States]
Serdox NNP12 [Servo B.V.]
Sterox NM [Monsanto] (12.5 EO)
Surfonic N-120 [Jefferson]
Synperonic NP12 [ICI Petrochem. Div.]
T-Det N-12 [Thompson-Hayward]
Teric GN12 [ICI Australia Ltd.] (tech. grade)
Teric N12 [ICI Australia Ltd.]

CATEGORY:

Dispersant, solubilizer, surfactant, detergent, wetting agent, penetrant, spreading agent, emulsifier, coemulsifier, stabilizer, intermediate, corrosion inhibitor

APPLICATIONS:

Cosmetic industry preparations: (Makon 12; Surfonic N-120); perfumery (Makon 12); shampoos (Makon 12)

Farm products: (Hyonic NP-120); agricultural oils/sprays (Makon 12; Sterox NM; Surfonic N-120; Teric N12); herbicides (Serdox NNP12); insecticides/pesticides

(Serdox NNP12)
Food applications: fruit/vegetable washing (Sterox NM); indirect food contact (Hyonic NP-120; Surfonic N-120)
Household detergents: (Alkasurf NP-12; Carsonon N-12; Hyonic NP-120; Igepal CO-720; Merpoxen NO120; Peganol NP12; Serdox NNP12; Sterox NM); detergent base (NP-55-120; T-Det N-12); heavy-duty cleaner (Peganol NP12; Teric N12); high-temperature detergents (Alkasurf NP-12; Hyonic NP-120); laundry detergent (Igepal CO-720; Sterox NM); liquid detergents (T-Det N-12); powdered detergents (Merpoxen NO120)
Industrial applications: (NP-55-120; Teric N12); dyes and pigments (Makon 12); industrial processing (Hyonic NP-120); lubricating/cutting oils (Surfonic N-120); metalworking (Igepal CO-720); paint mfg. (Igepal CO-720; Peganol NP12; Sterox NM; Teric N12); paper mfg. (Igepal CO-720; Makon 12; NP-55-120; Peganol NP12; Sterox NM); petroleum industry (Makon 12; Surfonic N-120); polymers/ polymerization (Polystep F-5); rubber (Surfonic N-120); textile/leather processing (Igepal CO-720; Makon 12; Merpoxen NO120; NP-55-120; Serdox NNP12; Surfonic N-120)
Industrial cleaners: (Alkasurf NP-12; Carsonon N-12; Igepal CO-720; Makon 12; Peganol NP12; Serdox NNP12; Sterox NM); dairy/food plant cleaners (Igepal CO-720; Sterox NM); institutional cleaners (Peganol NP12); lime soap dispersant (Surfonic N-120); metal processing surfactants (Makon 12; Sterox NM; T-Det N-12); sanitizers/germicides (Igepal CO-720; T-Det N-12); solvent cleaners (Teric N12); textile cleaning (Igepal CO-720; Makon 12; Peganol NP12; Serdox NNP12); wallpaper removal (Surfonic N-120)

PROPERTIES:

Form:

Liquid (Alkasurf NP-12; Carsonon N-12; Makon 12; Nissan Nonion NS-212; NP-55-120; Renex 682; Serdox NNP12; Surfonic N-120; Synperonic NP12; Teric GN12, N12)
Liquid to semisolid; clear liquid melt (Hyonic NP-120)
Clear to turbid liquid (Polystep F-5)
Clear liquid, haze may form during storage (Peganol NP12)
Clear, viscous liquid (Sterox NM)
Dispersed, opaque liquid (Igepal CO-720)
Opaque liquid (T-Det N-12)
Soft paste (Merpoxen NO120)

Color:

White (Merpoxen NO120)
Light straw (Makon 12)
Yellow (Sterox NM)
APHA 100 max. (Peganol NP12)
APHA 120 max. (Nissan Nonion NS-212)
Hazen 100 (Teric N12)

Hazen 150 max. (Synperonic NP12)
Hazen 250 (Teric GN12)

Odor:

Mild (Sterox NM)
Mild, aromatic (T-Det N-12)
Aromatic (Igepal CO-720)

Composition:

> 99% active (Hyonic NP-120; Synperonic NP12)
99.5% active (Sterox NM)
99.5% active min. (T-Det N-12)
100% active (Igepal CO-720; Makon 12; Merpoxen NO120; Polystep F-5; Surfonic N-120; Teric N12)
100% conc. (Alkasurf NP-12; Carsonon N-12; NP-55-120; Renex 682; Serdox NNP12)

Solubility:

Sol. in acetone (Surfonic N-120)
Sol. in alcohols (Synperonic NP12)
Sol. in benzene (Teric N12)
Sol. in butyl Celloslve (Igepal CO-720)
Sol. in carbon tetrachloride (Surfonic N-120)
Sol. in corn oil (Igepal CO-720)
Sol. in dibutyl phthalate (Igepal CO-720)
Sol. in diethylene glycol (Nissan Nonion NS-212)
Sol. in ethanol (Igepal CO-720; Teric N12)
Sol. in ether (Nissan Nonion NS-212)
Sol. in ethyl acetate (Teric N12)
Sol. in ethyl Icinol (Teric N12)
Sol. in ethylene dichloride (Igepal CO-720)
Sol. in glycol ethers (Synperonic NP12)
Sol. in methanol (Nissan Nonion NS-212; Surfonic N-120)
Sol. in heavy aromatic naphtha (Igepal CO-720)
Sol. in olein (Teric N12)
Sol. in perchloroethylene (Igepal CO-720; Peganol NP12; Teric N12)
Sol. in tetrachloromethan (Nissan Nonion NS-212)
Sol. in water (Carsonon N-12; Hyonic NP-120; Igepal CO-720; Makon 12; Nissan Nonion NS-212; Peganol NP12; Sterox NM; Surfonic N-120; Synperonic NP12; Teric N12)
Sol. in xylene (Igepal CO-720; Nissan Nonion NS-212; Peganol NP12; Surfonic N-120)

Ionic Nature:

Nonionic (Alkasurf NP-12; Carsonon N-12; Hyonic NP-120; Igepal CO-720; Makon 12; Merpoxen NO120; NP-55-120; Renex 682; Serdox NNP12; Sterox NM; Surfonic N-120; Synperonic NP12; T-Det N-12; Teric N12)

POE (12) nonyl phenyl ether (cont'd.)

M.W.:
748 (Surfonic N-120)
770 (Sterox NM)

Sp.gr.:
1.045 (50 C) (Teric N12)
1.06 (Igepal CO-720)
1.064 (Peganol NP12)
1.0674 (Sterox NM)
1.070 (T-Det N-12); (20/20 C) (Surfonic N-120)

Density:
1.062 g/ml (Synperonic NP12)
1.07 g/ml (Hyonic NP-120)
8.9 lb/gal (Makon 12; T-Det N-12); (20 C) (Surfonic N-120)

Visc.:
14.5 cs (100 C) (Peganol NP12)
67 cps (50 C) (Teric N12)
260–340 cps (Igepal CO-720)
265 cps (Synperonic NP12)
275 cps (Sterox NM)
300 cps (Makon 12)
560 SUS (100 F) (Surfonic N-120)

F.P.:
11 C (Synperonic NP12)
14 C (Surfonic N-120)

M.P.:
11 ± 2 C (Teric N12)

Pour Pt.:
12.2 C (Makon 12)
14 C (Synperonic NP12)
15 C (Hyonic NP-120)
17 C (Peganol NP12)
52 F (Sterox NM)
≈ 57 F (T-Det N-12)
62 ± 2 F (Igepal CO-720)

Solidification Pt.:
10 C (Makon 12)
57 ± 2 F (Igepal CO-720)

Flash Pt.:
> 200 C (Peganol NP12)
> 200 F (PMCC) (Igepal CO-720)
> 400 F (OC) (T-Det N-12)
525 F (OC) (Surfonic N-120)
535 F (Sterox NM)

Cloud Pt.:
73–84 C (1% aq.) (Nissan Nonion NS-212)
79–84 C (1% aq.) (Synperonic NP12)
80–83 C (Peganol NP12)
81 C (1% sol'n.) (Makon 12); (1% aq.) (Surfonic N-120)
82 ± 2 C (1% in hard water) (Teric N12)
83–85 C (Merpoxen NO120)
88–94 C (Sterox NM)
91 C (1% aq.) (Hyonic NP-120)
≈ 180 F (1%) (T-Det N-12)

HLB:
13.9 (Alkasurf NP-12; Renex 682; Synperonic NP12; Teric N12)
14.0 (Hyonic NP-120; NP-55-120; Serdox NNP12)
14.1 (Nissan Nonion NS-212; Surfonic N-120; T-Det N-12)
14.2 (Peganol NP12)
14.3 (Sterox NM)

Hydroxyl No.:
75 (Peganol NP12)
76 (Synperonic NP12)

Stability:
Good (Sterox NM)
Stable to acids, bases, salts (Surfonic N-120; T-Det N-12)
Good in hard or saline waters and in reasonable concs. of acids and alkalis (Teric N12)
Stable to acids, alkalis, dilute sol'ns. of many oxidizing and reducing agents (Igepal CO-720)

Ref. Index:
1.4860 (Sterox NM)
1.4869 (20 C) (Surfonic N-120)

pH:
6.0–7.5 (1%) (Sterox NM); (5% in distilled water) (Peganol NP12)
6.0–8.0 (1% aq.) (Synperonic NP12; Teric N12)
7.0 (1% aq.) (Hyonic NP-120); (1%) (T-Det N-12)
7.2 (1% sol'n.) (Makon 12)

Foam (Ross Miles):
115 mm initial, 105 mm after 5 min (0.05% sol'n.) (Hyonic NP-120)

Surface Tension:
32.3 dynes/cm (0.1%) (Surfonic N-120)
33.1 dynes/cm (Sterox NM)
34 dynes/cm (0.01% sol'n.) (Igepal CO-720)
35.2 dynes/cm (0.1%, 20 C) (Synperonic NP12; Teric N12)
36 dynes/cm (0.01% sol'n.) (Hyonic NP-120)

Wetting (Draves):
0.12% for a 25-s test (3-g hook) (Hyonic NP-120)

POE (12) nonyl phenyl ether (cont'd.)

Biodegradable: (Serdox NNP12); > 90% (Surfonic N-120)

TOXICITY/HANDLING:

May cause skin and eye irritation; spillages are slippery (Teric N12)
Causes eye burns, may cause skin irritation (Sterox NM)
Eye irritant; skin irritant on prolonged contact with conc. form (Synperonic NP12)

STORAGE/HANDLING:

Contact with conc. oxidizing or reducing agents may be explosive (Igepal CO-720; T-Det N-12)

STD. PKGS.:

Steel drums and bulk (Sterox NM)
200-kg net steel drums (Merpoxen NO120)
55-gal (200 l) steel drums, bulk (Hyonic NP-120)
460 lb net closed-head steel drums (T-Det N-12)

POE (13) nonyl phenyl ether

SYNONYMS:

Nonoxynol-13 (CTFA)
Nonyl phenol ethoxylate (13 moles EO)
Nonyl phenol polyglycol ether (13 moles EO)
Nonylphenoxy polyethoxy ethanol (13 moles EO)
Nonylphenoxy poly(ethyleneoxy) ethanol (13 moles EO)
PEG-13 nonyl phenyl ether
PEG (13) nonyl phenyl ether
POE (13) nonyl phenol

STRUCTURE:

$C_9H_{19}C_6H_4(OCH_2CH_2)_nOH$
where avg. $n = 13$

CAS No.:

9016-45-9 (generic); 26027-38-3 (generic); 37205-87-1 (generic)
RD No.: 977065-13-6

TRADENAME EQUIVALENTS:

Arkopal N-130 [Amer. Hoechst]
Lubrol N13 [ICI Ltd. Organics Div.]
Renex 679 [Atlas Chem. Industries NV]
Serdox NNP13 [Servo B.V.]
Synperonic NP13 [ICI Petrochem. Div.]
Tergitol NP-13 [Union Carbide]
Teric GN13 [ICI Australia Ltd.] (tech. grade)
Teric N13 [ICI Australia Ltd.]

CATEGORY:

Dispersant, solubilizer, detergent, wetting agent, emulsifier, scouring agent

APPLICATIONS:

Farm products: agricultural oils/sprays (Teric N13); herbicides (Serdox NNP13; Synperonic NP13); insecticides/pesticides (Lubrol N13; Serdox NNP13; Synperonic NP13)

Household detergents: (Serdox NNP13; Tergitol NP-13); heavy-duty cleaner (Teric N13); high-temperature detergents (Tergitol NP-13); liquid detergents (Tergitol NP-13); powdered detergents (Tergitol NP-13)

Industrial applications: (Arkopal N-130; Tergitol NP-13); construction (Teric N13); paint mfg. (Teric N13); solvents (Synperonic NP13); textile/leather processing (Serdox NNP13)

Industrial cleaners: (Serdox NNP13; Tergitol NP-13); solvent cleaners (Teric N13); textile scouring (Serdox NNP13)

PROPERTIES:

Form:

Liquid (Renex 679; Serdox NNP13; Teric N13)
Clear liquid (Tergitol NP-13)
Paste (Arkopal N-130; Lubrol N13; Synperonic NP13)

Color:

Slightly yellow (Tergitol NP-13)
Hazen 100 (Teric N13)
Hazen 150 max. (Synperonic NP13)
Hazen 250 (Teric GN13)

Odor:

Mild, characteristic (Tergitol NP-13)

Composition:

75% conc. (Lubrol N13)
99% active min. (Synperonic NP13)
100% active (Tergitol NP-13; Teric N13)
100% conc. (Arkopal N-130; Renex 679; Serdox NNP13)

Solubility:

Sol. in alcohols (Synperonic NP13)
Sol. in benzene (Teric N13)
Sol. in butyl acetate (Tergitol NP-13)
Sol. in butyl Cellosolve (Tergitol NP-13)
Sol. in ethanol (Teric N13)
Sol. in ethyl acetate (Teric N13)
Sol. in ethyl Icinol (Teric N13)
Sol. in ethylene glycol (Tergitol NP-13)
Sol. in glycol ethers (Synperonic NP13)
Sol. in isopropanol (anhyd.) (Tergitol NP-13)
Sol. in olein (Teric N13)

POE (13) nonyl phenyl ether *(cont'd.)*

Sol. in perchloroethylene (Teric N13)
Sol. in toluene (Tergitol NP-13)
Sol. in water (Synperonic NP13; Tergitol NP-13; Teric N13)

Ionic Nature:
Nonionic (Arkopal N-130; Lubrol N13; Renex 679; Serdox NNP13; Synperonic NP13; Tergitol NP-13; Teric N13)

M.W.:
792 (Tergitol NP-13)

Sp.gr.:
1.049 (50 C) (Teric N13)
1.068 (Synperonic NP13)
1.071 (20/20 C) (Tergitol NP-13)

Density:
8.90 lb/gal (20 C) (Tergitol NP-13)

Visc.:
75 cps (50 C) (Teric N13)
280 cps (Synperonic NP13)
410 cs (20 C) (Tergitol NP-13)

F.P.:
14 C (Synperonic NP13)

B.P.:
> 250 C (dec.) (Tergitol NP-13)

M.P.:
14 ± 2 C (Teric N13)

Pour Pt.:
17 C (Synperonic NP13)

Solidification Pt.:
16 C (Tergitol NP-13)

Cloud Pt.:
83 C (0.5% aq.) (Tergitol NP-13)
87–92 C (1% aq.) (Synperonic NP13)
89 ± 2 C (1% in hard water) (Teric N13)

HLB:
14.4 (Lubrol N13; Synperonic NP13; Tergitol NP-13; Teric N13)
14.5 (Serdox NNP13)

Hydroxyl No.:
71 (Synperonic NP13)

Stability:
Good in hard or saline waters and in reasonable concs. of acids and alkalis (Teric N13)

pH:
5.0–8.0 (10% aq.) (Tergitol NP-13)
6.0–8.0 (1% aq.) (Synperonic NP13; Teric N13)

Foam (Ross Miles):
185 mm initial, 13 mm after 5 min (50 C) (Tergitol NP-13)
Surface Tension:
34 dynes/cm (0.1% aq.) (Tergitol NP-13)
34.8 dynes/cm (0.1%, 20 C) (Synperonic NP13)
34.9 dynes/cm (0.1%, 20 C) (Teric N13)
Wetting (Draves):
16 s (0.1% conc.) (Tergitol NP-13)
Biodegradable: (Serdox NNP13)
TOXICITY/HANDLING:
Severe eye irritant; highly toxic to aquatic life; burning can produce carbon monoxide and/or carbon dioxide (Tergitol NP-13)
May cause skin and eye irritation; spillages are slippery (Teric N13)
Eye irritant; skin irritant on prolonged contact with conc. form (Synperonic NP13)
STORAGE/HANDLING:
Store at temps below 50 C (Tergitol NP-13)
STD. PKGS.:
55-gal (480 lb net) drums; 5-gal (45 lb net) pails; 1-gal (8.5 lb net) jugs (Tergitol NP-13)

POE (14) nonyl phenyl ether

SYNONYMS:
Nonoxynol-14 (CTFA)
Nonylphenol polyethyleneglycol ether (14 EO)
PEG-14 nonyl phenyl ether
PEG (14) nonyl phenyl ether
STRUCTURE:
$C_9H_{19}C_6H_4(OCH_2CH_2)_nOH$
where avg. $n = 14$
CAS No.:
9016-45-9 (generic); 26027-38-3 (generic); 37205-87-1 (generic)
RD No.: 977061-55-4
TRADENAME EQUIVALENTS:
Cremophor NP14 [BASF AG]
Lamacit 877 [Chemische Fabrik Grunau]
Makon 14 [Stepan]
CATEGORY:
Solubilizer, detergent, emulsifier, dyeing assistant
APPLICATIONS:
Cosmetic industry preparations: (Lamacit 877; Makon 14); perfumery (Cremophor

POE (14) nonyl phenyl ether (cont'd.)

NP14; Makon 14); shampoos (Makon 14)
Farm products: agricultural emulsions (Makon 14)
Food applications: essential oils (Cremophor NP14); flavors (Cremophor NP14)
Household detergents: detergent intermediate (Makon 14)
Industrial applications: chemical specialties (Makon 14); dyes and pigments (Makon 14); paper mfg. (Makon 14); petroleum industry (Makon 14); textile/leather processing (Makon 14)
Industrial cleaners: (Makon 14); metal processing surfactants (Makon 14); textile scouring (Makon 14)
Pharmaceutical applications: drugs (Lamacit 877)

PROPERTIES:

Form:
Liquid (Cremophor NP14; Lamacit 877; Makon 14)

Color:
Light straw (Makon 14)

Composition:
100% active (Makon 14)

Solubility:
Sol. in water (Makon 14)

Ionic Nature:
Nonionic (Makon 14)

Density:
8.9 lb/gal (Makon 14)

Visc.:
520 cps (Makon 14)

Pour Pt.:
18.9 C (Makon 14)

Solidification Pt.:
17.2 C (Makon 14)

Cloud Pt.:
94 C (1% sol'n.) (Makon 14)

pH:
7.2 (1% sol'n.) (Makon 14)

POE (15) nonyl phenyl ether

SYNONYMS:
Nonoxynol-15 (CTFA)
Nonyl phenol ethoxylate (15 moles EO)
Nonyl phenol 15 polyglycol ether
Nonylphenoxy polyethoxy ethanol (15 moles EO)

SYNONYMS *(cont'd.)*:
Nonylphenoxy poly(ethyleneoxy) ethanol (15 moles EO)
PEG-15 nonyl phenyl ether
PEG (15) nonyl phenyl ether
POE (15) nonyl phenol

STRUCTURE:
$C_9H_{19}C_6H_4(OCH_2CH_2)_nOH$
where avg. $n = 15$

CAS No.:
9016-45-9 (generic); 26027-38-3 (generic); 37205-87-1 (generic)
RD No.: 977006-97-5

TRADENAME EQUIVALENTS:
Ablunol NP15 [Taiwan Surfactant]
Alkasurf NP-15, NP-15 80% [Alkaril]
Arkopal N-150 [Amer. Hoechst]
Cedepal CO-730 [Domtar]
Chemax NP-15 [Chemax]
Conco NI-150 [Continental Chem.]
Iconol NP-15 [BASF Wyandotte]
Igepal CO-730 [GAF]
Merpoxen NO150 [Elektrochemische Fabrik Kempen]
Nikkol NP-15 [Nikko]
NP-55-150 [Hefti Ltd.]
Nutrol 640 [Clough Chem. Ltd.]
Peganol NP15 [Borg-Warner]
Polystep F-7 [Stepan]
Renex 678 [ICI United States]
Serdox NNP15 [Servo B.V.]
Siponic NP15 [Alcolac]
Surfonic N-150 [Jefferson]
Synperonic NP15 [ICI Petrochem. Div.]
Tergitol NP-15 [Union Carbide]
Teric GN15 [ICI Australia Ltd.] (tech. grade)
Teric N15 [ICI Australia Ltd.]
Triton N-150 [Rohm & Haas]

CATEGORY:
Dispersant, emulsifier, coemulsifier, solubilizer, detergent, wetting agent, rewetting agent, penetrant, stabilizer, surfactant, foaming agent, demulsifier

APPLICATIONS:
Cosmetic industry preparations: (Alkasurf NP-15; Conco NI-150; Nikkol NP-15; Surfonic N-150); perfumery (Teric N15)
Degreasers: (Surfonic N-150)

POE (15) nonyl phenyl ether (cont'd.)

Farm products: agricultural oils/sprays (Surfonic N-150); herbicides (Synperonic NP15); insecticides/pesticides (Conco NI-150; Nikkol NP-15; Synperonic NP15)

Food applications: indirect food additives (Surfonic N-150)

Household detergents: (Alkasurf NP-15; Cedepal CO-730; Conco NI-150; Igepal CO-730; Siponic NP15; Surfonic N-150; Tergitol NP-15; Teric N15); detergent base (NP-55-150); dishwashing (Renex 678); heavy-duty cleaner (Ablunol NP15; Alkasurf NP-15 80%); high-temperature detergents (Ablunol NP15; Igepal CO-730; NP-55-150; Peganol NP15; Renex 678); laundry detergent (Alkasurf NP-15); liquid detergents (Merpoxen NO150; Tergitol NP-15); paste detergents (Merpoxen NO150); powdered detergents (Merpoxen NO150; Tergitol NP-15)

Industrial applications: (Arkopal N-150; Nikkol NP-15; NP-55-150); construction (Alkasurf NP-15); dyes and pigments (Alkasurf NP-15); electroplating (Alkasurf NP-15); galvanic industries (NP-55-150); lubricating/cutting oils (Surfonic N-150); metalworking (Chemax NP-15); paint mfg. (Alkasurf NP-15; Chemax NP-15; Siponic NP15; Teric N15); paper mfg. (Chemax NP-15; Siponic NP15); petroleum industry (Conco NI-150; Surfonic N-150); polishes and waxes (Conco NI-150); polymers/polymerization (Conco NI-150; Polystep F-7; Siponic NP15); rubber (Surfonic N-150); solvents (Synperonic NP15); textile/leather processing (Alkasurf NP-15; Chemax NP-15; Conco NI-150; NP-55-150; Siponic NP15; Surfonic N-150)

Industrial cleaners: (Conco NI-150; Surfonic N-150; Tergitol NP-15); bottle washing (Ablunol NP15; Alkasurf NP-15 80%; Conco NI-150; Igepal CO-730; Peganol NP15); institutional cleaners (Alkasurf NP-15); janitorial cleaners (Alkasurf NP-15); metal processing surfactants (Ablunol NP15; Alkasurf NP-15, NP-15 80%; Igepal CO-730; Peganol NP15; Surfonic N-150); sanitizers/germicides (Teric N15); textile scouring (Ablunol NP15)

PROPERTIES:

Form:

Liquid (Alkasurf NP-15 80%; Cedepal CO-730; Chemax NP-15; Conco NI-150; Nikkol NP-15; NP-55-150; Nutrol 640; Serdox NNP15; Tergitol NP-15; Triton N-150)

Clear liquid (Renex 678)

Clear liquid, haze may form during storage (Peganol NP15)

Dispersed liquid (Igepal CO-730)

Opaque, viscous liquid (Iconol NP-15)

Turbid, viscous liquid (Polystep F-7)

Paste (Alkasurf NP-15; Arkopal N-150; Synperonic NP15; Teric N15)

Semisolid (Ablunol NP15; Siponic NP15; Surfonic N-150)

Soft wax (Merpoxen NO150)

Color:

Light (Alkasurf NP-15)

White (Merpoxen NO150; Surfonic N-150)

Yellow (Igepal CO-730)

APHA 100 max. (Iconol NP-15; Peganol NP15)
APHA 160 (Triton N-150)
Hazen 100 (Teric N15)
Hazen 150 max. (Synperonic NP15)
Hazen 250 (Teric GN15)

Odor:
Low (Alkasurf NP-15)
Aromatic (Igepal CO-730)

Composition:
80% active min. (Alkasurf NP-15 80%)
99% active min. (Synperonic NP15)
99% conc. min. (Nutrol 640)
100% active (Alkasurf NP-15; Conco NI-150; Igepal CO-730; Merpoxen NO150; Polystep F-7; Renex 678; Surfonic N-150; Teric N15; Triton N-150)
100% conc. (Ablunol NP15; Arkopal N-150; Cedepal CO-730; Nikkol NP-15; NP-55-150; Serdox NNP15; Siponic NP15; Tergitol NP-15)

Solubility:
Sol. in acetone (Surfonic N-150)
Sol. in alcohols (Synperonic NP15)
Sol. in benzene (Teric N15)
Sol. in butyl Cellosolve (Igepal CO-730)
Sol. in carbon tetrachloride (Surfonic N-150)
Sol. in dibutyl phthalate (Igepal CO-730)
Sol. in ethanol (Igepal CO-730; Teric N15)
Sol. in ethyl acetate (Teric N15)
Sol. in ethyl Icinol (Teric N15)
Sol. in ethylene dichloride (Igepal CO-730)
Sol. in ethylene glycol (Igepal CO-730)
Sol. in glycol ethers (Synperonic NP15)
Sol. in isopropanol (@ 10%) (Renex 678)
Sol. in methanol (Surfonic N-150)
Sol. in heavy aromatic naphtha (Igepal CO-730)
Sol. in olein (Teric N15)
Sol. in perchloroethylene (Igepal CO-730; Peganol NP15; Teric N15); (@ 10%) (Renex 678); disp. (@ 10%) (Alkasurf NP-15 80%)
Sol. in propylene glycol (@ 10%) (Renex 678)
Sol. in water (Alkasurf NP-15; Cedepal CO-730; Iconol NP-15; Igepal CO-730; Peganol NP15; Surfonic N-150; Synperonic NP15; Teric N15); (@ 10%) (Alkasurf NP-15 80%; Renex 678)
Sol. in xylene (Igepal CO-730; Peganol NP15; Surfonic N-150); (@ 10%) (Renex 678)

Ionic Nature:
Nonionic (Ablunol NP15; Alkasurf NP-15; Arkopal N-150; Cedepal CO-730; Conco NI-150; Iconol NP-15; Igepal CO-730; Merpoxen NO150; Nikkol NP-15; NP-55-

150; Nutrol 640; Peganol NP15; Serdox NNP15; Siponic NP15; Surfonic N-150; Synperonic NP15; Tergitol NP-15; Teric N15; Triton N-150)

M.W.:

875 (Iconol NP-15)
880 (Surfonic N-150)
880 avg. (Triton N-150)

Sp.gr.:

1.051 (50 C) (Teric N15)
1.065 (30/4 C) (Surfonic N-150)
1.07 (Alkasurf NP-15; Iconol NP-15; Igepal CO-730)
1.07–1.08 (Conco NI-150)
1.074 (Peganol NP15)
1.08 (Renex 678)

Density:

1.058 g/ml (40 C) (Synperonic NP15)
1.07 g/ml (Alkasurf NP-15 80%)
8.95 lb/gal (20 C) (Surfonic N-150)
9.0 lb/gal (Triton N-150)

Visc.:

17 cs (100 C) (Peganol NP15)
82 cps (50 C) (Teric N15)
125 cps (40 C) (Synperonic NP15)
319 cps (Renex 678)
450–550 cps (Igepal CO-730)
700 cps (Iconol NP-15)
4350 cps (Triton N-150)
89 SUS (210 F) (Surfonic N-150)

F.P.:

21 C (Synperonic NP15)
23 C (Surfonic N-150)

M.P.:

21 ± 2 C (Teric N15)

Pour Pt.:

21 C (Synperonic NP15)
23 C (Peganol NP15)
26 C (Iconol NP-15)
65 F (Triton N-150)
69 F (Renex 678)
71 F (Chemax NP-15)
71 ± 2 F (Igepal CO-730)

Solidification Pt.:

68 ± 2 F (Igepal CO-730)

Flash Pt.:
> 200 C (Peganol NP15)
> 200 F (PMCC) (Igepal CO-730)
> 300 F (Renex 678)
> 500 F (OC) (Surfonic N-150)

Cloud Pt.:
60–64 C (1% in 10% sodium chloride) (Alkasurf NP-15 80%)
94 C (1% aq.) (Surfonic N-150)
94–96 C (1% aq.) (Merpoxen NO150)
95 C (1%) (Triton N-150)
95 ± 2 C (1% in hard water) (Teric N15)
95–99 C (1% aq.) (Synperonic NP15)
95–100 C (1% aq.) (Iconol NP-15; Peganol NP15)
203–212 F (Conco NI-150); (1% sol'n.) (Igepal CO-730)
211 F (1% sol'n.) (Renex 678)

HLB:
14.9 (Siponic NP15)
15.0 (Alkasurf NP-15, NP-15 80%; Cedepal CO-730; Iconol NP-15; Igepal CO-730; NP-55-150; Nutrol 640; Peganol NP15; Renex 678; Serdox NNP15; Surfonic N-150; Synperonic NP15; Tergitol NP-15; Teric N15; Triton N-150)
16.0 (Ablunol NP15)
18.0 (Nikkol NP-15)

Hydroxyl No.:
64 (Peganol NP15; Synperonic NP15)

Stability:
Good (Alkasurf NP-15)
Stable to acids, bases, salts (Surfonic N-150)
Very good in acids, alkalis, and hard water (Renex 678)
Stable to acids, alkalis, hard water, and foam (Conco NI-150)
Stable to acids, alkalis, dilute sol'ns. of many oxidizing and reducing agents (Igepal CO-730)
Good in hard or saline waters and in reasonable concs. of acids and alkalis (Teric N15)

Ref. Index:
1.4815 (30 C) (Surfonic N-150)

pH:
5.0–8.0 (5% DW) (Alkasurf NP-15 80%)
6.0–7.5 (5% aq.) (Iconol NP-15; Peganol NP15)
6.0–8.0 (1% aq.) (Synperonic NP15; Teric N15)

Surface Tension:
33 dynes/cm (0.01% sol'n.) (Renex 678); (1%) (Triton N-150)
33.4 dynes/cm (0.1%, 20 C) (Synperonic NP15; Teric N15)
34.2 dynes/cm (0.1%) (Surfonic N-150)
36 dynes/cm (0.01% sol'n.) (Igepal CO-730)

POE (15) nonyl phenyl ether (cont'd.)

39 dynes/cm (0.1% aq.) (Iconol NP-15)

TOXICITY/HANDLING:

May cause skin and eye irritation; spillages are slippery (Teric N15)

Eye irritant; skin irritant on prolonged contact with conc. form (Synperonic NP15)

STORAGE/HANDLING:

Contact with conc. oxidizing or reducing agents may be explosive (Igepal CO-730)

STD. PKGS.:

200-kg net iron drums (Merpoxen NO150)

55-gal (450 lb net) steel drums (Iconol NP-15)

POE (20) nonyl phenyl ether

SYNONYMS:

Nonoxynol-20 (CTFA)

Nonyl phenol ethoxylate (20 moles EO)

Nonyl phenol 20 polyglycol ether

Nonylphenoxy polyethoxy ethanol (20 moles EO)

Nonylphenoxy poly(ethyleneoxy) ethanol (20 moles EO)

PEG-20 nonyl phenyl ether

PEG 1000 nonyl phenyl ether

POE (20) nonyl phenol

STRUCTURE:

$C_9H_{19}C_6H_4(OCH_2CH_2)_nOH$

where avg. $n = 20$

CAS No.:

9016-45-9 (generic); 26027-38-3 (generic); 37205-87-1 (generic)

RD No.: 977057-36-5

TRADENAME EQUIVALENTS:

Ablunol NP20 [Taiwan Surfactant]

Alkasurf NP-20, NP-20 70% [Alkaril]

Chemax NP-20 [Chemax]

Chemcol NPE-200 [Chemform]

Conco NI-185 [Continental Chem.]

Ethylan 20 [Lankro Chem. Ltd.]

Iconol NP-20 [BASF Wyandotte]

Igepal CO-850 [GAF]

Macol NP-20 [Mazer]

Makon 20 [Stepan]

Merpoxen NO200 [Elektrochemische Fabrik Kempen]

Peganol NP20 [Borg-Warner]

Polystep F-8 [Stepan]

TRADENAME EQUIVALENTS *(cont'd.)*:

Renex 649 [Atlas Chem. Industries NV]
Rexol 25/20 [Hart Chem. Ltd.]
Serdox NNP20/70 [Servo B.V.]
Sermul EN20/70 [Servo B.V.]
Surfonic N-200 [Jefferson]
Synperonic NP20 [ICI Petrochem. Div.]
T-Det N-20 [Thompson-Hayward]
Teric N20 [ICI Australia Ltd.]
Trycol NP-20 [Emery]

CATEGORY:

Dispersant, solubilizer, detergent, emulsifier, coemulsifier, demulsifier, wetting agent, rewetting agent, stabilizer, intermediate, penetrant, surfactant, foaming agent

APPLICATIONS:

Cosmetic industry preparations: (Conco NI-185; Makon 20); perfumery (Ethylan 20; Makon 20; Teric N20); shampoos (Makon 20)

Farm products: agricultural oils/sprays (Chemcol NPE-200; Makon 20; Surfonic N-200); insecticides/pesticides (Conco NI-185; Ethylan 20)

Food applications: indirect food additives (Surfonic N-200)

Household detergents: (Chemcol NPE-200; Conco NI-185; Ethylan 20; Teric N20); heavy-duty detergents (Ablunol NP20); high-temperature detergents (Igepal CO-850; Peganol NP20; T-Det N-20); liquid detergents (Merpoxen NO200); paste detergents (Merpoxen NO200); powdered detergents (Merpoxen NO200; Teric N20)

Industrial applications: glass processing (Igepal CO-850; Peganol NP20); latex applications (Ethylan 20; Igepal CO-850; Peganol NP20; Serdox NNP20/70); lubricating/cutting oils (Makon 20; Surfonic N-200); paint mfg. (Chemax NP-20; Teric N20); paper mfg. (Chemax NP-20; Makon 20); petroleum industry (Alkasurf NP-20, NP-20 70%; Chemcol NPE-200; Conco NI-185; Igepal CO-850; Makon 20; Peganol NP20; Surfonic N-200; T-Det N-20); polishes and waxes (Conco NI-185); polymers/polymerization (Conco NI-185; Sermul EN20/70; T-Det N-20; Trycol NP-20); rubber (Surfonic N-200); textile/leather processing (Chemax NP-20; Chemcol NPE-200; Conco NI-185; Makon 20; Surfonic N-200)

Industrial cleaners: (Conco NI-185; Makon 20); bottle washing (Ablunol NP20; Conco NI-185); metal processing surfactants (Ablunol NP20; Chemax NP-20; Makon 20); sanitizers/germicides (Teric N20); textile scouring (Ablunol NP20; Alkasurf NP-20, NP-20 70%; Makon 20)

PROPERTIES:

Form:

Liquid (Alkasurf NP-20 70%)
Clear to hazy liquid (Polystep F-8)
Gel (Polystep F-8)

Semisolid (Ablunol NP20; Surfonic N-200)
Opaque solid (T-Det N-20)
Solid (Alkasurf NP-20; Chemax NP-20; Chemcol NPE-200; Renex 649; Rexol 25/20; Serdox NNP20/70; Sermul EN20/70; Synperonic NP20; Teric N20; Trycol NP-20)
Soft wax (Peganol NP20)
Wax (Conco NI-185; Iconol NP-20; Igepal CO-850; Merpoxen NO200)
Waxy solid (Ethylan 20; Macol NP-20; Makon 20)

Color:
White (Alkasurf NP-20; Ethylan 20; Iconol NP-20; Merpoxen NO200; Surfonic N-200)
Off-white (Chemcol NPE-200; Makon 20)
Pale yellow (Igepal CO-850)
APHA 100 max. (50 C) (Peganol NP20)
Gardner 1 (Trycol NP-20)
Hazen 100 (Teric N20)
Hazen 150 max. (Synperonic NP20)

Odor:
Aromatic (Igepal CO-850)
Negligible (Ethylan 20)

Composition:
70% active (Polystep F-8)
70% active min. (Alkasurf NP-20 70%)
70% conc. (Serdox NNP20/70; Sermul EN20/70)
99% active min (Alkasurf NP-20; Synperonic NP20)
99.5% active min. (T-Det N-20)
100% active (Chemcol NPE-200; Conco NI-185; Ethylan 20; Igepal CO-850; Makon 20; Merpoxen NO200; Rexol 25/20; Surfonic N-200; Teric N20)
100% conc. (Ablunol NP20; Macol NP-20; Renex 649)

Solubility:
Sol. in acetone (Surfonic N-200)
Sol. in alcohols (Synperonic NP20)
Sol. in benzene (Teric N20)
Sol. in butyl Cellosolve (Igepal CO-850)
Sol. in carbon tetrachloride (Surfonic N-200)
Sol. in dibutyl phthalate (Igepal CO-850)
Sol. in ethanol (Igepal CO-850; Teric N20)
Sol. in ethyl acetate (Teric N20)
Sol. in ethyl Icinol (Teric N20)
Sol. in ethylene dichloride (Igepal CO-850)
Sol. in ethylene glycol (Igepal CO-850)
Sol. in glycol ethers (Synperonic NP20)
Sol. in methanol (Surfonic N-200)
Sol. in heavy aromatic naphtha (Igepal CO-850)

Sol. in oil in the boiling range (T-Det N-20)
Sol. in olein (Teric N20)
Sol. in perchloroethylene (Igepal CO-850; Peganol NP20; Teric N20)
Sol. in water (Ethylan 20; Iconol NP-20; Igepal CO-850; Macol NP-20; Makon 20; Peganol NP20; Surfonic N-200; Synperonic NP20; T-Det N-20; Teric N20); (@ 10%) (Alkasurf NP-20, NP-20 70%); (@ 5%) (Trycol NP-20)
Sol. in xylene (Igepal CO-850; Peganol NP20; Surfonic N-200); (@ 5%) (Trycol NP-20)

Ionic Nature:
Nonionic (Ablunol NP20; Chemcol NPE-200; Conco NI-185; Ethylan 20; Igepal CO-850; Macol NP-20; Merpoxen NO200; Peganol NP20; Rexol 25/20; Serdox NNP20/70; Sermul EN20/70; Surfonic N-200; Synperonic NP20; T-Det N-20; Teric N20)

M.W.:
1095 (Iconol NP-20)
1100 (Surfonic N-200)

Sp.gr.:
1.052 (Peganol NP20)
1.061 (50 C) (Teric N20)
1.065 (40 C) (Ethylan 20)
1.08 (50 C) (Igepal CO-850); (50/25 C) (Iconol NP-20); (105 F) (T-Det N-20)

Density:
1.073 g/ml (40 C) (Synperonic NP20)
1.08 g/ml (Alkasurf NP-20, NP-20 70%)
9.0 lb/gal (Makon 20); (20 C) (Surfonic N-200); (105 F) (T-Det N-20)

Visc.:
20 cs (100 C) (Peganol NP20)
100 cps (50 C) (Teric N20)
160 cs (20 C) (Ethylan 20)
168 cps (40 C) (Synperonic NP20)
105 SUS (210 F) (Surfonic N-200)

F.P.:
30 C (Synperonic NP20)
34 C (Surfonic N-200)

M.P.:
30 C (Iconol NP-20)
30 ± 2 C (Teric N20)
34 C (Trycol NP-20)

Pour Pt.:
30 C (Ethylan 20; Makon 20; Synperonic NP20)
33 C (Peganol NP20)
≈ 36 C (Chemcol NPE-200)
90 F (T-Det N-20)

POE (20) nonyl phenyl ether (cont'd.)

91 F (Chemax NP-20)
91 ± 2 F (Igepal CO-850)

Solidification Pt.:
30 C (Makon 20)
86 ± 2 F (Igepal CO-850)

Flash Pt.:
> 200 C (Peganol NP20)
> 200 F (PMCC) (Igepal CO-850)
> 400 F (COC) (Ethylan 20)
> 450 F (TOC) (Chemcol NPE-200)
> 500 F (OC) (Surfonic N-200); (PMCC) (T-Det N-20)
515 F (Trycol NP-20)

Cloud Pt.:
63–73 C (1% in 10% NaCl) (Chemcol NPE-200)
71 C (1% in 10% NaCl) (Alkasurf NP-20, NP-20 70%)
71–73 C (1 g in 100 ml 10% NaCl sol'n.) (Merpoxen NO200)
72–74 C (1% in 10% NaCl) (Rexol 25/20)
88 C (5% saline) (Trycol NP-20)
> 100 C (Peganol NP20); (1% aq.) (Ethylan 20; Iconol NP-20; Surfonic N-200; Synperonic NP20); (1% in hard water) (Teric N20)
Clear at 100 C (Makon 20)
Clear at 212 F (1% sol'n.) (Igepal CO-850)
> 212 F (Conco NI-185); (1% sol'n.) (T-Det N-20)

HLB:
15.8 (Surfonic N-200)
16.0 (Ablunol NP20; Alkasurf NP-20, NP-20 70%; Chemcol NPE-200; Ethylan 20; Iconol NP-20; Igepal CO-850; Macol NP-20; Peganol NP20; Rexol 25/20; Serdox NNP20/70; Synperonic NP20; T-Det N-20; Teric N20; Trycol NP-20)

Hydroxyl No.:
49 (Synperonic NP20)
51 (Peganol NP20)

Stability:
Stable to high concs. of electrolytes (Peganol NP20)
Stable to acids, bases, salts (Surfonic N-200; T-Det N-20)
Very stable against hydrolysis by acids and alkalis (Trycol NP-20)
Stable to acids, alkalis, hard water, and foam (Conco NI-185)
Stable to acids, alkalis, dilute sol'ns. of many oxidizing and reducing agents (Igepal CO-850)
Good in hard or saline waters and in reasonable concs. of acids and alkalis (Teric N20)
Stable to electrolytes, hard water, extremes of pH and temperature (Ethylan 20)

Ref. Index:
1.4720 (50 C) (Surfonic N-200)

pH:

3.5–4.5 (10% sol'n.) (Chemcol NPE-200)
5.0–7.0 (1% aq.) (T-Det N-20)
6.0–7.5 (5% aq.) (Iconol NP-20); (5% in distilled water) (Peganol NP20)
6.0–8.0 (1% aq.) (Ethylan 20; Synperonic NP20; Teric N20)
9.0 (1% sol'n.) (Makon 20)

Surface Tension:

39 dynes/cm (0.01% sol'n.) (Igepal CO-850)
40 dynes/cm (0.1% aq.) (Iconol NP-20)
41.7 dynes/cm (0.1%, 20 C) (Synperonic NP20; Teric N20)

TOXICITY/HANDLING:

Nonhazardous; however, considered to possess low acute oral and skin penetratin toxicity (T-Det N-20)
May cause skin and eye irritation; spillages are slippery (Teric N20)
Eye irritant; skin irritant on prolonged contact with conc. form (Synperonic NP20)
Avoid prolonged contact with conc. form; spillages may be slippery (Ethylan 20)

STORAGE/HANDLING:

Contact with conc. oxidizing or reducing agents may be explosive (Ethylan 20; Igepal CO-850; T-Det N-20)

STD. PKGS.:

200-kg net iron drums (Merpoxen NO200)
200-kg net mild-steel drums or bulk (Ethylan 20)
55-gal (450 lb net) steel drums (Iconol NP-20)
460 lb drums (T-Det N-20)

POE (1) octyl phenyl ether

SYNONYMS:

Ethylene glycol octyl phenyl ether
Octoxynol-1 (CTFA)
Octyl phenol ethoxylate (1 EO)
Octyl phenol polyethoxy ethanol (1 EO)
Octyl phenol polyglycol ether (1 EO)
Octylphenoxy ethoxy ethanol (1 EO)
PEG-1 octyl phenyl ether

STRUCTURE:

$C_8H_{17}C_6H_4OCH_2CH_2OH$

CAS No.:

9002-93-1 (generic); 9004-87-9 (generic); 9036-19-5 (generic)
RD No.: 977054-43-5

POE (1) octyl phenyl ether (cont'd.)

TRADENAME EQUIVALENTS:

Alkasurf OP-1 [Alkaril]
Cedepal CA-210 [Domtar] (1.5 EO)
Igepal CA-210 [GAF] (1.5 EO)
Rexol 45/1 [Hart Chem. Ltd.]
Siponic OP1.5 [Alcolac] (1.5 EO)
Triton X-15 [Rohm & Haas]

CATEGORY:

Dispersant, emulsifier, coemulsifier, coupling agent, detergent, retardant, wetting agent

APPLICATIONS:

Cosmetic industry preparations: hair preparations (Alkasurf OP-1)
Farm products: insecticides/pesticides (Igepal CA-210)
Household detergents: (Rexol 45/1)
Industrial applications: petroleum industry (Alkasurf OP-1); polishes and waxes (Igepal CA-210); solvents (Igepal CA-210; Rexol 45/1)
Industrial cleaners: drycleaning compositions (Igepal CA-210); solvent cleaners (Igepal CA-210)

PROPERTIES:

Form:

Liquid (Alkasurf OP-1; Cedepal CA-210; Igepal CA-210; Rexol 45/1; Siponic OP1.5; Triton X-15)

Color:

Yellow (Igepal CA-210)
APHA 250 (Triton X-15)

Odor:

Aromatic (Igepal CA-210)

Composition:

100% active (Alkasurf OP-1; Igepal CA-210; Rexol 45/1; Triton X-15)
100% conc. (Cedepal CA-210; Siponic OP1.5)

Solubility:

Sol. in aromatic solvent (@ 10%) (Alkasurf OP-1)
Sol. in butyl Cellosolve (Igepal CA-210)
Sol. in ethanol (Igepal CA-210)
Sol. in deodorized kerosene (Igepal CA-210)
Sol. in min. oil (@ 10%) (Alkasurf OP-1); sol. in low-visc. white min. oil (Igepal CA-210)
Sol. in min. spirits (@ 10%) (Alkasurf OP-1)
Sol. in oil (Siponic OP1.5)
Sol. in perchloroethylene (Igepal CA-210); (@ 10%) (Alkasurf OP-1)
Sol. in Stoddard solvent (Igepal CA-210)
Insol. in water (Triton X-15)
Sol. in xylene (Igepal CA-210)

Ionic Nature:
Nonionic (Cedepal CA-210; Rexol 45/1; Siponic OP1.5; Triton X-15)
M.W.:
250 avg. (Triton X-15)
Sp.gr.:
0.99 (Igepal CA-210)
Density:
0.99 g/ml (Alkasurf OP-1)
8.2 lb/gal (Triton X-15)
Visc.:
500–800 cps (Igepal CA-210)
790 cps (Triton X-15)
Pour Pt.:
15 F (Triton X-15)
23 ± 2 F (Igepal CA-210)
Solidification Pt.:
18 ± 2 F (Igepal CA-210)
Flash Pt.:
> 200 F (PMCC) (Igepal CA-210)
> 300 F (TOC) (Triton X-15)
Cloud Pt.:
Insol. (1% sol'n.) (Igepal CA-210)
HLB:
3.5 (Cedepal CA-210)
3.6 (Alkasurf OP-1; Rexol 45/1; Triton X-15)
4.7 (Siponic OP1.5)
4.8 (Igepal CA-210)
Stability:
Stable to acids, alkalis and dilute sol'ns. of many oxidizing and reducing agents (Igepal CA-210)
pH:
5.0–8.0 (5% DW) (Alkasurf OP-1)
Surface Tension:
Insol. (0.01% sol'n.) (Igepal CA-210)
TOXICITY/HANDLING:
Severe eye irritant (Igepal CA-210)
STORAGE/HANDLING:
Contact with conc. oxidizing or reducing agents may be hazardous (explosive) (Igepal CA-210)

POE (3) octyl phenyl ether

SYNONYMS:

Octoxynol-3 (CTFA)
Octyl phenol ethoxylate (3 EO)
Octyl phenol polyethoxy ethanol (3 EO)
Octyl phenol polyglycol ether (3 EO)
Octylphenoxy polyethoxy ethanol (3 EO)
Octylphenoxypoly (ethyleneoxy) ethanol (3 EO)
PEG-3 octyl phenyl ether
PEG (3) octyl phenyl ether
POE (3) octyl phenol

STRUCTURE:

$C_8H_{17}C_6H_4(OCH_2CH_2)_nOH$
where avg. $n = 3$

CAS No.:

9002-93-1 (generic); 9004-87-9 (generic); 9010-43-9; 9036-19-5 (generic)
RD No.: 977057-42-3

TRADENAME EQUIVALENTS:

Chemax OP-3 [Chemax]
Igepal CA-420 [GAF]
Nikkol OP-3 [Nikko]
Peganol OP3 [GAF]
Rexol 45/3 [Hart Chem. Ltd.]
T-Det O-4 [Thompson-Hayward] (3–4 moles EO)
Triton X-35 [Rohm & Haas]

CATEGORY:

Dispersant, emulsifier, detergent, coupling agent, solubilizer, coemulsifier, detergent, wetting agent, surfactant

APPLICATIONS:

Cosmetic industry preparations: (Nikkol OP-3)
Degreasers: (T-Det O-4)
Farm products: agricultural oils/sprays (T-Det O-4); insecticides/pesticides (Chemax OP-3; Igepal CA-420; Nikkol OP-3)
Household detergents: (T-Det O-4)
Industrial applications: (Nikkol OP-3); metalworking (T-Det O-4); paint mfg. (T-Det O-4); polishes and waxes (Chemax OP-3; Igepal CA-420); solvents (Chemax OP-3; Igepal CA-420; Rexol 45/3); textile/leather processing (T-Det O-4)
Industrial cleaners: drycleaning compositions (Igepal CA-420; T-Det O-4); solvent cleaners (Igepal CA-420)

PROPERTIES:

Form:

Liquid (Chemax OP-3; Igepal CA-420; Nikkol OP-3; Peganol OP3; Rexol 45/3; Triton X-35)
Clear liquid (T-Det O-4)

Color:
Pale yellow (Igepal CA-420)
APHA 125 (Triton X-35)
Odor:
Aromatic (Igepal CA-420)
Composition:
99.5% active min. (T-Det O-4)
100% active (Igepal CA-420; Rexol 45/3; Triton X-35)
100% conc. (Nikkol OP-3; Peganol OP3)
Solubility:
Sol. in butyl Cellosolve (Igepal CA-420)
Sol. in ethanol (Igepal CA-420)
Sol. in deodorized kerosene (Igepal CA-420)
Sol. in perchloroethylene (Igepal CA-420)
Sol. in Stoddard solvent (Igepal CA-420)
Disp. in water (T-Det O-4)
Sol. in xylene (Igepal CA-420)
Ionic Nature:
Nonionic (Igepal CA-420; Nikkol OP-3; Peganol OP3; Rexol 45/3; T-Det O-4; Triton X-35)
M.W.:
338 avg. (Triton X-35)
Sp.gr.:
1.02 (Igepal CA-420)
1.03 (68 F) (T-Det O-4)
Density:
8.5 lb/gal (Triton X-35)
8.6 lb/gal (68 F) (T-Det O-4)
Visc.:
370 cps (Triton X-35)
Pour Pt.:
–10 F (Triton X-35)
–10 ± 2 F (Igepal CA-420)
–9 F (Chemax OP-3)
< 0 F (T-Det O-4)
Solidification Pt.:
–15 ± 2 F (Igepal CA-420)
Flash Pt.:
> 200 F (Igepal CA-420)
> 300 F (TOC) (Triton X-35)
> 500 F (PMCC) (T-Det O-4)
Cloud Pt.:
Insol. (1% sol'n.) (Igepal CA-420)

POE (3) octyl phenyl ether (cont'd.)

HLB:
6.0 (Nikkol OP-3)
7.8 (Peganol OP3; Rexol 45/3; Triton X-35)
8.0 (Igepal CA-420)
8.5 (T-Det O-4)
Stability:
Stable to acids, bases, salts (T-Det O-4)
Stable to acids, alkalis, dilute sol'ns. of many oxidizing and reducing agents (Igepal CA-420)
pH:
6.0–8.0 (1% in 15% IPA/water) (T-Det O-4)
Surface Tension:
Insol. (0.01% sol'n.) (Igepal CA-420)
29 dynes/cm (0.01%) (Triton X-35)
TOXICITY/HANDLING:
Severe eye irritant (Igepal CA-420)
Nonhazardous; however, considered to possess low acute oral and skin penetration toxicity (T-Det O-4)
STORAGE/HANDLING:
Contact with conc. oxidizing or reducing agents may be explosive (Igepal CA-420; T-Det O-4)
STD. PKGS.:
460-lb drums (T-Det O-4)

POE (5) octyl phenyl ether

SYNONYMS:
Octoxynol-5 (CTFA)
Octyl phenol ethoxylate (5 EO)
Octyl phenol polyethoxy ethanol (5 EO)
Octyl phenol polyglycol ether (5 EO)
Octylphenoxy polyethoxy ethanol (5 EO)
Octylphenoxypoly (ethyleneoxy) ethanol (5 EO)
PEG-5 octyl phenyl ether
PEG (5) octyl phenyl ether
POE (5) octyl phenol
STRUCTURE:
$C_8H_{17}C_6H_4(OCH_2CH_2)_nOH$
where avg. $n = 5$

CAS No.:

9002-93-1 (generic); 9004-87-9 (generic); 9036-19-5 (generic)
RD No.: 977054-44-6

TRADENAME EQUIVALENTS:

Alkasurf OP-5 [Alkaril]
Cedepal CA-520 [Domtar]
Chemax OP-5 [Chemax]
Iconol OP-5 [BASF Wyandotte]
Igepal CA-520 [GAF]
Rexol 45/5 [Hart Chem. Ltd.]
Teric X5 [ICI Australia Ltd.]
Triton X-45 [Rohm & Haas]

CATEGORY:

Dispersant, emulsifier, coemulsifier, detergent, surfactant, wetting agent

APPLICATIONS:

Farm products: agricultural powders (Teric X5); insecticides/pesticides (Chemax OP-5; Igepal CA-520; Teric X5; Triton X-45)

Household detergents: (Teric X5)

Industrial applications: dyes and pigments (Teric X5); industrial processing (Triton X-45); metalworking (Teric X5); petroleum industry (Igepal CA-520); polishes and waxes (Chemax OP-5; Igepal CA-520; Teric X5); solvent systems (Alkasurf OP-5; Chemax OP-5; Igepal CA-520; Teric X5); textile/leather processing (Teric X5)

Industrial cleaners: alkaline cleaners (Teric X5); drycleaning compositions (Igepal CA-520; Triton X-45); metal processing surfactants (Teric X5); sanitizers/germicides (Teric X5); solvent cleaners (Alkasurf OP-5; Igepal CA-520; Teric X5)

PROPERTIES:

Form:

Liquid (Alkasurf OP-5; Cedepal CA-520; Chemax OP-5; Iconol OP-5; Igepal CA-520; Rexol 45/5; Teric X5; Triton X-45)

Color:

Light (Alkasurf OP-5)
Pale yellow (Igepal CA-520)
APHA 100 (Triton X-45)
APHA 100 max. (Iconol OP-5)
Hazen 100 (Teric X5)

Odor:

Characteristic (Alkasurf OP-5)
Aromatic (Igepal CA-520)

Composition:

99% active (Iconol OP-5)
100% active (Alkasurf OP-5; Igepal CA-520; Rexol 45/5; Teric X5; Triton X-45)
100% conc. (Cedepal CA-520)

POE (5) octyl phenyl ether (cont'd.)

Solubility:

Sol. in benzene (Teric X5)
Sol. in butyl Cellosolve (Igepal CA-520)
Sol. in ethanol (Igepal CA-520; Teric X5)
Sol. in ethyl acetate (Teric X5)
Sol. in ethyl Icinol (Teric X5)
Sol. in kerosene (Teric X5)
Partly sol. in min. oil (Teric X5)
Sol. in oil (Alkasurf OP-5; Cedepal CA-520)
Sol. in olein (Teric X5)
Sol. in paraffin oil (Teric X5)
Sol. in perchloroethylene (Igepal CA-520; Teric X5)
Sol. in Stoddard solvent (Igepal CA-520)
Sol. in veg. oil (Teric X5)
Disp. in water (Iconol OP-5; Igepal CA-520; Teric X5)
Sol. in xylene (Igepal CA-520)

Ionic Nature:

Nonionic (Alkasurf OP-5; Cedepal CA-520; Iconol OP-5; Igepal CA-520; Rexol 45/5; Teric X5; Triton X-45)

M.W.:

425 (Iconol OP-5)
426 avg. (Triton X-45)

Sp.gr.:

1.04 (Iconol OP-5; Igepal CA-520)
1.046 (20 C) (Teric X5)
1.05 (15/15 C) (Alkasurf OP-5)

Density:

8.7 lb/gal (Triton X-45)

Visc.:

250–280 cps (Igepal CA-520)
290 cps (Triton X-45)
300 cps (Iconol OP-5)
465 cps (20 C) (Teric X5)

M.P.:

< 0 ± 2 C (Teric X5)

Pour Pt.:

< 0 C (Iconol OP-5)
–15 F (Chemax OP-5; Triton X-45)
–15 ± 2 F (Igepal CA-520)

Solidification Pt.:

–20 ± 2 F (Igepal CA-520)

Flash Pt.:

> 200 F (PMCC) (Igepal CA-520)

> 300 F (TOC) (Triton X-45)

Cloud Pt.:

Insol. (1% sol'n.) (Igepal CA-520)
< 0 C (1%) (Triton X-45)
< 25 C (1% aq.) (Iconol OP-5)
63–66 C (10% in 25% diethylene glycol butyl ether) (Rexol 45/5)

HLB:

10.0 (Cedepal CA-520; Iconol OP-5; Igepal CA-520)
10.4 (Alkasurf OP-5; Rexol 45/5; Teric X5; Triton X-45)

Stability:

Stable to acids, alkalis, dilute sol'ns. of many oxidizing and reducing agents (Igepal CA-520)
Good in hard or saline waters and in reasonable concs. of acids and alkalis (Teric X5)

pH:

6.0–7.5 (5% aq.) (Iconol OP-5)
6.0–8.0 (1% aq.) (Teric X5)

Surface Tension:

28 dynes/cm (1%) (Triton X-45); (0.1% aq.) (Iconol OP-5)
30 dynes/cm (0.01% sol'n.) (Igepal CA-520)

TOXICITY/HANDLING:

Severe eye irritant (Igepal CA-520)
May cause skin and eye irritation; spillages are slippery (Teric X5)

STORAGE/HANDLING:

Contact with conc. oxidizing or reducing agents may be explosive (Igepal CA-520)

STD. PKGS.:

55-gal (450 lb net) steel drums (Iconol OP-5)

POE (6) octyl phenyl ether

SYNONYMS:

Octyl phenol ethoxylate (6 EO)
Octyl phenol polyethoxy ethanol (6 EO)
Octyl phenol polyglycol ether (6 EO)
Octylphenoxy polyethoxy ethanol (6 EO)
Octylphenoxypoly (ethyleneoxy) ethanol (6 EO)
PEG-6 octyl phenyl ether
PEG (6) octyl phenyl ether
POE (6) octyl phenol

STRUCTURE:

$C_8H_{17}C_6H_4(OCH_2CH_2)_nOH$
where avg. $n = 6$

POE (6) octyl phenyl ether (cont'd.)

TRADENAME EQUIVALENTS:

Alkasurf OP-6 [Alkaril]
Peganol OP6 [GAF]
T-Det O-6 [Thompson-Hayward] (6–7 moles EO)

CATEGORY:

Coemulsifier, emulsifier, dispersant, wetting agent, detergent, surfactant

APPLICATIONS:

Farm products: agricultural oils/sprays (T-Det O-6)
Household detergents: (T-Det O-6)
Industrial applications: textile/leather processing (T-Det O-6)
Industrial cleaners: drycleaning compositions (Alkasurf OP-6); solvent cleaners (Alkasurf OP-6)

PROPERTIES:

Form:

Liquid (Alkasurf OP-6; Peganol OP6)
Clear liquid (T-Det O-6)

Composition:

99.5% active min. (T-Det O-6)
100% conc. (Alkasurf OP-6; Peganol OP6)

Solubility:

Disp. in water (T-Det O-6)

Ionic Nature:

Nonionic (Alkasurf OP-6; Peganol OP6; T-Det O-6)

Sp.gr.:

1.05 (68 F) (T-Det O-6)

Density:

8.7 lb/gal (68 F) (T-Det O-6)

Pour Pt.:

< 0 F (T-Det O-6)

Flash Pt.:

> 500 F (PMCC) (T-Det O-6)

HLB:

10.0 (Peganol OP6)
11.4 (Alkasurf OP-6)
11.6 (T-Det O-6)

Stability:

Stable to acids, bases, salts (T-Det O-6)

pH:

6.0–8.0 (1% in 15% IPA/water) (T-Det O-6)

TOXICITY/HANDLING:

Nonhazardous; however, considered to possess low acute oral and skin penetration toxicity (T-Det O-6)

STORAGE/HANDLING:
Do not mix with concentrated oxidizing or reducing agents—potentially explosive (T-Det O-6)
STD. PKGS.:
460-lb drums (T-Det O-6)

POE (7) octyl phenyl ether

SYNONYMS:
Octoxynol-7 (CTFA)
Octyl phenol ethoxylate (7 EO)
Octyl phenol polyethoxy ethanol (7 EO)
Octyl phenol polyglycol ether (7 EO)
Octylphenoxy polyethoxy ethanol (7 EO)
Octylphenoxypoly (ethyleneoxy) ethanol (7 EO)
PEG-7 octyl phenyl ether
PEG (7) octyl phenyl ether
POE (7) octyl phenol
STRUCTURE:
$C_8H_{17}C_6H_4(OCH_2CH_2)_nOH$
where avg. $n = 7$
CAS No.:
9002-93-1 (generic); 9004-87-9 (generic); 9036-19-5 (generic)
RD No.: 977057-42-3
TRADENAME EQUIVALENTS:
Chemax OP-7 [Chemax]
Hyonic OP-70 [Diamond Shamrock]
Iconol OP-7 [BASF Wyandotte]
Igepal CA-620 [GAF]
Rexol 45/7 [Hart Chem. Ltd.]
Siponic OP7 [Alcolac]
Teric X7 [ICI Australia Ltd.] (7.5 moles EO)
CATEGORY:
Surfactant, dispersant, emulsifier, wetting agent, intermediate, detergent, stabilizer
APPLICATIONS:
Cleansers: waterless hand cleaner (Chemax OP-7; Igepal CA-620)
Farm products: agricultural oils/sprays (Hyonic OP-70); agricultural powders (Teric X7); insecticides/pesticides (Teric X7)
Household detergents: (Chemax OP-7; Igepal CA-620; Teric X7); hard surface cleaner (Igepal CA-620; Rexol 45/7)
Industrial applications: dyes and pigments (Teric X7); paint mfg. (Hyonic OP-70);

paper mfg. (Igepal CA-620); polishes and waxes (Teric X7); polymers/polymerization (Hyonic OP-70); textile/leather processing (Hyonic OP-70; Igepal CA-620; Teric X7)

Industrial cleaners: (Igepal CA-620; Teric X7); acid cleaners (Chemax OP-7); alkaline cleaners (Teric X7); metal processing surfactants (Chemax OP-7; Igepal CA-620; Teric X7); sanitizers/germicides (Igepal CA-620; Teric X7); solvent cleaners (Teric X7)

PROPERTIES:

Form:

Liquid (Chemax OP-7; Iconol OP-7; Igepal CA-620; Rexol 45/7; Siponic OP7; Teric X7)

Clear liquid (Hyonic OP-70)

Color:

Pale yellow (Igepal CA-620)

APHA 100 max. (Iconol OP-7)

Hazen 100 (Teric X7)

Odor:

Aromatic (Igepal CA-620)

Composition:

99% active min. (Hyonic OP-70; Iconol OP-7)

100% active (Igepal CA-620; Rexol 45/7; Teric X7)

100% conc. (Siponic OP7)

Solubility:

Sol. in benzene (Teric X7)

Sol. in butyl Cellosolve (Igepal CA-620)

Sol. in ethanol (Igepal CA-620; Teric X7)

Sol. in ethyl acetate (Teric X7)

Sol. in ethyl Icinol (Teric X7)

Sol. in ethylene glycol (Igepal CA-620)

Limited sol. in oil (Siponic OP7)

Sol. in olein (Teric X7)

Sol. in perchloroethylene (Igepal CA-620; Teric X7)

Sol. in veg. oil (Teric X7)

Sol. in water (Iconol OP-7; Igepal CA-620; Teric X7); limited sol. (Siponic OP7); disp. (Hyonic OP-70)

Sol. in xylene (Igepal CA-620)

Ionic Nature:

Nonionic (Iconol OP-7; Igepal CA-620; Rexol 45/7; Siponic OP7; Teric X7)

M.W.:

515 (Iconol OP-7)

Sp.gr.:

1.05 (Iconol OP-7; Igepal CA-620)

1.059 (20 C) (Teric X7)

Density:
1.06 g/ml (Hyonic OP-70)
Visc.:
240–260 cps (Igepal CA-620)
300 cps (Iconol OP-7)
420 cps (20 C) (Teric X7)
M.P.:
< 0 ± 2 C (Teric X7)
Pour Pt.:
–11 C (Hyonic OP-70)
–9 C (Iconol OP-7)
15 F (Chemax OP-7)
15 ± 2 F (Igepal CA-620)
Solidification Pt.:
10 ± 2 F (Igepal CA-620)
Flash Pt.:
> 200 F (PMCC) (Igepal CA-620)
Cloud Pt.:
21–25 C (1% aq.) (Iconol OP-7; Rexol 45/7)
23 C (1% aq. sol'n.) (Hyonic OP-70)
28 ± 2 C (1% in hard water) (Teric X7)
70–75 F (1% sol'n.) (Igepal CA-620)
HLB:
11.5 (Siponic OP7)
12.0 (Hyonic OP-70; Igepal CA-620)
12.3 (Teric X7)
12.4 (Iconol OP-7; Rexol 45/7)
Stability:
Stable to acids, alkalis, dilute sol'ns. of many oxidizing and reducing agents (Igepal CA-620)
Good in hard or saline waters and in reasonable concs. of acids and alkalis (Teric X7)
pH:
6.0–7.5 (5% aq.) (Iconol OP-7)
6.0–8.0 (1% aq.) (Teric X7)
7.0 (1% aq.) (Hyonic OP-70)
Foam (Ross Miles):
14 mm initial, 14 mm after 5 min (0.05% in DW) (Hyonic OP-70)
Surface Tension:
27 dynes/cm (0.01%) (Hyonic OP-70)
29 dynes/cm (0.1% aq.) (Iconol OP-7)
29.5 dynes/cm (*0.1%, 20 C) (Teric X7)
30 dynes/cm (0.01% sol'n.) (Igepal CA-620)

POE (7) octyl phenyl ether (cont'd.)

Wetting (Draves):
0.05% for a 25-s test (3-g hook) (Hyonic OP-70)

TOXICITY/HANDLING:
May cause skin and eye irritation; spillages are slippery (Teric X7)
Severe eye irritant (Igepal CA-620)

STORAGE/HANDLING:
Contact with conc. oxidizing or reducing agents may be explosive (Igepal CA-620)

STD. PKGS.:
55-gal (450 lb net) steel drums (Iconol OP-7)

POE (8) octyl phenyl ether

SYNONYMS:
Octoxynol-8 (CTFA)
Octyl phenol ethoxylate (8 EO)
Octyl phenol polyethoxy ethanol (8 EO)
Octyl phenol polyglycol ether (8 EO)
Octylphenoxy polyethoxy ethanol (8 EO)
Octylphenoxypoly (ethyleneoxy) ethanol (8 EO)
PEG-8 octyl phenyl ether
PEG 400 octyl phenyl ether
POE (8) octyl phenol

STRUCTURE:
$C_8H_{17}C_6H_4(OCH_2CH_2)_nOH$
where avg. $n = 8$

CAS No.:
9004-87-9 (generic); 9036-19-5 (generic)

TRADENAME EQUIVALENTS:
Alkasurf OP-8 [Alkaril] (7–8 moles EO)
Hyonic OP-80 [Diamond Shamrock]
Peganol OP8 [GAF]
T-Det O-8 [Thompson-Hayward] (7–8 moles EO)
Teric X8 [ICI Australia Ltd.] (8.5 moles EO)
Triton X-114 [Rohm & Haas] (7–8 moles EO)
Triton X-114SB [Rohm & Haas]

CATEGORY:
Dispersant, emulsifier, coemulsifier, detergent, wetting agent

APPLICATIONS:
Farm products: agricultural formulations (T-Det O-8; Teric X8); insecticides/pesticides (Alkasurf OP-8; Teric X8)
Household detergents: (Alkasurf OP-8; T-Det O-8; Teric X8); laundry detergent

(Alkasurf OP-8; Triton X-114, X-114SB)

Industrial applications: construction (Alkasurf OP-8); dyes and pigments (Alkasurf OP-8; Teric X8); electroplating (Alkasurf OP-8); paint mfg. (Alkasurf OP-8; T-Det O-8); paper mfg. (Alkasurf OP-8); petroleum industry (T-Det O-8); polishes and waxes (Teric X8); textile/leather processing (Alkasurf OP-8; T-Det O-8; Teric X8)

Industrial cleaners: (Alkasurf OP-8; Triton X-114, X-114SB); alkaline cleaners (Teric X8); metal processing surfactants (Alkasurf OP-8; Teric X8); sanitizers/germicides (Teric X8); solvent cleaners (Teric X8)

PROPERTIES:

Form:

Liquid (Alkasurf OP-8; Hyonic OP-80; Peganol OP8; Teric X8; Triton X-114, X-114SB)

Clear liquid (T-Det O-8)

Color:

Light (Alkasurf OP-8)

APHA 100 (Triton X-114, X-114SB)

Hazen 100 (Teric X8)

Odor:

Low (Alkasurf OP-8)

Composition:

80% active in isopropyl alcohol (Triton X-114SB)

99.5% active min. (T-Det O-8)

100% active (Alkasurf OP-8; Teric X8; Triton X-114)

100% conc. (Hyonic OP-80; Peganol OP8)

Solubility:

Sol. in benzene (Teric X8)

Sol. in ethanol (Teric X8)

Sol. in ethyl acetate (Teric X8)

Sol. in ethyl Icinol (Teric X8)

Sol. in olein (Teric X8)

Sol. in perchloroethylene (Teric X8)

Partly sol. in veg. oil (Teric X8)

Sol. in water (Alkasurf OP-8; T-Det O-8; Teric X8)

Ionic Nature:

Nonionic (Alkasurf OP-8; Hyonic OP-80; Peganol OP8; T-Det O-8; Teric X8; Triton X-114, X-114SB)

M.W.:

536 avg. (Triton X-114, X-114SB)

Sp.gr.:

1.04 (68 F) (T-Det O-8)

1.06 (15/15 C) (Alkasurf OP-8); (20 C) (Teric X8)

Density:

8.2 lb/gal (Triton X-114SB)

POE (8) octyl phenyl ether (*cont'd.*)

8.7 lb/gal (68 F) (T-Det O-8)
8.8 lb/gal (Triton X-114)

Visc.:
260 cps (Triton X-114)
402 cps (20 C) (Teric X8)

M.P.:
< 0 ± 2 C (Teric X8)

Pour Pt.:
–5 F (Triton X-114SB)
15 F (T-Det O-8; Triton X-114)

Flash Pt.:
61 F (Setaflash CC) (Triton X-114SB)
> 300 F (TOC) (Triton X-114)
> 500 F (PMCC) (T-Det O-8)

Cloud Pt.:
22 C (1%) (Triton X-114)
47 ± 2 C (1% in hard water) (Teric X8)
70 F (1%) (T-Det O-8)

HLB:
12.4 (Alkasurf OP-8; T-Det O-8; Triton X-114, X-114SB)
12.6 (Teric X8)
12.8 (Peganol OP8)

Stability:
Good (Alkasurf OP-8)
Stable in acids, bases, salts (T-Det O-8)
Good in hard or saline waters and in reasonable concs. of acids and alkalis (Teric X8)

pH:
5.0–7.0 (1% in 15% IPA/water) (T-Det O-8)
6.0–8.0 (1% aq.) (Teric X8)

Surface Tension:
29 dynes/cm (1%) (Triton X-114)
30.2 dynes/cm (0.1%, 20 C) (Teric X8)

TOXICITY/HANDLING:
May cause skin and eye irritation; spillages are slippery (Teric X8)
Nonhazardous; however, considered to possess low acute oral and skin penetration toxicity (T-Det O-8)

STORAGE/HANDLING:
Do not mix with conc. oxidizing or reducing agents—potentially explosive (T-Det O-8)

STD. PKGS.:
460-lb drums (T-Det O-8)

SYNONYMS:
Octoxynol-9 (CTFA)
Octyl phenol ethoxylate (9 EO)
Octyl phenol polyethoxy ethanol (9 EO)
Octyl phenol polyglycol ether (9 EO)
Octylphenoxy polyethoxy ethanol (9 EO)
Octylphenoxypoly (ethyleneoxy) ethanol (9 EO)
PEG-9 octyl phenyl ether
PEG (9) octyl phenyl ether
POE (9) octyl phenol
Polyethoxylated octylphenol (9 EO)

STRUCTURE:
$C_8H_{17}C_6H_4(OCH_2CH_2)_nOH$
where avg. $n = 9$

CAS No.:
9002-93-1 (generic); 9004-87-9 (generic); 9010-43-9; 9036-19-5 (generic)

TRADENAME EQUIVALENTS:
Cedepal CA-630 [Domtar]
Chemax OP-9 [Chemax]
Hyonic OP-100 [Diamond Shamrock]
Igepal CA-630 [GAF]
Makon OP-9 [Stepan] (9–10 moles EO)
Nutrol 100 [Clough Chem. Co. Ltd.]
Serdox NOP9 [Servo B.V.]
Siponic F-90 [Alcolac]
T-Det O-9 [Thompson-Hayward]
Triton X-100, X-120 [Rohm & Haas] (9–10 EO)

CATEGORY:
Dispersant, detergent, wetting agent, emulsifier, coemulsifier, penetrant, surfactant, solubilizer

APPLICATIONS:
Cleansers: waterless hand cleaners (Chemax OP-9; Igepal CA-630)
Cosmetic industry preparations: perfumery (Siponic F-90)
Degreasers: (Hyonic OP-100)
Farm products: agricultural oils/sprays (Hyonic OP-100); agricultural wettable powders (Triton X-120); herbicides (Triton X-100); insecticides/pesticides (Nutrol 100; Triton X-100)
Household detergents: (Chemax OP-9; Hyonic OP-100; Igepal CA-630; T-Det O-9; Triton X-100); hard surface cleaner (Igepal CA-630; Makon OP-9)
Industrial applications: paper mfg. (Hyonic OP-100; Igepal CA-630; Makon OP-9; Serdox NOP9); textile/leather processing (Hyonic OP-100; Igepal CA-630; Makon OP-9; Serdox NOP9; Triton X-100)
Industrial cleaners: (Hyonic OP-100; Triton X-100); acid cleaners (Chemax OP-9;

POE (9) octyl phenyl ether (cont'd.)

Hyonic OP-100); alkaline cleaners (Hyonic OP-100); metal processing surfactants (Chemax OP-9; Igepal CA-630; Nutrol 100); sanitizers/germicides (Igepal CA-630); textile scouring (Triton X-100)

PROPERTIES:

Form:

Liquid (Cedepal CA-630; Chemax OP-9; Igepal CA-630; Nutrol 100; Serdox NOP9; Triton X-100)
Clear liquid (Hyonic OP-100; Makon OP-9; T-Det O-9)
Solid (Triton X-120)

Color:

White (Triton X-120)
Pale straw (Makon OP-9)
Pale yellow (Igepal CA-630)
APHA 100 (Triton X-100)

Odor:

Aromatic (Igepal CA-630)

Composition:

40% active (Triton X-120)
> 99% active (Hyonic OP-100; Nutrol 100)
99.5% active min. (T-Det O-9)
100% active (Igepal CA-630; Makon OP-9; Triton X-100)
100% conc. (Cedepal CA-630; Serdox NOP9)

Solubility:

Sol. in butyl Cellosolve (Igepal CA-630)
Sol. in ethanol (Igepal CA-630)
Sol. in ethylene glycol (Igepal CA-630)
Sol. in perchloroethylene (Igepal CA-630)
Sol. in water (Cedepal CA-630; Hyonic OP-100; Igepal CA-630; Makon OP-9; T-Det O-9)
Sol. in xylene (Igepal CA-630)

Ionic Nature:

Nonionic (Cedepal CA-630; Hyonic OP-100; Igepal CA-630; Makon OP-9; Nutrol 100; Serdox NOP9; T-Det O-9; Triton X-100, X-120)

M.W.:

628 avg. (Triton X-100, X-120)
628 ± 10 (Makon OP-9)

Sp.gr.:

1.06 (Igepal CA-630); (68 F) (T-Det O-9)

Density:

1.07 g/ml (Hyonic OP-100)
8.8 lb/gal (68 F) (T-Det O-9)
8.9 lb/gal (Triton X-100)

Visc.:
200–300 cps (Makon OP-9)
230–260 cps (Igepal CA-630)
240 cps (Triton X-100)
Pour Pt.:
10 C (Hyonic OP-100)
45 F (Chemax OP-9; Makon OP-9; T-Det O-9; Triton X-100)
45 ± 2 F (Igepal CA-630)
Solidification Pt.:
40 ± 2 F (Igepal CA-630)
Flash Pt.:
200 F (PMCC) (Igepal CA-630)
> 300 F (TOC) (Triton X-100)
> 500 F (PMCC) (T-Det O-9)
Cloud Pt.:
65 C (1%) (Triton X-100); (1% aq.) (Hyonic OP-100)
65 ± 2 (1% aq.) (Makon OP-9)
146–153 F (1%) (Igepal CA-630)
150 F (1%) (T-Det O-9)
HLB:
13.0 (Cedepal CA-630; Hyonic OP-100; Igepal CA-630; Nutrol 100; Serdox NOP9)
13.5 (Makon OP-9; T-Det O-9; Triton X-100)
Hydroxyl No.:
89 ± 5 (Makon OP-9)
Stability:
Stable to acids, bases, salts (T-Det O-9)
Stable to acids, alkalis, dilute sol'ns. of many oxidizing and reducing agents (Igepal CA-630)
pH:
5.0–7.0 (1% aq.) (T-Det O-9)
7.0 (1% aq.) (Hyonic OP-100)
7.5–8.5 (5% in 1:1 water:IPA) (Makon OP-9)
Foam (Ross Miles):
55 mm initial, 50 mm after 5 min (0.05%) (Hyonic OP-100)
Surface Tension:
30 dynes/cm (0.01%) (Hyonic OP-100); (1%) (Triton X-100)
31 dynes/cm (0.01% sol'n.) (Igepal CA-630)
Wetting (Draves):
0.04% for a 25-s test (3-g hook) (Hyonic OP-100)
TOXICITY/HANDLING:
Nonhazardous; however, considered to possess low acute oral and skin penetration toxicity (T-Det O-9)
Severe eye irritant (Igepal CA-630)

POE (9) octyl phenyl ether (cont'd.)

STORAGE/HANDLING:

Do not mix with concentrated oxidizing or reducing agents—potentially explosive (T-Det O-9)

Contact with conc. oxidizing or reducing agents may be explosive (Igepal CA-630)

STD. PKGS.:

55-gal (200 l) steel drums (Hyonic OP-100)

460-lb drums (T-Det O-9)

POE (10) octyl phenyl ether

SYNONYMS:

Octoxynol-10 (CTFA)
Octyl phenol ethoxylate (10 EO)
Octyl phenol polyethoxy ethanol (10 EO)
Octyl phenol polyglycol ether (10 EO)
Octylphenoxy polyethoxy ethanol (10 EO)
Octylphenoxypoly (ethyleneoxy) ethanol (10 EO)
PEG-10 octyl phenyl ether
PEG 500 octyl phenyl ether
POE (10) octyl phenol

STRUCTURE:

$C_8H_{17}C_6H_4(OCH_2CH_2)_nOH$
where avg. $n = 10$

CAS No.:

9002-93-1 (generic); 9004-87-9 (generic); 9036-19-5 (generic)

TRADENAME EQUIVALENTS:

Alkasurf OP-10 [Alkaril]
Iconol OP-10 [BASF Wyandotte]
Nikkol OP-10 [Nikko]
Octapol 100 [Sanyo]
Peganol OP10 [GAF]
Renex 750 [ICI Specialty Chem.]
Rexol 45/10 [Hart Chem. Ltd.]
Siponic OP10 [Alcolac]
Synperonic OP10 [ICI Petrochem. Div.]
Teric X10 [ICI Australia Ltd.]

CATEGORY:

Dispersant, emulsifier, coemulsifier, wetting agent, detergent, surfactant, solubilizer

APPLICATIONS:

Cosmetic industry preparations: (Nikkol OP-10)

Farm products: agricultural formulations (Rexol 45/10; Teric X10); insecticides/

pesticides (Nikkol OP-10; Teric X10)
Household detergents: (Alkasurf OP-10; Synperonic OP10; Teric X10); detergent base (Octapol 100; Rexol 45/10); laundry detergent (Alkasurf OP-10)
Industrial applications: (Nikkol OP-10); dyes and pigments (Alkasurf OP-10; Teric X10); electroplating (Alkasurf OP-10); paint mfg. (Alkasurf OP-10; Rexol 45/10); paper mfg. (Rexol 45/10); plastics (Rexol 45/10; Siponic OP10); polishes and waxes (Teric X10); polymers/polymerization (Siponic OP10); printing inks (Alkasurf OP-10); textile/leather processing (Alkasurf OP-10; Rexol 45/10; Teric X10); wood pulping (Rexol 45/10)
Industrial cleaners: (Alkasurf OP-10; Rexol 45/10); acid cleaners (Alkasurf OP-10); caustic cleaners (Alkasurf OP-10; Teric X10); metal processing surfactants (Alkasurf OP-10; Teric X10); sanitizers/germicides (Teric X10); solvent cleaners (Teric X10)

PROPERTIES:

Form:
Liquid (Alkasurf OP-10; Iconol OP-10; Nikkol OP-10; Octapol 100; Peganol OP10; Renex 750; Rexol 45/10; Siponic OP10; Synperonic OP10; Teric X10)

Color:
Light (Alkasurf OP-10)
APHA 100 max. (Iconol OP-10)
Hazen 100 (Teric X10)
Hazen 150 max. (Synperonic OP10)

Odor:
Low (Alkasurf OP-10)

Composition:
99% active min. (Iconol OP-10; Synperonic OP10)
100% active (Alkasurf OP-10; Rexol 45/10; Teric X10)
100% conc. (Nikkol OP-10; Octapol 100; Peganol OP10; Renex 750; Siponic OP10)

Solubility:
Sol. in alcohols (Synperonic OP10)
Sol. in benzene (Teric X10)
Sol. in ethanol (Teric X10)
Sol. in ethyl acetate (Teric X10)
Sol. in ethyl Icinol (Teric X10)
Sol. in glycol ethers (Synperonic OP10)
Sol. in olein (Teric X10)
Sol. in perchloroethylene (Teric X10)
Partly sol. in veg. oil (Teric X10)
Sol. in water (Alkasurf OP-10; Iconol OP-10; Siponic OP10; Synperonic OP10; Teric X10)

Ionic Nature:
Nonionic (Alkasurf OP-10; Iconol OP-10; Nikkol OP-10; Octapol 100; Peganol OP10; Renex 750; Rexol 45/10; Siponic OP10; Synperonic OP10; Teric X10)

POE (10) octyl phenyl ether (cont'd.)

M.W.:
650 (Iconol OP-10)

Sp.gr.:
1.06 (Iconol OP-10)
1.062 (20 C) (Teric X10)
1.1 (15/15 C) (Alkasurf OP-10)

Density:
1.062 g/ml (20 C) (Synperonic OP10)

Visc.:
250 cps (Iconol OP-10)
393 cps (20 C) (Synperonic OP10; Teric X10)

F.P.:
7 C (Synperonic OP10)

M.P.:
7 ± 2 C (Teric X10)

Pour Pt.:
7 C (Iconol OP-10; Synperonic OP10)

Cloud Pt.:
63–67 C (1% aq.) (Iconol OP-10; Synperonic OP10)
64–67 C (1% DW) (Rexol 45/10)
65 ± 2 C (1% in hard water) (Teric X10)

HLB:
11.5 (Nikkol OP-10)
13.5 (Alkasurf OP-10; Iconol OP-10; Peganol OP10; Rexol 45/10)
13.6 (Octapol 100; Renex 750; Synperonic OP10; Teric X10)
14.4 (Siponic OP10)

Hydroxyl No.:
90 (Synperonic OP10)

Stability:
Good (Alkasurf OP-10)
Good in hard or saline waters and in reasonable concs. of acids and alkalis (Teric X10)

pH:
6.0–8.0 (1% aq.) (Synperonic OP10; Teric X10)

Surface Tension:
30 dynes/cm (0.1% aq.) (Iconol OP-10)
31.8 dynes/cm (0.1%, 20 C) (Synperonic OP10; Teric X10)

TOXICITY/HANDLING:
Eye irritant; skin irritant on prolonged contact with conc. form (Synperonic OP10)
May cause skin and eye irritation; spillages are slippery (Teric X10)

STD. PKGS.:
55-gal (450 lb net) steel drums (Iconol OP-10)

POE (11) octyl phenyl ether

SYNONYMS:

Octoxynol-11 (CTFA)
Octyl phenol ethoxylate (11 EO)
Octyl phenol polyethoxy ethanol (11 EO)
Octyl phenol polyglycol ether (11 EO)
Octylphenoxy polyethoxy ethanol (11 EO)
Octylphenoxypoly (ethyleneoxy) ethanol (11 EO)
PEG-11 octyl phenyl ether
PEG (11) octyl phenyl ether
POE (11) octyl phenol

STRUCTURE:

$C_8H_{17}C_6H_4(OCH_2CH_2)_nOH$
where avg. $n = 11$

CAS No.:

9004-87-9 (generic); 9036-19-5 (generic)

TRADENAME EQUIVALENTS:

Synperonic OP11 [ICI Petrochem. Div.]
Teric X11 [ICI Australia Ltd.]

CATEGORY:

Dispersant, detergent, wetting agent, emulsifier

APPLICATIONS:

Farm products: agricultural powders (Teric X11); insecticides/pesticides (Teric X11)
Household detergents: (Synperonic OP11; Teric X11)
Industrial applications: dyes and pigments (Teric X11); polishes and waxes (Teric X11); textile/leather processing (Teric X11)
Industrial cleaners: alkaline cleaners (Teric X11); metal processing surfactants (Teric X11); sanitizers/germicides (Teric X11); solvent cleaners (Teric X11)

PROPERTIES:

Form:

Liquid (Synperonic OP11; Teric X11)

Color:

Hazen 100 (Teric X11)
Hazen 150 max. (Synperonic OP11)

Composition:

99% active min. (Synperonic OP11)
100% active (Teric X11)

Solubility:

Sol. in alcohols (Synperonic OP11)
Sol. in benzene (Teric X11)
Sol. in ethanol (Teric X11)
Sol. in ethyl acetate (Teric X11)
Sol. in ethyl Icinol (Teric X11)
Sol. in glycol ethers (Synperonic OP11)

POE (11) octyl phenyl ether (cont'd.)

Sol. in olein (Teric X11)
Sol. in perchloroethylene (Teric X11)
Partly sol. in veg. oil (Teric X11)
Sol. in water (Synperonic OP11; Teric X11)

Ionic Nature:
Nonionic (Synperonic OP11; Teric X11)

Sp.gr.:
1.067 (20 C) (Teric X11)

Density:
1.067 g/ml (20 C) (Synperonic OP11)

Visc.:
371 cps (20 C) (Synperonic OP11; Teric X11)

F.P.:
8 C (Synperonic OP11)

M.P.:
8 ± 2 C (Teric X11)

Pour Pt.:
8 C (Synperonic OP11)

Cloud Pt.:
79–82 C (1% aq.) (Synperonic OP11)
80 ± 2 C (1% in hard water) (Teric X11)

HLB:
14.0 (Synperonic OP11; Teric X11)

Hydroxyl No.:
87 (Synperonic OP11)

Stability:
Good in hard or saline waters and in reasonable concs. of acids and alkalis (Teric X11)

pH:
6.0–8.0 (1% aq.) (Synperonic OP11)

Surface Tension:
31 dynes/cm (0.1%, 20 C) (Synperonic OP11; Teric X11)

TOXICITY/HANDLING:
Eye irritant; skin irritant on prolonged contact with conc. form (Synperonic OP11)
May cause skin and eye irritation; spillages are slippery (Teric X11)

POE (12) octyl phenyl ether

SYNONYMS:
Octyl phenol ethoxylate (12 EO)
Octyl phenol polyethoxy ethanol (12 EO)
Octyl phenol polyglycol ether (12 EO)

SYNONYMS *(cont'd.)*:
Octylphenoxy polyethoxy ethanol (12 EO)
Octylphenoxypoly (ethyleneoxy) ethanol (12 EO)
PEG-12 octyl phenyl ether
PEG (12) octyl phenyl ether
POE (12) octyl phenol

STRUCTURE:
$C_8H_{17}C_6H_4(OCH_2CH_2)_nOH$
where avg. $n = 12$

TRADENAME EQUIVALENTS:
Alkasurf OP-12 [Alkaril] (12–13 moles EO)
Cedepal CA-720 [Domtar] (12.5 EO)
Igepal CA-720 [GAF] (12.5 EO)
Rexol 45/12 [Hart Chem. Ltd.]
Triton X-102 [Rohm & Haas] (12–13 EO)

CATEGORY:
Dispersant, detergent, emulsifier, wetting agent

APPLICATIONS:
Household detergents: (Alkasurf OP-12; Rexol 45/12; Triton X-102); hard surface cleaner (Igepal CA-720); laundry detergent (Alkasurf OP-12)
Industrial applications: dyes and pigments (Alkasurf OP-12); electroplating (Alkasurf OP-12); paint mfg. (Alkasurf OP-12); textile/leather processing (Alkasurf OP-12)
Industrial cleaners: (Alkasurf OP-12; Rexol 45/12; Triton X-102); acid cleaners (Alkasurf OP-12); caustic cleaners (Alkasurf OP-12); metal cleaning/pickling (Alkasurf OP-12; Igepal CA-720; Rexol 45/12); steam cleaning (Igepal CA-720)

PROPERTIES:

Form:
Liquid (Alkasurf OP-12; Cedepal CA-720; Rexol 45/12; Triton X-102)
Dispersed, opaque liquid (Igepal CA-720)

Color:
Light (Alkasurf OP-12)
APHA 100 (Triton X-102)

Odor:
Low (Alkasurf OP-12)
Aromatic (Igepal CA-720)

Composition:
100% active (Alkasurf OP-12; Cedepal CA-720; Igepal CA-720; Rexol 45/12; Triton X-102)

Solubility:
Sol. in butyl Cellosolve (Igepal CA-720)
Sol. in ethanol (Igepal CA-720)
Sol. in ethylene glycol (Igepal CA-720)
Sol. in perchloroethylene (Igepal CA-720)

POE (12) octyl phenyl ether (*cont'd.*)

Sol. in water (Alkasurf OP-12; Igepal CA-720)
Sol. in xylene (Igepal CA-720)

Ionic Nature:
Nonionic (Alkasurf OP-12; Cedepal CA-720; Igepal CA-720; Rexol 45/12; Triton X-102)

M.W.:
756 avg. (Triton X-102)

Sp.gr.:
1.07 (Alkasurf OP-12; Igepal CA-720)

Density:
8.9 lb/gal (Triton X-102)

Visc.:
70–76 cps (50 C) (Igepal CA-720)
330 cps (Triton X-102)

Pour Pt.:
60 F (Triton X-102)
69 ± 2 F (Igepal CA-720)

Solidification Pt.:
64 ± 2 F (Igepal CA-720)

Flash Pt.:
> 200 F (PMCC) (Igepal CA-720)
> 300 F (TOC) (Triton X-102)

Cloud Pt.:
86–90 C (1% DW) (Rexol 45/12)
88 C (1%) (Triton X-102)
187–194 F (1% sol'n.) (Igepal CA-720)

HLB:
14.5 (Rexol 45/12)
14.6 (Alkasurf OP-12; Cedepal CA-720; Igepal CA-720; Triton X-102)

Stability:
Good (Alkasurf OP-12)
Stable to acids and alkalis (Cedepal CA-720)
Stable to acids, alkalis, dilute sol'ns. of many oxidizing and reducing agents (Igepal CA-720)

Surface Tension:
32 dynes/cm (0.01% sol'n.) (Igepal CA-720); (1%) (Triton X-102)

TOXICITY/HANDLING:
Severe eye irritant (Igepal CA-720)

STORAGE/HANDLING:
Contact with conc. oxidizing or reducing agents may be explosive (Igepal CA-720)

POE (30) octyl phenyl ether

SYNONYMS:

Octoxynol-30 (CTFA)
Octyl phenol ethoxylate (30 EO)
Octyl phenol polyethoxy ethanol (30 EO)
Octyl phenol polyglycol ether (30 EO)
Octylphenoxy polyethoxy ethanol (30 EO)
Octylphenoxypoly (ethyleneoxy) ethanol (30 EO)
PEG-30 octyl phenyl ether
PEG (30) octyl phenyl ether
POE (30) octyl phenol
POE (30) octyl phenyl alcohol

STRUCTURE:

$C_8H_{17}C_6H_4(OCH_2CH_2)_nOH$
where avg. $n = 30$

CAS No.:

9004-87-9 (generic); 9036-19-5 (generic)

TRADENAME EQUIVALENTS:

Alkasurf OP-30, OP-30 70% [Alkaril]
Iconol OP-30 [BASF Wyandotte]
Igepal CA-887 [GAF]
Nikkol OP-30 [Nikko]
Octapol 300 [Sanyo]
Peganol OP30, OP 30 70% [GAF]
Rexol 45/307 [Hart Chem. Ltd.]
Serdox NOP30/70 [Servo B.V.]
Siponic F-300 [Alcolac]
Siponic OP30 [Alcolac]
Triton X-305 [Rohm & Haas]

CATEGORY:

Dispersant, surfactant, solubilizer, coupling agent, emulsifier, coemulsifier, stabilizer, detergent, lubricant, wetting agent, dyeing assistant

APPLICATIONS:

Cosmetic industry preparations: (Nikkol OP-30; Siponic F-300)
Farm products: insecticides/pesticides (Nikkol OP-30)
Household detergents: (Siponic F-300); detergent base (Octapol 300)
Industrial applications: (Nikkol OP-30); dyes and pigments (Igepal CA-887; Siponic F-300); latex applications (Alkasurf OP-30 70%; Igepal CA-887; Siponic OP30); plastics (Alkasurf OP-30; Igepal CA-887; Rexol 45/307; Siponic F-300, OP30); polymers/polymerization (Alkasurf OP-30; Igepal CA-887; Rexol 45/307; Serdox NOP30/70; Siponic F-300)
Industrial cleaners: (Siponic F-300); metal processing surfactants (Siponic F-300); textile scouring (Siponic F-300)

POE (30) octyl phenyl ether (cont'd.)

PROPERTIES:

Form:

Liquid (Alkasurf OP-30 70%; Igepal CA-887; Peganol OP30 70%; Rexol 45/307; Serdox NOP30/70; Siponic F-300, OP30; Triton X-305)
Paste (Nikkol OP-30)
Solid (Alkasurf OP-30; Iconol OP-30; Octapol 300; Peganol OP30)

Color:

Light (Alkasurf OP-30 70%)
White (Alkasurf OP-30)
Pale yellow (Igepal CA-887)
APHA 150 (Triton X-305)

Odor:

Low (Alkasurf OP-30 70%)
Aromatic (Igepal CA-887)

Composition:

70% active (Alkasurf OP-30 70%; Igepal CA-887; Peganol OP30 70%; Rexol 45/307; Serdox NOP30/70; Siponic F-300, OP30)
70% active in water (Triton X-305)
100% active (Alkasurf OP-30)
100% conc. (Iconol OP-30; Nikkol OP-30; Octapol 300; Peganol OP30)

Solubility:

High sol. in water (Alkasurf OP-30 70%; Siponic OP30); sol. (Triton X-305); (@ 10%) (Alkasurf OP-30)

Ionic Nature:

Nonionic (Iconol OP-30; Igepal CA-887; Nikkol OP-30; Octapol 300; Peganol OP30, OP30 70%; Rexol 45/307; Serdox NOP30/70; Siponic F-300, OP70; Triton X-305)

M.W.:

1526 avg. (Triton X-305)

Sp.gr.:

1.1 (Alkasurf OP-30 70%; Igepal CA-887)

Density:

1.10 g/ml (Alkasurf OP-30)
9.1 lb/gal (Triton X-305)

Visc.:

470 cps (Triton X-305)

Pour Pt.:

35 F (Triton X-305)
36 ± 2 F (Igepal CA-887)

Solidification Pt.:

31 ± 2 F (Igepal CA-887)

Flash Pt.:

> 200 F (PMCC) (Igepal CA-887)
> 300 F (TOC) (Triton X-305)

Cloud Pt.:
73–77 C (1% in 10% sodium chloride) (Rexol 45/307)
75 C (1% in 10% sodium chloride) (Alkasurf OP-30)
> 100 C (1%) (Triton X-305)
Clear at 212 F (1% sol'n.) (Igepal CA-887)

HLB:
17.0 (Iconol OP-30; Nikkol OP-30)
17.3 (Alkasurf OP-30 70%; Octapol 300; Peganol OP30, OP30 70%; Siponic F-300, OP30; Triton X-305)
17.4 (Igepal CA-887; Rexol 45/307)
17.5 (Alkasurf OP-30; Serdox NOP30/70)

Stability:
Stable to acids, alkalis, dilute sol'ns. of many oxidizing and reducing agents (Igepal CA-887)

pH:
5.0–8.0 (5% DW) (Alkasurf OP-30)

Surface Tension:
39 dynes/cm (0.01% sol'n.) (Igepal CA-887)
41 dynes/cm (1%) (Triton X-305)

TOXICITY/HANDLING:
Severe eye irritant (Igepal CA-887)

STORAGE/HANDLING:
Contact with conc. oxidizing or reducing agents may be explosive (Igepal CA-887)

POE (40) octyl phenyl ether

SYNONYMS:
Octoxynol-40 (CTFA)
Octyl phenol ethoxylate (40 EO)
Octyl phenol polyethoxy ethanol (40 EO)
Octyl phenol polyglycol ether (40 EO)
Octylphenoxy polyethoxy ethanol (40 EO)
Octylphenoxypoly (ethyleneoxy) ethanol (40 EO)
PEG-40 octyl phenyl ether
PEG 2000 octyl phenyl ether
POE (40) octyl phenol
POE (40) octyl phenyl alcohol
Polyethoxylated octyl phenol (40 EO)

STRUCTURE:
$C_8H_{17}C_6H_4(OCH_2CH_2)_nOH$
where avg. $n = 40$

POE (40) octyl phenyl ether (cont'd.)

CAS No.:
9002-93-1 (generic); 9004-87-9 (generic); 9036-19-5 (generic)
RD No.: 977066-20-5

TRADENAME EQUIVALENTS:
Alkasurf OP-40 [Alkaril]
Cedepal CA-890, CA-897 [Domtar]
Chemax OP-40/70 [Chemax]
Hyonic OP-407 [Diamond Shamrock]
Iconol OP-40, O-40-70% [BASF Wyandotte]
Igepal CA-890, CA-897 [GAF]
Nissan Nonion HS-240 [Nippon Oil & Fats]
Octapol 400 [Sanyo]
Peganol OP40, OP40 70% [GAF]
Rexol 45/407 [Hart Chem. Ltd.]
Serdox NOP40/70 [Servo B.V.]
Siponic F-400, OP40 [Alcolac]
T-Det O-40, O-407 [Thompson-Hayward]
Teric X40, X40L [ICI Australia Ltd.]
Triton X-405 [Rohm & Haas]
Trycol OP-407 [Emery]

CATEGORY:
Dispersant, detergent, emulsifier, coemulsifier wetting agent, stabilizer, surfactant, dyeing assistant

APPLICATIONS:
Cosmetic industry preparations: (Siponic F-400)
Farm products: agricultural formulations (Hyonic OP-407; T-Det O-40; Teric X40, X40L); insecticides/pesticides (Teric X40)
Household detergents: (Alkasurf OP-40; Siponic F-400; Teric X40, X40L); detergent base (Octapol 400); high-temperature detergents (Hyonic OP-407; T-Det O-40, O-407); laundry detergent (Alkasurf OP-40)
Industrial applications: (T-Det O-407); dyes and pigments (Alkasurf OP-40; Igepal CA-890, CA-897; Siponic F-400; Teric X40); electroplating (Alkasurf OP-40); latex applications (Alkasurf OP-40; Hyonic OP-407; Igepal CA-890, CA-897; Siponic OP40); plastics (Cedepal CA-890; Chemax OP-40/70; Igepal CA-890, CA-897; Rexol 45/407; Siponic F-400, OP40; Trycol OP-407); polishes and waxes (Teric X40); polymers/polymerization (Cedepal CA-890; Chemax OP-40/70; Igepal CA-890, CA-897; Rexol 45/407; Serdox NOP40/70; Siponic F-400; T-Det O-40; Trycol OP-407); textile/leather processing (Alkasurf OP-40; T-Det O-40; Teric X40, X40L)
Industrial cleaners: (Alkasurf OP-40; Siponic F-400; Teric X40, X40L); alkaline cleaners (Teric X40); metal processing surfactants (Alkasurf OP-40; Teric X40); sanitizers/germicides (Teric X40); solvent cleaners (Teric X40); textile scouring (Siponic F-400)

PROPERTIES:

Form:

Liquid (Alkasurf OP-40; Cedepal CA-897; Chemax OP-40/70; Iconol O-40-70%; Igepal CA-897; Peganol OP40 70%; Rexol 45/407; Serdox NOP40/70; Siponic F-400, OP40; T-Det O-407; Teric X40L; Triton X-405; Trycol OP-407)
Clear liquid (Hyonic OP-407)
Semisolid (Nissan Nonion HS-240)
Solid (Octapol 400; Peganol OP40; T-Det O-40; Teric X40)
Cast solid (Iconol OP-40)
Wax (Cedepal CA-890; Igepal CA-890)

Color:

Light (Alkasurf OP-40)
White (T-Det O-40)
Off-white (Igepal CA-890)
Pale yellow (Igepal CA-897)
APHA 100 max. (Iconol O-40-70%; T-Det O-407)
APHA 250 (Triton X-405)
Gardner 1 (Trycol OP-407); 1 max. (Iconol OP-40)
Hazen 100 (Teric X40)

Odor:

Low (Alkasurf OP-40)
Aromatic (Igepal CA-890, CA-897)

Composition:

68.5–71.5% active in water (Iconol O-40-70%)
70% active (Cedepal CA-897; Chemax OP-40/70; Hyonic OP-407; Igepal CA-897; Peganol OP40 70%; Rexol 45/407; Serdox NOP40/70; Siponic F-400, OP40; Teric X40L)
70% active in water (Alkasurf OP-40; Triton X-405; Trycol OP-407)
70 ± 0.5% active (T-Det O-407)
99% active min. (Iconol OP-40)
99.5% active min. (T-Det O-40)
100% active (Cedepal CA-890; Igepal CA-890; Nissan Nonion HS-240; Octapol 400; Peganol OP40; Teric X40)

Solubility:

Sol. in benzene (Teric X40)
Sol. in butyl Cellosolve (Igepal CA-890)
Sol. in ethanol (Igepal CA-890; Teric X40)
Sol. in ethyl acetate (Teric X40)
Sol. in ethyl Icinol (Teric X40)
Sol. in perchloroethylene (Teric X40)
High sol. in water (Alkasurf OP-40; T-Det O-407); sol. (Hyonic OP-407; Iconol OP-40, O-40-70%; Igepal CA-890; Siponic OP40; T-Det O-40; Teric X40; Trycol OP-407)

POE (40) octyl phenyl ether (cont'd.)

Ionic Nature:

Nonionic (Alkasurf OP-40; Cedepal CA-890, CA-897; Iconol OP-40, O-40-70%; Igepal CA-890, CA-897; Nissan Nonion HS-240; Octapol 400; Peganol OP40, OP40 70%; Rexol 45/407; Serdox NOP40/70; Siponic F-400, OP40; T-Det O-40, O-407; Teric X40, X40L; Triton X-405)

M.W.:

1966 avg. (Triton X-405)
1970 (Iconol OP-40); (of active) (Iconol O-40-70%)

Sp.gr.:

1.06 (150 F) (T-Det O-40)
1.08 (50 C) (Igepal CA-890)
1.082 (50 C) (Teric X40)
1.09 (50/25 C) (Iconol OP-40)
1.1 (Alkasurf OP-40; Iconol O-40-70%; Igepal CA-897)

Density:

0.91 g/ml (Hyonic OP-407)
8.8 lb/gal (150 F) (T-Det O-40)
9.0 lb/gal (Trycol OP-407)
≈ 9.1 lb/gal (T-Det O-407)
9.2 lb/gal (Triton X-405)

Visc.:

40 cps (100 C) (Iconol OP-40)
245 cps (50 C) (Teric X40)
490 cps (Triton X-405)
500 cps (Iconol O-40-70%)
600 cps (T-Det O-407)
220 cSt (100 F) (Trycol OP-407)

M.P.:

44 ± 2 C (Teric X40)
50 C (Iconol OP-40)

Pour Pt.:

–2 C (Hyonic OP-407)
–4 C (Iconol O-40-70%)
13 C (Trycol OP-407)
25 F (Chemax OP-40/70; Triton X-405)
25 ± 2 F (Igepal CA-897)
≈ 30 F (T-Det O-407)
115 ± 2 F (Igepal CA-890)
120 F (T-Det O-40)

Solidification Pt.:

20 ± 2 F (Igepal CA-897)
110 ± 2 F (Igepal CA-890)

Flash Pt.:
> 100 C (TOC) (T-Det O-407)
> 200 F (PMCC) (Igepal CA-890, CA-897)
> 212 F (TOC) (Triton X-405)
> 500 F (PMCC) (T-Det O-40)

Cloud Pt.:
73 C (1% aq.) (Hyonic OP-407)
73–77 C (1% in 10% sodium chloride) (Rexol 45/407)
74 C (10% saline) (Trycol OP-407)
> 100 C (1%) (T-Det O-407; Triton X-405); (1% aq.) (Iconol OP-40, O-40-70%); (1% in hard water) (Teric X40)
Clear at 212 F (1% sol'n.) (Igepal CA-890, CA-897)
> 212 F (1% sol'n.) (T-Det O-40)

HLB:
17.6 (Teric X40)
17.9 (Iconol OP-40, O-40-70%; Nissan Nonion HS-240; Octapol 400; Peganol OP40, OP40 70%; Siponic F-400, OP40; T-Det O-40; Triton X-405)
18.0 (Alkasurf OP-40; Cedepal CA-890, CA-897; Hyonic OP-407; Igepal CA-890, CA-897; Rexol 45/407; Serdox NOP40/70)

Stability:
Good (Alkasurf OP-40)
Very stable against hydrolysis by acids and alkalis (Trycol OP-407)
Stable to acids, bases, and salts (T-Det O-40, O-407)
Good in hard or saline waters and in reasonable concs. of acids and alkalis (Teric X40)
Stable to acids, alklais, dilute sol'ns. of many oxidizing and reducing agents (Igepal CA-890, CA-897)

pH:
5.0–7.0 (1% aq.) (T-Det O-40)
6.0–7.5 (5% aq.) (Iconol OP-40, O-40-70%)
6.0–8.0 (1% aq.) (Teric X40)
7.0 (1% sol'n.) (T-Det O-407); (1% aq.) (Hyonic OP-407)

Foam (Ross Miles):
120 mm initial, 115 mm after 5 min (0.05% sol'n.) (Hyonic OP-407)

Surface Tension:
36.4 dynes/cm (0.1%, 20 C) (Teric X40)
42 dynes/cm (0.1% aq.) (Iconol OP-40); (0.01%) (Igepal CA-890)
44 dynes/cm (1%) (Triton X-405)
48 dynes/cm (0.01% sol'n.) (Hyonic OP-407; Igepal CA-897)

TOXICITY/HANDLING:
Nonhazardous; however, considered to possess low acute oral and skin penetration toxicity (T-Det O-40)
May cause skin and eye irritation; spillages are slippery (Teric X40)
Severe eye irritant (Igepal CA-890)

POE (40) octyl phenyl ether (cont'd.)

STORAGE/HANDLING:

Contact with conc. oxidizing or reducing agents may be explosive (Igepal CA-890, CA-897; T-Det O-40, O-407)

STD. PKGS.:

55-gal (200 l) steel drums, bulk (Hyonic OP-407)
55-gal (450 lb net) steel drums (Iconol OP-40)
55-gal (450 lb net) lined fiber drums (Iconol O-40-70%)
480-lb net closed-head steel drums (T-Det O-407)

POE (70) octyl phenyl ether

SYNONYMS:

Octoxynol-70 (CTFA)
Octyl phenol ethoxylate (70 EO)
Octyl phenol polyethoxy ethanol (70 EO)
Octyl phenol polyglycol ether (70 EO)
Octylphenoxy polyethoxy ethanol (70 EO)
Octylphenoxypoly (ethyleneoxy) ethanol (70 EO)
PEG-70 octyl phenyl ether
PEG (70) octyl phenyl ether
POE (70) octyl phenol
Polyethoxylated octyl phenol (70 EO)

STRUCTURE:

$C_8H_{17}C_6H_4(OCH_2CH_2)_nOH$
where avg. $n = 70$

CAS No.:

9004-87-9 (generic); 9036-19-5 (generic)

TRADENAME EQUIVALENTS:

Alkasurf OP-70-50% [Alkaril]
Hyonic OP-705 [Diamond Shamrock]
Nissan Nonion HS-270 [Nippon Oil & Fats]
Peganol OP70, OP70 70% [GAF]
Siponic F-707 [Alcolac]
Triton X-705, X-705-50% [Rohm & Haas]

CATEGORY:

Stabilizer, wetting agent, emulsifier, coemulsifier, detergent, surfactant

APPLICATIONS:

Cosmetic industry preparations: (Siponic F-707)
Farm products: agricultural formulations (Hyonic OP-705)
Household detergents: (Siponic F-707); high-temperature detergents (Hyonic OP-705)

Industrial applications: dyes and pigments (Siponic F-707); latex applications (Hyonic OP-705); plastics (Alkasurf OP-70-50%; Siponic F-707); polymers/polymerization (Alkasurf OP-70-50%; Siponic F-707)
Industrial cleaners: (Siponic F-707); metal processing surfactants (Siponic F-707); textile scouring (Siponic F-707)

PROPERTIES:

Form:
Liquid (Alkasurf OP-70-50%; Peganol OP70 70%; Siponic F-707; Triton X-705-50%)
Clear liquid (Hyonic OP-705)
Semisolid (Nissan Nonion HS-270)
Solid (Peganol OP70; Triton X-705)

Color:
White (Triton X-705)
APHA 200 (Triton X-705-50%)

Composition:
50% active (Alkasurf OP-70-50%; Hyonic OP-705; Triton X-705-50%)
70% active (Peganol OP70 70%; Siponic F-707)
100% active (Nissan Nonion HS-270; Peganol OP70; Triton X-705)

Solubility:
Sol. in water (Hyonic OP-705); (@ 10%) (Alkasurf OP-70-50%)

Ionic Nature:
Nonionic (Alkasurf OP-70-50%; Hyonic OP-705; Nissan Nonion HS-270; Peganol OP70, OP70 70%; Siponic F-707; Triton X-705, X-705-50%)

M.W.:
3286 avg. (Triton X-705, X-705-50%)

Density:
1.08 g/ml (Alkasurf OP-70-50%)
1.10 g/ml (Hyonic OP-705)
9.0 lb/gal (Triton X-705-50%)
9.4 lb/gal (Triton X-705)

Pour Pt.:
–12 C (Hyonic OP-705)

Flash Pt.:
> 212 F (TOC) (Triton X-705-50%)
> 500 F (COC) (Triton X-705)

Cloud Pt.:
76 C (1% in 10% sodium chloride) (Alkasurf OP-70-50%)
91 C (1% aq.) (Hyonic OP-705)
> 100 C (1%) (Triton X-705, X-705-50%)

HLB:
18.5 (Hyonic OP-705)
18.7 (Alkasurf OP-70-50%; Nissan Nonion HS-270; Peganol OP70, OP70 70%; Siponic F-707; Triton X-705, X-705-50%)

POE (70) octyl phenyl ether (cont'd.)

pH:
5.0–8.0 (5% DW) (Alkasurf OP-70-50%)
7.0 (1% aq.) (Hyonic OP-705)
Foam (Ross Miles):
125 mm initial, 125 mm after 5 min (0.05% sol'n.) (Hyonic OP-705)
Surface Tension:
41 dynes/cm (1%) (Triton X-705, X-705-50%)
49 dynes/cm (0.01%) (Hyonic OP-705)
STD. PKGS.:
55-gal (200 l) steel drums, bulk (Hyonic OP-705)

POE (5) oleyl amide

SYNONYMS:
PEG-5 oleamide (CTFA)
PEG (5) oleyl amide
POE (5) oleamide
STRUCTURE:

$$CH_3(CH_2)_7CH{=}CH(CH_2)_7\overset{\displaystyle O}{\overset{\|}{C}}{-}NH{-}(CH_2CH_2O)_nH$$

where avg. $n = 5$
CAS No.:
RD No.: 977065-67-0
TRADENAME EQUIVALENTS:
Ethomid O/15 [Armak]
CATEGORY:
Emulsifier, dispersant, detergent, leveling agent, lubricant
APPLICATIONS:
Industrial applications: dyes and pigments (Ethomid O/15); silicone finishing agents (Ethomid O/15); textile/leather processing (Ethomid O/15)
PROPERTIES:
Form:
Liquid (Ethomid O/15)
Color:
Gardner 8 max. (Ethomid O/15)
Solubility:
Sol. in carbon tetrachloride (< 25 C) (Ethomid O/15)
Sol. in dioxane (< 25 C) (Ethomid O/15)
Sol. in isopropanol (< 50 C) (Ethomid O/15)
Forms gel in water @ 50–80 C (Ethomid O/15)

Sp.gr.:
1.0 (Ethomid O/15)
Flash Pt.:
225 F (PM) (Ethomid O/15)
Hydroxyl No.:
100–120 (Ethomid O/15)
Stability:
Stable in acid or alkaline sol'ns. (Ethomid O/15)
pH:
Neutral (Ethomid O/15)
Surface Tension:
35 dynes/cm (0.1%) (Ethomid O/15)

POE (2) oleyl amine

SYNONYMS:
PEG-2 oleamine (CTFA)
PEG 100 oleyl amine
STRUCTURE:

$$\begin{array}{lll} CH(CH_2)_7CH_3 & & (CH_2CH_2O)_xH \\ \| & & | \\ CH(CH_2)_7CH_2 & \text{——} & N\text{——}(CH_2CH_2O)_yH \end{array}$$

where avg. $(x + y) = 2$
CAS No.:
RD No.: 977065-39-6
TRADENAME EQUIVALENTS:
Crodamet 1.02 [Croda] (primary amine)
Ethomeen O/12 [Armak] (tertiary amine)
Hetoxamine O2 [Heterene]
CATEGORY:
Emulsifier, dispersant, desizing agent, softener, antistatic agent, water repellent
APPLICATIONS:
Farm products: agricultural oils/sprays (Hetoxamine O2)
Industrial applications: textile/leather processing (Ethomeen O/12; Hetoxamine O2); waxes and oils (Hetoxamine O2)
Industrial cleaners: metal processing surfactants (Hetoxamine O2)
PROPERTIES:
Form:
Liquid (Hetoxamine O2)
Clear liquid (Ethomeen O/12)

POE (2) oleyl amine (cont'd.)

Color:
Gardner 8 max. (Ethomeen O/12)
Composition:
96% min. tertiary amine (Ethomeen O/12)
Solubility:
Sol. in isopropanol (Hetoxamine O2)
Sol. in min. oil (Hetoxamine O2)
Ionic Nature:
Cationic (Crodamet 1.02; Ethomeen O/12)
M.W.:
350 (Hetoxamine O2)
Sp.gr.:
0.916 (Ethomeen O/12)
Flash Pt.:
470 F (COC) (Ethomeen O/12)
Surface Tension:
31.5 dynes/cm (0.1%) (Ethomeen O/12)
TOXICITY/HANDLING:
Skin irritant, severe eye irritant (Ethomeen O/12)

POE (5) oleyl amine

SYNONYMS:
PEG-5 oleamine (CTFA)
PEG (5) oleyl amine
STRUCTURE:
$CH(CH_2)_7CH_3$ $(CH_2CH_2O)_xH$
|| |
$CH(CH_2)_7CH_2$——N——$(CH_2CH_2O)_yH$
where avg. $(x + y) = 5$
CAS No.:
RD No.: 977065-68-1
TRADENAME EQUIVALENTS:
Crodamet 1.05 [Croda] (primary amine)
Ethomeen O/15 [Armak] (tertiary amine)
Hetoxamine O5 [Heterene]
CATEGORY:
Emulsifier, dispersant, desizing agent, softener, antistatic agent, water repellent
APPLICATIONS:
Farm products: agricultural oils/sprays (Hetoxamine O5)
Industrial applications: textile/leather processing (Ethomeen O/15; Hetoxamine O5);

waxes and oils (Hetoxamine O5)
Industrial cleaners: metal processing surfactants (Hetoxamine O5)

PROPERTIES:

Form:
Liquid (Hetoxamine O5)
Clear liquid (Ethomeen O/15)

Color:
Gardner 8 max. (Ethomeen O/15)

Solubility:
Sol. in isopropanol (Hetoxamine O5)
Sol. in min. oil (Hetoxamine O5)

Ionic Nature:
Cationic (Crodamet 1.05; Ethomeen O/15)

M.W.:
492 (Hetoxamine O5)

Sp.gr.:
0.960 (Ethomeen O/15)

Flash Pt.:
540 F (COC) (Ethomeen O/15)

Surface Tension:
35.3 dynes/cm (0.1%) (Ethomeen O/15)

TOXICITY/HANDLING:
Skin irritant, severe eye irritant (Ethomeen O/15)

POE (15) oleyl amine

SYNONYMS:
PEG-15 oleamine (CTFA)
PEG (15) oleyl amine

STRUCTURE:

$$\begin{array}{lll} CH(CH_2)_7CH_3 & & (CH_2CH_2O)_xH \\ \| & & | \\ CH(CH_2)_7CH_2 & — & N — (CH_2CH_2O)_yH \end{array}$$

where avg. $(x + y) = 15$

CAS No.:
RD No.: 977063-33-4

TRADENAME EQUIVALENTS:
Crodamet 1.015 [Croda] (primary amine)
Ethomeen O/25 [Armak] (tertiary amine)
Hetoxamine O15 [Heterene]

POE (15) oleyl amine (*cont'd.*)

CATEGORY:
Emulsifier, dispersant, desizing agent, softener, antistatic agent, water repellent
APPLICATIONS:
Farm products: agricultural oils/sprays (Hetoxamine O15)
Industrial applications: textile/leather processing (Ethomeen O/25; Hetoxamine O15); waxes and oils (Hetoxamine O15)
Industrial cleaners: metal processing surfactants (Hetoxamine O15)
PROPERTIES:
Form:
Liquid (Hetoxamine O15)
Clear liquid (Ethomeen O/25)
Color:
Gardner 8 max. (Ethomeen O/25)
Solubility:
Sol. in isopropanol (Hetoxamine O15)
Sol. in water (Hetoxamine O15)
Ionic Nature:
Cationic (Crodamet 1.015; Ethomeen O/25)
M.W.:
930 (Hetoxamine O15)
Sp.gr.:
1.04 (Ethomeen O/25)
Flash Pt.:
380 F (PM) (Ethomeen O/25)
TOXICITY/HANDLING:
Skin irritant, severe eye irritant (Ethomeen O/25)

POE (5) oleyl ether

SYNONYMS:
Oleth-5 (CTFA)
PEG-5 oleyl ether
PEG (5) oleyl ether
STRUCTURE:
$CH_3(CH_2)_7CH=CH(CH_2)_7CH_2(OCH_2CH_2)_nOH$
where avg. $n = 5$
CAS No.:
9004-98-2 (generic); 25190-05-0 (generic)
RD No.: 977057-51-4

TRADENAME EQUIVALENTS:

Eumulgin O5 [Henkel]
Hetoxol OL5 [Heterene]
Hostacerin O-5 [Amer. Hoechst]
Lipocol O-2 [Lipo]
Macol OA-5 [Mazer]
Standamul O-5 [Henkel]
Volpo 5 [Croda]
Volpo O5 [Croda Chem. Ltd.]

CATEGORY:

Solubilizer, emulsifier, wetting agent, dispersant, detergent, scouring agent, coupling agent, emollient, superfatting agent, conditioning agent, lubricant, spreading agent, leveling agent, bodying agent, antistat, plasticizer

APPLICATIONS:

Automobile cleaners: car shampoo (Volpo O5)

Bath products: bubble bath (Volpo 5); bath oils (Hetoxol OL5; Volpo 5)

Cosmetic industry preparations: (Eumulgin O5; Macol OA-5; Standamul O-5; Volpo 5, O5); astringent creams and lotions (Volpo 5); creams and lotions (Hetoxol OL5; Hostacerin O-5; Lipocol O-2); hair straighteners, cold waves (Volpo 5); makeup (Volpo 5); perfumery (Volpo 5, O5); shampoos (Hetoxol OL5; Volpo 5)

Degreasers: (Volpo O5)

Farm products: herbicides (Volpo O5); insecticides/pesticides (Eumulgin O5; Volpo O5)

Household formulations: (Hetoxol OL5; Macol OA-5); detergent base/intermediate (Volpo O5); floor polish (Eumulgin O5); hard surface cleaner (Volpo O5); liquid detergents (Volpo O5)

Industrial applications: (Macol OA-5); dyes and pigments (Hetoxol OL5; Lipocol O-2; Volpo O5); latexes (Volpo O5); lubricating/cutting oils (Macol OA-5; Volpo O5); metalworking (Macol OA-5); paint mfg. (Eumulgin O5); petroleum industry (Volpo O5); polishes and waxes (Volpo O5); polymers/polymerization (Volpo O5); silicone products (Hetoxol OL5); textile/leather processing (Hetoxol OL5; Macol OA-5; Volpo O5); waxes and oils (Hetoxol OL5)

Industrial cleaners: dairy cleaners (Volpo O5); glass cleaners (Volpo O5); lime soap dispersant (Volpo O5); metal processing surfactants (Volpo O5); solvent cleaners (Volpo O5)

Pharmaceutical applications: (Volpo O5); antiperspirant/deodorant (Lipocol O-2); depilatories (Lipocol O-2; Volpo 5)

PROPERTIES:

Form:

Liquid (Hetoxol OL5; Macol OA-5; Volpo O5)
Clear liquid (Eumulgin O5)
Hazy liquid (Volpo 5)
Slightly cloudy liquid (Hostacerin O-5)

POE (5) oleyl ether (cont'd.)

Viscous liquid (Standamul O-5)

Color:

Colorless (Macol OA-5)
Pale straw (Volpo O5)
Pale yellow (Eumulgin O5)
Yellowish (Hostacerin O-5)

Odor:

Very low (Volpo 5)

Composition:

97% active min. (Volpo O5)
100% active (Hostacerin O-5; Lipocol O-2)

Solubility:

Sol. in alcohols (Standamul O-5; Volpo 5)
Almost sol. in arachis oil (Volpo O5)
Sol. in most aromatic solvents (Volpo 5)
Partly sol. in butyl stearate (Volpo O5)
Sol. in most chlorinated solvents (Volpo 5)
Sol. in ethanol (Volpo O5)
Sol. in fatty acids (Hostacerin O-5)
Sol. in fatty alcohols (Hostacerin O-5)
Sol. in fatty esters (Hostacerin O-5)
Sol. in glycols (Volpo 5)
Sol. in hydrocarbons (Hostacerin O-5; Standamul O-5)
Sol. in isopropanol (Hetoxol OL5); sol. (@ 1%) (Macol OA-5)
Partly sol. in kerosene (Volpo O5)
Sol. in ketones (Volpo 5)
Sol. in min. oil (Volpo 5); sol. (@ 1%) (Macol OA-5); partly sol. (Volpo O5)
Sol. in nonpolar oils (Volpo 5)
Sol. in oleic acid (Volpo O5)
Partly sol. in oleyl alcohol (Volpo O5)
Sol. in most organic solvents (Hostacerin O-5; Standamul O-5)
Sol. in trichlorethylene (Volpo O5)
Sol. in water (Hetoxol OL5); dissolves to cloudy sol'n. (Hostacerin O-5); partly sol. (Volpo O5); insol. (Volpo 5); insol. (@ 1%) (Macol OA-5)

Ionic Nature:

Nonionic (Eumulgin O5; Hetoxol OL5; Hostacerin O-5; Lipocol O-2; Macol OA-5; Standamul O-5; Volpo 5, O5)

Sp.gr.:

0.95 (20 C) (Eumulgin O5)
0.953 ± 0.005 (20 C) (Hostacerin O-5)

Visc.:

67 ± 3 cps (20 C) (Hostacerin O-5)

Solidification Pt.:
50–68 F (Eumulgin O5)
Flash Pt.:
240 C (Marcusson) (Hostacerin O-5)
HLB:
8.7 (Macol OA-5)
8.8 (Volpo 5)
9.0 (Volpo O5)
9.1 (Hetoxol OL5)
Acid No.:
1.0 (Hostacerin O-5)
1.0 max. (Volpo O5)
2.0 max. (Volpo 5)
Iodine No.:
40–52 (Wijs) (Volpo 5)
42–48 (Volpo O5)
Saponification No.:
1.0 (Hostacerin O-5)
Hydroxyl No.:
110–120 (Hetoxol OL5)
115–125 (Volpo O5)
120–130 (Eumulgin O5)
120–135 (Volpo 5)
125 (Macol OA-5)
Stability:
Excellent stability; unaffected by acids and alkalies of medium conc. (Eumulgin O5)
Stable to hydrolysis by strong acids and alkalies (Macol OA-5)
Stable in sol'ns. of metallic ions and to many acids and alkalies (Volpo 5)
Stable to many alkalies and acids under extreme pH conditions (Standamul O-5)
Stable to alkalies and strong mineral acids (Volpo O5)
Acid and alkaline stable (Lipocol O-2)
Storage Stability:
Stable under normal storage conditions (Eumulgin O5)
Ref. Index:
1.463 (20 C) (Hostacerin O-5)
pH:
5.0–7.0 (3% aq. sol'n.) (Volpo 5)
6.0–7.5 (3%) (Volpo O5)
Neutral (Hostacerin O-5)
Surface Tension:
31.0 dynes/cm (0.1% aq.) (Volpo O5)
TOXICITY/HANDLING:
Observe normal safety precautions (Eumulgin O5)

POE (5) oleyl ether (*cont' d.*)

STORAGE/HANDLING:
Store at temps. above 50 F to avoid solidification (Eumulgin O5)
STD. PKGS.:
55-gal (396 lb net) steel drums (Eumulgin O5)

POE (20) sorbitan monoisostearate

SYNONYMS:
PEG-20 sorbitan isostearate (CTFA)
PEG-20 sorbitan monoisostearate
PEG 1000 sorbitan monoisostearate
Sorbitan monoisostearate, polyethoxylated (20)
CAS No.:
RD No.: 977055-33-6
TRADENAME EQUIVALENTS:
Crillet 6 [Croda]
Montanox 70 [Seppic]
Nikkol TI-10 [Nikko]
T-Maz 67 [Mazer]
CATEGORY:
Solubilizer, emulsifier, wetting agent, surfactant, antistat, stabilizer, dispersant, viscosity modifier
APPLICATIONS:
Cosmetic industry preparations: (Crillet 6; Nikkol TI-10; T-Maz 67)
Farm products: herbicides (Crillet 6); insecticides/pesticides (Crillet 6)
Food applications: (Crillet 6; T-Maz 67); food packaging (Crillet 6)
Household detergents: (Crillet 6)
Industrial applications: metalworking (Crillet 6; T-Maz 67); paint mfg. (Crillet 6); polishes and waxes (Crillet 6); polymers/polymerization (Crillet 6); printing inks (Crillet 6); textile/leather processing (Crillet 6; T-Maz 67)
Pharmaceutical applications: (Crillet 6; T-Maz 67)
PROPERTIES:
Form:
Liquid (Montanox 70; Nikkol TI-10)
Clear liquid (Crillet 6)
Color:
Yellow (Crillet 6; T-Maz 67)
Composition:
97% active min. (T-Maz 67)
100% conc. (Montanox 70; Nikkol TI-10)

Solubility:
Miscible with acetone in certain proportions (T-Maz 67)
Sol. in ethanol (Crillet 6); miscible in certain proportions (T-Maz 67)
Miscible with naphtha in certain proportions (T-Maz 67)
Sol. in oleic acid (Crillet 6)
Sol. in oleyl alcohol (Crillet 6)
Sol. in toluol (T-Maz 67)
Sol. in trichlorethylene (Crillet 6)
Sol. in water (Crillet 6; T-Maz 67)
Sol. in xylene (Crillet 6)
Ionic Nature:
Nonionic (Crillet 6; Montanox 70; Nikkol TI-10; T-Maz 67)
Sp.gr.:
1.0 (T-Maz 67)
Visc.:
400 cps (T-Maz 67)
HLB:
14.8 (T-Maz 67)
14.9 (Crillet 6)
15.0 (Nikkol TI-10)
Acid No.:
2 max. (T-Maz 67)
Saponification No.:
40–50 (Crillet 6)
45–55 (T-Maz 67)
Hydroxyl No.:
80–95 (T-Maz 67)
Surface Tension:
38.6 dynes/cm (0.1%) (Crillet 6)

POE (40) sorbitan peroleate

SYNONYMS:
PEG-40 sorbitan peroleate (CTFA)
PEG 2000 sorbitan peroleate
POE (40) sorbitol septaoleate
CAS No.:
RD No.: 977063-07-2
TRADENAME EQUIVALENTS:
Arlatone T [ICI United States]

POE (40) sorbitan peroleate (cont'd.)

CATEGORY:
Solubilizer, emulsifier, antistat, lubricant, surfactant
APPLICATIONS:
Industrial applications: textile/leather processing (Arlatone T)
PROPERTIES:
Form:
Liquid (Arlatone T)
Color:
Yellow (Arlatone T)
Composition:
100% active (Arlatone T)
Solubility:
Sol. in isopropyl myristate (Arlatone T)
Sol. in isopropyl palmitate (Arlatone T)
Sol. in min. oil (Arlatone T)
Sol. in veg. oil (Arlatone T)
Disp. in water (Arlatone T)
Sol. in xylene
Ionic Nature:
Nonionic (Arlatone T)
Sp.gr.:
≈ 1.0 (Arlatone T)
Visc.:
≈ 175 cps (Arlatone T)
Flash Pt.:
> 300 F (Arlatone T)
Fire Pt.:
> 300 F (Arlatone T)
HLB:
9.0 (Arlatone T)

POE (10) stearyl amine

SYNONYMS:
PEG-10 stearamine (CTFA)
PEG 500 stearyl amine
POE (10) octadecylamine
STRUCTURE:

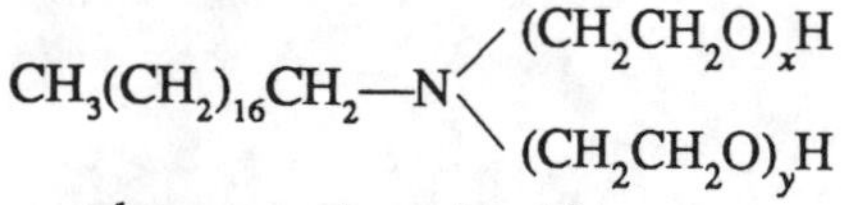

$$CH_3(CH_2)_{16}CH_2—N\begin{cases}(CH_2CH_2O)_xH \\ (CH_2CH_2O)_yH\end{cases}$$

where avg. $(x + y) = 10$

CAS No.:
RD No.: 977065-32-9
TRADENAME EQUIVALENTS:
Ethomeen 18/20 [Armak]
CATEGORY:
Emulsifier, dispersant
APPLICATIONS:
Industrial applications: textile/leather processing (Ethomeen 18/20)
PROPERTIES:
Form:
Liquid to paste (Ethomeen 18/20)
Color:
Gardner 8 max. (Ethomeen 18/20)
Ionic Nature:
Cationic (Ethomeen 18/20)
Sp.gr.:
1.020 (Ethomeen 18/20)
Flash Pt.:
540 F (COC) (Ethomeen 18/20)
Surface Tension:
40.0 dynes/cm (0.1%) (Ethomeen 18/20)
TOXICITY/HANDLING:
Skin irritant, severe eye irritant (Ethomeen 18/20)

POE (15) stearyl amine

SYNONYMS:
PEG-15 stearamine (CTFA)
PEG (15) stearyl amine
POE (15) octadecylamine
STRUCTURE:

$$CH_3(CH_2)_{16}CH_2—N \begin{matrix} / (CH_2CH_2O)_xH \\ \backslash (CH_2CH_2O)_yH \end{matrix}$$

where avg. $(x + y) = 15$
CAS No.:
RD No.: 977063-35-6
TRADENAME EQUIVALENTS:
Ethomeen 18/25 [Armak] (tertiary amine)
Hetoxamine ST-15 [Heterene] (tertiary amine)

POE (15) stearyl amine (*cont'd.*)

CATEGORY:

Emulsifier, dispersant, desizing agent, antistat, water repellent

APPLICATIONS:

Farm applications: agricultural oils/sprays (Hetoxamine ST-15)

Industrial applications: textile/leather processing (Ethomeen 18/20; Hetoxamine ST-15); waxes and oils (Hetoxamine ST-15)

Industrial cleaners: metal processing surfactants (Hetoxamine ST-15)

PROPERTIES:

Form:

Liquid to paste (Ethomeen 18/25)

Solid (Hetoxamine ST-15)

Color:

Gardner 8 max. (Ethomeen 18/25; Hetoxamine ST-15)

Composition:

95.0% min. tertiary amine (Hetoxamine ST-15)

Solubility:

Sol. in isopropanol (Hetoxamine ST-15)

Insol. in min. oil (Hetoxamine ST-15)

Sol. in water (Hetoxamine ST-15)

Ionic Nature:

Cationic (Ethomeen 18/25)

Sp.gr.:

1.015 (Ethomeen 18/25)

Flash Pt.:

560 F (COC) (Ethomeen 18/25)

Equivalent Wt.:

870–940 (Hetoxamine ST-15)

Surface Tension:

43.0 dynes/cm (0.1%) (Ethomeen 18/25)

POE (2) stearyl ether

SYNONYMS:

PEG-2 stearyl ether

PEG 100 stearyl ether

POE (2) stearyl alcohol

Steareth-2 (CTFA)

STRUCTURE:

$CH_3(CH_2)_{16}CH_2(OCH_2CH_2)_nOH$

where avg. $n = 2$

CAS No.:
9005-00-9 (generic)
RD No.: 977055-79-0

TRADENAME EQUIVALENTS:
Alkasurf SA-2 [Alkaril]
Brij 72 [ICI United States]
Hetoxol STA2 [Heterene]
Lipocol S-2 [Lipo]
Macol SA-2 [Mazer]
Simulsol 72 [Seppic]
ST-55-2 [Hefti Ltd.]
Volpo S.2 [Croda]

CATEGORY:
Dispersant, solubilizer, coupling agent, detergent, wetting agent, emulsifier, surfactant, defoamer, conditioning agent, leveling agent, intermediate, stabilizer

APPLICATIONS:
Bath products: bath oils (Hetoxol STA2)
Cosmetic industry preparations: (Macol SA-2; ST-55-2; Volpo S.2); creams and lotions (Hetoxol STA2; Lipocol S-2); shampoos (Hetoxol STA2; Lipocol S-2); topical products (Alkasurf SA-2)
Household products: (Hetoxol STA2; Macol SA-2); bleaches/dyes (Lipocol S-2); detergents (Lipocol S-2)
Industrial applications: (Macol SA-2); dyes and pigments (Hetoxol STA2; Lipocol S-2); latex emulsion polymerization (ST-55-2); metalworking lubricants (Macol SA-2); silicone products (Hetoxol STA2); textile/leather processing (Hetoxol STA2; Macol SA-2; ST-55-2)
Pharmaceutical applications: (ST-55-2); antiperspirant/deodorant (Lipocol S-2); depilatories (Lipocol S-2)

PROPERTIES:

Form:
Solid (Alkasurf SA-2; Hetoxol STA2; Macol SA-2; Simulsol 72; ST-55-2; Volpo S.2)
Waxy solid (Brij 72; Lipocol S-2)

Color:
White (Alkasurf SA-2; Brij 72; Lipocol S-2; Macol SA-2)

Composition:
100% conc. (Macol SA-2; Simulsol 72; ST-55-2; Volpo S.2)

Solubility:
Sol. in alcohols (Brij 72)
Sol. in aromatic solvent (@ 10%) (Alkasurf SA-2)
Sol. in cottonseed oil (Brij 72)
Sol. in isopropanol (Hetoxol STA2; Macol SA-2)
Sol. in perchloroethylene (@ 10%) (Alkasurf SA-2)

POE (2) stearyl ether (cont'd.)

Ionic Nature:
Nonionic (Brij 72; Hetoxol STA2; Lipocol S-2; Macol SA-2; Simulsol 72; ST-55-2; Volpo S.2)
M.P.:
43 C (Macol SA-2)
Pour Pt.:
43 C (Brij 72)
Flash Pt.:
> 300 F (Brij 72)
Fire Pt.:
> 300 F (Brij 72)
Cloud Pt.:
57–61 C (10% in 25% butyl Carbitol) (Alkasurf SA-2)
HLB:
4.9 (Alkasurf SA-2; Brij 72; Hetoxol STA2; Macol SA-2; Simulsol 72)
4.9 ± 1 (Lipocol S-2)
5.0 (ST-55-2)
5.6 (Volpo S.2)
Acid No.:
1.0 max. (Lipocol S-2)
Hydroxyl No.:
145–165 (Hetoxol STA2; Macol SA-2)
155–165 (Lipocol S-2)
Stability:
Stable to hydrolysis by strong acids and alkalies (Macol SA-2)
Acid and alkaline stable (Lipocol S-2)
Stable at high and low pH; stable in presence of ionic matter (Volpo S.2)
STD. PKGS.:
55-gal open-head steel drums (Brij 72)

POE (10) stearyl ether

SYNONYMS:
PEG-10 stearyl ether
PEG 500 stearyl ether
POE (10) stearyl alcohol
Steareth-10 (CTFA)
STRUCTURE:
$CH_3(CH_2)_{16}CH_2(OCH_2CH_2)_nOH$
where avg. $n = 10$

CAS No.:
9005-00-9 (generic)
RD No.: 977055-80-3

TRADENAME EQUIVALENTS:
Brij 76 [ICI United States] (with BHA and citric acid)
Hetoxol STA-10 [Heterene]
Lipocol S-10 [Lipo]
Macol SA-10 [Mazer]
Simulsol 76 [Seppic]
Volpo S10 [Croda]

CATEGORY:
Solubilizer, surfactant, dispersant, coupling agent, emulsifier, defoamer, wetting agent, conditioner, detergent, leveling agent, intermediate, dye assistant

APPLICATIONS:
Bath products: bath oils (Hetoxol STA-10)
Cosmetic industry preparations: (Macol SA-10; Volpo S10); creams and lotions (Hetoxol STA-10; Lipocol S-10); perfumery (Brij 76); shampoos (Hetoxol STA-10)
Household formulations: (Hetoxol STA-10; Macol SA-10)
Industrial applications: (Macol SA-10); dyes and pigments (Hetoxol STA-10; Lipocol S-10); metalworking lubricants (Macol SA-10); textile/leather processing (Hetoxol STA-10; Macol SA-10)
Pharmaceutical applications: antiperspirant/deodorant (Lipocol S-10); depilatories (Lipocol S-10)

PROPERTIES:

Form:
Solid (Hetoxol STA-10; Macol SA-10; Simulsol 76; Volpo S10)
Waxy solid (Brij 76; Lipocol S-10)

Color:
White (Brij 76; Lipocol S-10; Macol SA-10)
Gardner 1 max. (Hetoxol STA-10)

Composition:
99% conc. min. (Hetoxol STA-10)
100% conc. (Lipocol S-10; Macol SA-10; Simulsol 76; Volpo S10)

Solubility:
Sol. in ethanol (Brij 76)
Sol. in isopropanol (Hetoxol STA-10; Macol SA-10)
Insol. in min. oil (Hetoxol STA-10)
Sol. in propylene glycol (Brij 76)
Sol. in water (Hetoxol STA-10)

Ionic Nature:
Nonionic (Brij 76; Hetoxol STA-10; Lipocol S-10; Macol SA-10; Simulsol 76; Volpo S10)

POE (10) stearyl ether (cont'd.)

M.P.:
38 C (Macol SA-10)
Pour Pt.:
38 C (Brij 76)
Flash Pt.:
> 300 F (Brij 76)
Fire Pt.:
> 300 F (Brij 76)
HLB:
12.4 (Brij 76; Macol SA-10; Simulsol 76; Volpo S10)
12.4 ± 1 (Hetoxol STA-10; Lipocol S-10)
Acid No.:
1.0 max. (Hetoxol STA-10; Lipocol S-10)
Hydroxyl No.:
45–60 (Hetoxol STA-10)
75–90 (Lipocol S-10; Macol SA-10)
Stability:
Acid and alkali stable (Lipocol S-10)
Stable to hydrolysis by strong acids and alkalis (Macol SA-10)
Stable over a wide pH range; stable in presence of ionic matter (Volpo S10)

POE (15) stearyl ether

SYNONYMS:
PEG-15 stearyl ether
PEG (15) stearyl ether
POE (15) stearyl alcohol
Steareth-15 (CTFA)
STRUCTURE:
$CH_3(CH_2)_{16}CH_2(OCH_2CH_2)_nOH$
where avg. $n = 15$
CAS No.:
9005-00-9 (generic)
TRADENAME EQUIVALENTS:
Macol SA-15 [Mazer]
CATEGORY:
Dispersant, solubilizer, coupling agent, detergent, wetting agent, emulsifier
APPLICATIONS:
Cosmetic industry preparations: (Macol SA-15)
Household products: (Macol SA-15)
Industrial applications: (Macol SA-15); metalworking lubricants (Macol SA-15);

textile/leather processing (Macol SA-15)

PROPERTIES:

Form:
Solid (Macol SA-15)

Color:
White (Macol SA-15)

Composition:
100% conc. (Macol SA-15)

Solubility:
Sol. in isopropanol (Macol SA-15)
Sol. in water (Macol SA-15)

Ionic Nature:
Nonionic (Macol SA-15)

M.P.:
38 C (Macol SA-15)

HLB:
14.2 (Macol SA-15)

Hydroxyl No.:
57–72 (Macol SA-15)

Stability:
Stable to hydrolysis by strong acids and alkalies (Macol SA-15)

POE (20) stearyl ether

SYNONYMS:
PEG-20 stearyl ether
PEG 1000 stearyl ether
POE (20) stearyl alcohol
Steareth-20 (CTFA)

STRUCTURE:
$CH_3(CH_2)_{16}CH_2(OCH_2CH_2)_nOH$
where avg. $n = 20$

CAS No.:
9005-00-9 (generic)
RD No.: 977053-41-0

TRADENAME EQUIVALENTS:
Alkasurf SA-20 [Alkaril]
Brij 78 [ICI United States] (with BHA and citric acid)
Hetoxol STA-20 [Heterene]
Lipal 20SA [PVO]
Lipocol S-20 [Lipo]

POE (20) stearyl ether (cont'd.)

TRADENAME EQUIVALENTS *(cont'd.):*
Macol SA-20 [Mazer]
Simulsol 78 [Seppic]
ST-55-20 [Hefti Ltd.]
Trycol SAL-20 [Emery]
Volpo S20 [Croda]

CATEGORY:
Dispersant, detergent, lubricant, suspending aid, solubilizer, coupling agent, surfactant, emulsifier, coemulsifier, leveling agent, intermediate, wetting agent, defoamer, conditioning agent, stabilizer

APPLICATIONS:
Bath products: bath oils (Hetoxol STA-20)
Cleansers: germicidal skin cleanser (Lipal 20SA)
Cosmetic industry preparations: (Macol SA-20; ST-55-20; Volpo S20); creams and lotions (Hetoxol STA-20; Lipal 20SA; Lipocol S-20); hair preparations (Lipal 20SA); makeup (Lipal 20SA); perfumery (Brij 78); shampoos (Hetoxol STA-20); shaving preparations (Lipal 20SA)
Food applications: citrus fruit coatings (Trycol SAL-20)
Household formulations: (Hetoxol STA-20; Macol SA-20)
Industrial applications: (Macol SA-20); dyes and pigments (Hetoxol STA-20; Lipocol S-20; ST-55-20; Trycol SAL-20); latex applications (Trycol SAL-20); metalworking lubricants (Macol SA-20); polishes and waxes (ST-55-20); silicone products (Hetoxol STA-20; ST-55-20); textile/leather processing (Alkasurf SA-20; Hetoxol STA-20; Macol SA-20; ST-55-20; Trycol SAL-20)
Industrial cleaners: textile scouring (Alkasurf SA-20)
Pharmaceutical applications: (ST-55-20); antiperspirant/deodorant (Alkasurf SA-20; Lipal 20SA; Lipocol S-20); depilatories (Lipocol S-20); ointments (Lipal 20SA)

PROPERTIES:

Form:
Solid (Alkasurf SA-20; Hetoxol STA-20; Lipal 20SA; Macol SA-20; Simulsol 78; Trycol SAL-20; Volpo S20)
Flakes (ST-55-20)
Waxy solid (Brij 78; Lipocol S-20)

Color:
White (Alkasurf SA-20; Brij 78; Lipocol S-20; Macol SA-20)
Gardner 1 (Lipal 20SA; Trycol SAL-20)
Gardner 1 max. (Hetoxol STA-20)

Composition:
99+% active (Hetoxol STA-20)
100% active (Lipal 20SA; Lipocol S-20)
100% conc. (Macol SA-20; Simulsol 78; ST-55-20; Volpo S20)

Solubility:
Sol. in alcohols (Brij 78)

Disp. in aromatic solvent (@ 10%) (Alkasurf SA-20)
Disp. in butyl stearate (@ 5%) (Trycol SAL-20)
Disp. in glycerol trioleate (@ 5%) (Trycol SAL-20)
Sol. in isopropanol (Hetoxol STA-20; Lipal 20SA; Macol SA-20)
Disp. in min. oil (@ 5%) (Trycol SAL-20); insol. (Hetoxol STA-20)
Disp. in perchloroethylene (@ 10%) (Alkasurf SA-20)
Sol. in propylene glycol (Lipal 20SA)
Disp. in Stoddard solvent (@ 5%) (Trycol SAL-20)
Sol. in water (Hetoxol STA-20; Lipal 20SA; Macol SA-20); sol. (@ 10%) (Alkasurf SA-20); sol. (@ 5%) (Trycol SAL-20)
Sol. in xylene (@ 5%) (Trycol SAL-20)

Ionic Nature:
Nonionic (Brij 78; Hetoxol STA-20; Lipal 20SA; Lipocol S-20; Macol SA-20; Simulsol 78; ST-55-20; Trycol SAL-20; Volpo S20)

M.P.:
38 C (Macol SA-20)
40 C (Trycol SAL-20)

Pour Pt.:
38 C (Brij 78)

Flash Pt.:
> 300 F (Brij 78)
560 F (Trycol SAL-20)

Fire Pt.:
> 300 F (Brij 78)

Cloud Pt.:
70–75 C (1% sol'n.) (Lipal 20SA)
73–77 C (1% in 10% sodium chloride) (Alkasurf SA-20)
91 C (5% saline) (Trycol SAL-20)

HLB:
15.0 (ST-55-20)
15.3 (Alkasurf SA-20; Brij 78; Hetoxol STA-20; Macol SA-20; Simulsol 78; Trycol SAL-20)
15.3 ± 1 (Lipal 20SA; Lipocol S-20)
15.5 (Volpo S20)

Acid No.:
1.0 max. (Lipocol S-20)
2.0 max. (Hetoxol STA-20)

Hydroxyl No.:
45–60 (Hetoxol STA-20; Lipocol S-20; Macol SA-20)
50–70 (Lipal 20SA)

Stability:
Stable over a wide pH range (Lipal 20SA)
Stable over a wide pH range and in presence of ionic matter (Volpo S20)

POE (20) stearyl ether (cont'd.)

Acid and alkaline stable (Lipocol S-20)
Stable to hydrolysis by strong acids and alkalis (Macol SA-20)

POE (21) stearyl ether

SYNONYMS:
PEG-21 stearyl ether
PEG (21) stearyl ether
POE (21) stearyl alcohol
Steareth-21 (CTFA)
STRUCTURE:
$CH_3(CH_2)_{16}CH_2(OCH_2CH_2)_nOH$
where avg. $n = 21$
TRADENAME EQUIVALENTS:
Brij 721 [ICI United States] (with BHA and citric acid)
CATEGORY:
Solubilizer, emulsifier
APPLICATIONS:
Cosmetic industry preparations: (Brij 721); creams and lotions (Brij 721); shaving preparations (Brij 721)
Industrial applications: aerosols (Brij 721)
Pharmaceutical applications: antiperspirant/deodorant (Brij 721)
PROPERTIES:
Form:
Waxy solid (Brij 721)
Color:
Almost colorless (Brij 721)
Odor:
Slight, waxy (Brij 721)
Composition:
98% conc. min. (Brij 721)
Solubility:
Insol. in cottonseed oil (@ 1% and 10%) (Brij 721)
Insol. in ethanol (@ 1% and 10%) (Brij 721)
Insol. in glycerin (@ 1% and 10%) (Brij 721)
Insol. in isopropanol (@ 1% and 10%) (Brij 721)
Insol. in min. oil (@ 1% and 10%) (Brij 721)
Insol. in propylene gycol (@ 1% and 10%) (Brij 721)
Insol. in dist. water (@ 1% and 10%) (Brij 721)
Ionic Nature:
Nonionic (Brij 721)

Pour Pt.:
45 C (Brij 721)
Flash Pt.:
> 230 F (PMCC) (Brij 721)
HLB:
15.5 (Brij 721)
Acid No.:
2.0 max. (Brij 721)
Hydroxyl No.:
44–61 (Brij 721)
Stability:
Stable to alkali (Brij 721)
TOXICITY/HANDLING:
Classed as mild eye irritant (Brij 721)

POE (40) stearyl ether

SYNONYMS:
PEG-40 stearyl ether
PEG 2000 stearyl ether
POE (40) stearyl alcohol
Steareth-40 (CTFA)
STRUCTURE:
$CH_3(CH_2)_{16}CH_2(OCH_2CH_2)_nOH$
where avg. $n = 40$
CAS No.:
9005-00-9 (generic)
TRADENAME EQUIVALENTS:
Macol SA-40 [Mazer]
CATEGORY:
Dispersant, solubilizer, coupling agent, detergent, wetting agent, emulsifier
APPLICATIONS:
Cosmetic industry preparations: (Macol SA-40)
Household products: (Macol SA-40)
Industrial applications: (Macol SA-40); metalworking lubricants (Macol SA-40); textile/leather processing (Macol SA-40)
PROPERTIES:
Form:
Solid (Macol SA-40)
Color:
White (Macol SA-40)

POE (40) stearyl ether (cont'd.)

Composition:
100% conc. (Macol SA-40)
Solubility:
Sol. in isopropanol (Macol SA-40)
Sol. in water (Macol SA-40)
Ionic Nature:
Nonionic (Macol SA-40)
M.P.:
40 C (Macol SA-40)
HLB:
17.4 (Macol SA-40)
Hydroxyl No.:
25–40 (Macol SA-40)
Stability:
Stable to hydrolysis by strong acids and alkalies (Macol SA-40)

POE (50) tallow amine

SYNONYMS:
PEG-50 tallow amine (CTFA)
PEG (50) tallow amine
PEG (50) tallow amine, hydrogenated
STRUCTURE:

$$R{-}N\begin{cases}(CH_2CH_2O)_xH \\ (CH_2CH_2O)_yH\end{cases}$$

where R represents the tallow radical and avg. $(x + y) = 50$
CAS No.:
61791-26-2 (generic)
RD No.: 977069-79-6
TRADENAME EQUIVALENTS:
Chemeen HT-50 [Chemax]
Ethomeen T/60 [Armak]
Varonic U-250 [Sherex]
CATEGORY:
Dispersant, emulsifier, antistat, lubricant, dye intermediate, plasticizer, leveling agent
APPLICATIONS:
Industrial applications: dyes and pigments (Varonic U-250); fiberglass (Chemeen HT-50); lacquer (Varonic U-250); metal buffing (Chemeen HT-50); plastics/resins (Varonic U-250); rubbers (Chemeen HT-50); textile/leather processing (Chemeen HT-50; Ethomeen T/60)

PROPERTIES:
Form:
Paste to solid (Ethomeen T/60)
Solid (Chemeen HT-50; Varonic U-250)
Color:
Gardner 7 (Varonic U-250)
Gardner 10 max. (Ethomeen T/60)
Composition:
97% ethoxylated amine (Varonic U-250)
Solubility:
Emulsifiable in benzene (Varonic U-250)
Sol. in carbon tetrachloride (Varonic U-250)
Sol. in isopropanol (Varonic U-250)
Emulsifiable-sol. in MEK (Varonic U-250)
Emulsifiable in Stoddard solvent (Varonic U-250)
Sol. in water (Varonic U-250)
M.W.:
2362–2562 (Ethomeen T/60)
2470 (Chemeen HT-50)
Sp.gr.:
1.111 (Varonic U-250)
1.115 (60 F) (Ethomeen T/60)
Flash Pt.:
> 400 F (PM) (Ethomeen T/60)
Neutral. Equiv.:
2470 (Varonic U-250)

POE (3) tallow aminopropylamine

SYNONYMS:
PEG-3 tallow aminopropylamine (CTFA)
PEG (3) tallow aminopropylamine
STRUCTURE:

$$R—N(CH_2CH_2O)_zH—(CH_2)_3—N[(CH_2CH_2O)_xH][(CH_2CH_2O)_yH]$$

where R represents the tallow radical and
avg. $(x + y + z) = 3$

POE (3) tallow aminopropylamine (cont'd.)

CAS No.:
977066-69-5
TRADENAME EQUIVALENTS:
Ethoduomeen T/13 [Akzo]
CATEGORY:
Dispersant, emulsifier
APPLICATIONS:
Industrial applications: bitumen emulsions (Ethoduomeen T/13); waxes (Ethoduomeen T/13)
PROPERTIES:
Form:
Liquid (Ethoduomeen T/13)
Color:
Gardner 18 max. (Ethoduomeen T/13)
Composition:
95% active min. (Ethoduomeen T/13)
Ionic Nature:
Cationic (Ethoduomeen T/13)
Sp.gr.:
0.95 (Ethoduomeen T/13)
F.P.:
20 C (Ethoduomeen T/13)
B.P.:
150 C min. (Ethoduomeen T/13)
M.P.:
17 C (Ethoduomeen T/13)
Flash Pt.:
204 C (COC) (Ethoduomeen T/13)
HLB:
10.1 (Ethoduomeen T/13)
Surface Tension:
34.5 dynes/cm (0.1%) (Ethoduomeen T/13)
TOXICITY/HANDLING:
Skin irritant, severe eye irritant (Ethoduomeen T/13)
STORAGE/HANDLING:
Avoid contact with strong oxidizing agents (Ethoduomeen T/13)

POE (9) tridecyl ether

SYNONYMS:

PEG-9 tridecyl ether
PEG 450 tridecyl ether
POE (9) tridecyl alcohol
Trideceth-9 (CTFA)
Tridecyl alcohol, ethoxylated (9 EO)
Tridecyl alcohol ethylene oxide adduct (9 EO)

STRUCTURE:

$C_{13}H_{27}(OCH_2CH_2)_nOH$
where avg. $n = 9$

CAS No.:

24938-91-8 (generic)
RD No.: 977062-57-9

TRADENAME EQUIVALENTS:

Chemal TDA-9 [Chemax]
Hetoxol TD-9 [Heterene]
Iconol TDA-9 [BASF Wyandotte]
Siponic TD-9-90 [Alcolac]
Teric 13A9 [ICI Australia Ltd.]
Trycol TDA-9 [Emery]

CATEGORY:

Dispersant, wetting agent, rewetting agent, detergent, emulsifier, coemulsifier, foam stabilizer, solubilizer, surfactant, emollient, lubricant

APPLICATIONS:

Cosmetic industry preparations: creams and lotions (Siponic TD-9-90); skin preparations (Siponic TD-9-90)
Farm products: agricultural applications (Teric 13A9)
Household detergents:; high-foaming cleaner (Trycol TDA-9); heavy-duty cleaner (Trycol TDA-9); light-duty cleaners (Trycol TDA-9)
Industrial applications: dyes and pigments (Teric 13A9); industrial processing (Teric 13A9; Trycol TDA-9); paper towels (Trycol TDA-9)
Industrial cleaners: (Hetoxol TD-9); solvent cleaners (Teric 13A9); textile scouring (Trycol TDA-9)

PROPERTIES:

Form:

Liquid (Chemal TDA-9; Hetoxol TD-9; Iconol TDA-9; Siponic TD-9-90; Teric 13A9; Trycol TDA-9)

Color:

APHA 70 max. (Iconol TDA-9)
Gardner 1 max. (Hetoxol TD-9; Trycol TDA-9)
Hazen 150 (Teric 13A9)

Composition:

90% active (Siponic TD-9-90; Teric 13A9)

POE (9) tridecyl ether (cont'd.)

99.5% active min. (Hetoxol TD-9)
100% active (Iconol TDA-9; Trycol TDA-9)

Solubility:
Disp. in butyl stearate (@ 5%) (Trycol TDA-9)
Sol. in ethanol (Teric 13A9)
Sol. in ethyl acetate (Teric 13A9)
Sol. in ethyl Icinol (Teric 13A9)
Sol. in glycerol trioleate (@ 5%) (Trycol TDA-9)
Sol. in isopropanol (Hetoxol TD-9)
Disp. in min. oil (@ 5%) (Trycol TDA-9); insol. (Hetoxol TD-9)
Sol. in olein (Teric 13A9)
Disp. in perchloroethylene (@ 5%) (Trycol TDA-9)
Sol. in Stoddard solvent (@ 5%) (Trycol TDA-9)
Sol. in water (Hetoxol TD-9; Iconol TDA-9; Teric 13A9); sol. (@ 5%) (Trycol TDA-9)

Ionic Nature:
Nonionic (Iconol TDA-9; Siponic TD-9-90; Teric 13A9; Trycol TDA-9)

M.W.:
590 (Iconol TDA-9)

Sp.gr.:
1.0 (Iconol TDA-9)
1.018 (20 C) (Teric 13A9)

Density:
8.3 lb/gal (Trycol TDA-9)

Visc.:
75 cps (Iconol TDA-9)
75 cs (Trycol TDA-9)
142 cps (20 C) (Teric 13A9)

M.P.:
< 0 C (Teric 13A9)

Pour Pt.:
20 C (Iconol TDA-9)

Cloud Pt.:
55 C (Trycol TDA-9)
55–62 C (1% aq.) (Iconol TDA-9)
63 ± 2 C (1% in hard water) (Teric 13A9)

HLB:
12.9 (Siponic TD-9-90)
13.2 (Hetoxol TD-9; Iconol TDA-9; Trycol TDA-9)
13.3 (Teric 13A9)

Acid No.:
2.0 max. (Hetoxol TD-9)

Hydroxyl No.:
88–102 (Chemal TDA-9)

90–100 (Hetoxol TD-9)

Stability:

Stable to hydrolysis by acids and alkalies; conc. mineral acids will react chemically with this product (Trycol TDA-9)

pH:

6.0–7.5 (5% aq.) (Iconol TDA-9)

6.0–8.0 (1% aq.) (Teric 13A9)

Surface Tension:

27.9 dynes/cm (0.1%, 20 C) (Teric 13A9)

Biodegradable: Nonbiodegradable (Teric 13A9)

TOXICITY/HANDLING:

May cause skin and eye irritation; spillages are slippery (Teric 13A9)

STD. PKGS.:

55-gal (420 lb net) steel drums (Iconol TDA-9)

POE (10) tridecyl ether

SYNONYMS:

Ethoxylated tridecyl alcohol (10 EO)

PEG-10 tridecyl ether

PEG 500 tridecyl ether

Trideceth-10 (CTFA)

Tridecyloxypoly (ethyleneoxy) ethanol (10 EO)

STRUCTURE:

$C_{13}H_{27}(OCH_2CH_2)_nOH$

where avg. $n = 10$

CAS No.:

24938-91-8 (generic)

RD No.: 977058-50-6

TRADENAME EQUIVALENTS:

Carsonon TD-10 [Carson]

Emulphogene BC-720 [GAF]

Iconol TDA-10 [BASF Wyandotte]

Lipal 10TD [PVO]

Macol TD-10 [Mazer]

Volpo T10 [Croda]

CATEGORY:

Emulsifier, solubilizer, dispersant, detergent, wetting agent, rewetting agent, foam builder, foam stabilizer, surfactant

APPLICATIONS:

Cleansers: germicidal skin cleanser (Lipal 10TD)

POE (10) tridecyl ether (cont'd.)

Cosmetic industry preparations: (Macol TD-10); creams and lotions (Lipal 10TD); hair preparations (Lipal 10TD); makeup (Lipal 10TD); shaving preparations (Lipal 10TD)

Household detergents: (Carsonon TD-10; Emulphogene BC-720); heavy-duty cleaner (Emulphogene BC-720); light-duty cleaners (Emulphogene BC-720)

Industrial applications: (Macol TD-10); corrosion inhibitors (Carsonon TD-10; Emulphogene BC-720); industrial processing (Carsonon TD-10; Emulphogene BC-720); paper mfg. (Carsonon TD-10; Emulphogene BC-720); textile/leather processing (Carsonon TD-10)

Industrial cleaners: (Carsonon TD-10); textile scouring (Emulphogene BC-720)

Pharmaceutical applications: (Macol TD-10); antiperspirant/deodorant (Lipal 10TD); ointments (Lipal 10TD)

PROPERTIES:

Form:

Liquid (Carsonon TD-10)
Opaque, viscous, pourable liquid (Emulphogene BC-720)
Liquid/paste (Iconol TDA-10)
Soft paste (Volpo T10)
Semisolid (Macol TD-10)
Solid (Lipal 10TD)

Color:

Off-white (Volpo T10)
APHA 70 max. (Iconol TDA-10)
Gardner 1 (Carsonon TD-10; Lipal 10TD)
VCS 2 max. (@ 50 C) (Emulphogene BC-720)

Odor:

Mild, pleasant (Emulphogene BC-720)

Composition:

100% active (Carsonon TD-10; Emulphogene BC-720; Iconol TDA-10; Lipal 10TD; Macol TD-10)

Solubility:

Sol. in acetone (Emulphogene BC-720)
Sol. in alcohols (Emulphogene BC-720)
Partly sol. in arachis oil (Volpo T10)
Partly sol. in butyl stearate (Volpo T10)
Sol. in ethanol (Volpo T10)
Sol. in isopropanol (Lipal 10TD)
Partly sol. in kerosene (Volpo T10)
Partly sol. in oleic acid (Volpo T10)
Partly sol. in oleyl alcohol (Volpo T10)
Sol. in propylene glycol (Lipal 10TD)
Sol. in toluene (Emulphogene BC-720)
Partly sol. in tetrachlorethylene (Emulphogene BC-720)

Sol. in trichloroethylene (Volpo T10)
Sol. in water (Carsonon TD-10; Emulphogene BC-720; Iconol TDA-10; Lipal 10TD; Volpo T10)

Ionic Nature:
Nonionic (Carsonon TD-10; Emulphogene BC-720; Iconol TDA-10; Lipal 10TD; Macol TD-10)

M.W.:
640 (Iconol TDA-10)

Sp.gr.:
1.00–1.03 (40 C) (Emulphogene BC-720)
1.03 (Iconol TDA-10)

Density:
8.5 lb/gal (Carsonon TD-10)

Pour Pt.:
20 C (Emulphogene BC-720; Iconol TDA-10)
67 F (Carsonon TD-10)

Flash Pt.:
> 200 F (PMCC) (Emulphogene BC-720)
> 300 F (COC) (Carsonon TD-10)

Fire Pt.:
> 165 C (Emulphogene BC-720)

Cloud Pt.:
58–62 C (1% sol'n.) (Emulphogene BC-720)
61 C (1% aq.) (Volpo T10)
75–85 C (1% sol'n.) (Lipal 10TD)
77–88 C (1% aq.) (Iconol TDA-10)
162–166 F (1% in distilled water) (Carsonon TD-10)

HLB:
13.7 (Carsonon TD-10; Iconol TDA-10; Macol TD-10; Volpo T10)
13.8 ± 1 (Lipal 10TD)

Acid No.:
1.0 (Volpo T10)

Iodine No.:
1.0 max. (Volpo T10)

Hydroxyl No.:
80–95 (Lipal 10TD)
90–100 (Volpo T10)

Stability:
Stable to sulfuric acid, alkalies, metallic ions (Emulphogene BC-720)
Stable over a wide pH range (Lipal 10TD)
Stable to alkalies and strong mineral acids (Volpo T10)

pH:
5.0–7.0 (3% aq.) (Carsonon TD-10)

POE (10) tridecyl ether (cont'd.)

6.0–7.5 (5% aq.) (Iconol TDA-10); (3%) (Volpo T10)
6.0–8.0 (10% sol'n.) (Emulphogene BC-720)

Surface Tension:
29.5 dynes/cm (0.1% aq.) (Volpo T10)

Biodegradable: (Volpo T10)

STD. PKGS.:
55-gal (420 lb net) steel drums (Iconol TDA-10)

POE (11) tridecyl ether

SYNONYMS:
PEG-11 tridecyl ether
PEG (11) tridecyl ether
Trideceth-11 (CTFA)

STRUCTURE:
$C_{13}H_{27}(OCH_2CH_2)_nOH$
where avg. $n = 11$

CAS No.:
24938-91-8 (generic)
RD No.: 977067-68-7

TRADENAME EQUIVALENTS:
Carsonon TD-11 [Carson]
Trycol TDA-11 [Emery]

CATEGORY:
Detergent, dispersant, emulsifier, wetting agent, foam builder

APPLICATIONS:
Household detergents: (Carsonon TD-11); heavy-duty cleaners (Trycol TDA-11); light-duty cleaners (Trycol TDA-11)
Industrial applications: corrosion inhibitors (Carsonon TD-11); industrial processing (Carsonon TD-11); paper mfg. (Carsonon TD-11); textile/leather processing (Carsonon TD-11)
Industrial cleaners: (Carsonon TD-11)

PROPERTIES:

Form:
Liquid (Trycol TDA-11)
Waxy solid (Carsonon TD-11)

Color:
< Gardner 1 (Trycol TDA-11)
Gardner 1 (Carsonon TD-11)

Composition:
Avail. as 100% active or 70% aq. sol'n. (Carsonon TD-11)
100% active (Trycol TDA-11)

Solubility:
Disp. in butyl stearate (@ 5%) (Trycol TDA-11)
Sol. in glycerol trioleate (@ 5%) (Trycol TDA-11)
Disp. in min. oil (@ 5%) (Trycol TDA-11)
Disp. in perchloroethylene (@ 5%) (Trycol TDA-11)
Sol. in Stoddard solvent (@ 5%) (Trycol TDA-11)
Sol. in water (@ 5%) (Trycol TDA-11)
Ionic Nature:
Nonionic (Carsonon TD-11; Trycol TDA-11)
Density:
8.4 lb/gal (Trycol TDA-11)
8.5 lb/gal (Carsonon TD-11)
Pour Pt.:
92 F (Carsonon TD-11)
Flash Pt.:
> 300 F (COC) (Carsonon TD-11)
Cloud Pt.:
74 C (Trycol TDA-11)
137–141 F (1% in 10% NaCl sol'n.) (Carsonon TD-11)
HLB:
13.7 (Trycol TDA-11)
14.6 (Carsonon TD-11)
Stability:
Compatible with electrolytes, strong alkalies (Carsonon TD-11)
Stable to hydrolysis by acids and alkalies; conc. mineral acids will react chemically with product (Trycol TDA-11)
pH:
5.0–7.0 (3% aq.) (Carsonon TD-11)

POE (15) tridecyl ether

SYNONYMS:
PEG-15 tridecyl ether
PEG (15) tridecyl ether
Trideceth-15 (CTFA)
Tridecyloxypoly (ethyleneoxy) ethanol (15 EO)
STRUCTURE:
$C_{13}H_{27}(OCH_2CH_2)_nOH$
where avg. $n = 15$
CAS No.:
24938-91-8 (generic)
RD No.: 977067-69-8

POE (15) tridecyl ether (cont'd.)

TRADENAME EQUIVALENTS:

Alkasurf TDA-15 [Alkaril]
Emulphogene BC-840 [GAF]
Renex 31 [ICI United States]
Volpo T15 [Croda]

CATEGORY:

Emulsifier, solubilizer, detergent, coemulsifier, leveling agent, detergent, foam stabilizer, wetting agent

APPLICATIONS:

Household detergents: (Emulphogene BC-840); dishwashing (Renex 31); heavy-duty cleaner (Emulphogene BC-840); high-temperature detergents (Alkasurf TDA-15; Emulphogene BC-840); light-duty cleaners (Emulphogene BC-840)

Industrial applications: industrial processing (Emulphogene BC-840); latexes (Emulphogene BC-840)

Industrial cleaners: metal processing surfactants (Renex 31)

PROPERTIES:

Form:

Soft paste (Renex 31; Volpo T15)
Paste (Emulphogene BC-840)
Solid (Alkasurf TDA-15)

Color:

White (Alkasurf TDA-15; Renex 31)
Off-white (Volpo T15)
VCS 2 max. (50 C) (Emulphogene BC-840)

Odor:

Mild (Alkasurf TDA-15)
Mild, pleasant (Emulphogene BC-840)

Composition:

100% active (Alkasurf TDA-15; Emulphogene BC-840; Renex 31)

Solubility:

Sol. in acetone (Emulphogene BC-840)
Sol. in alcohols (Emulphogene BC-840); sol. in lower alcohols (Renex 31)
Partly sol. in arachis oil (Volpo T15)
Sol. in butyl Cellosolve (Renex 31)
Partly sol. in butyl stearate (Volpo T15)
Sol. in carbon tetrachloride (Renex 31)
Sol. in ethanol (Volpo T15)
Partly sol. in oleic acid (Volpo T15)
Partly sol. in oleyl alcohol (Volpo T15)
Slightly sol. in perchloroethylene (Renex 31)
Sol. in propylene glycol (Renex 31)
Sol. in toluene (Emulphogene BC-840)
Partly sol. in tetrachlorethylene (Emulphogene BC-840)

Sol. in trichloroethylene (Volpo T15)
Completely sol. in water (Alkasurf TDA-15); sol. (Emulphogene BC-840; Renex 31; Volpo T15)
Sol. in xylene (Renex 31)

Ionic Nature:
Nonionic (Alkasurf TDA-15; Emulphogene BC-840; Renex 31)

Sp.gr.:
1.0 (Renex 31)
1.01–1.04 (40 C) (Emulphogene BC-840)

Visc.:
130 cps (Renex 31)

M.P.:
30 C (Alkasurf TDA-15)

Pour Pt.:
33–40 C (Emulphogene BC-840)
61 F (Renex 31)

Flash Pt.:
> 200 F (PMCC) (Emulphogene BC-840)
> 300 F (Renex 31)

Fire Pt.:
> 165 C (Emulphogene BC-840)
> 300 F (Renex 31)

Cloud Pt.:
94 C (1% aq.) (Volpo T15)
> 95 C (1% sol'n.) (Emulphogene BC-840)
210 F (1% aq.) (Renex 31)

HLB:
15.4 (Renex 31)
15.5 (Volpo T15)

Acid No.:
1.0 max. (Renex 31; Volpo T15)

Iodine No.:
1.0 max. (Volpo T15)

Hydroxyl No.:
60–74 (Renex 31)
65–75 (Volpo T15)

Stability:
Stable to sulfuric acid, alkalies, metallic ions (Emulphogene BC-840)
Stable to alkalies and strong mineral acids (Volpo T15)

pH:
6.0–7.5 (3%) (Volpo T15)
6.0–8.0 (10% sol'n.) (Emulphogene BC-840)

POE (15) tridecyl ether (cont'd.)

Surface Tension:

32.0 dynes/cm (0.1% aq.) (Volpo T15)

34 dynes/cm (0.01% sol'n.) (Renex 31)

Biodegradable: (Volpo T15)

STD. PKGS.:

55-gal open-head steel drums (Renex 31)

POE (6) tridecyl ether phosphate

SYNONYMS:

PEG-6 tridecyl ether phosphate

PEG 300 tridecyl ether phosphate

Trideceth-6 phosphate (CTFA)

CAS No.:

9046-01-9 (generic)

TRADENAME EQUIVALENTS:

Gafac RS-610 [GAF]

CATEGORY:

Emulsifier, dispersant, detergent, wetting agent, antistat, lubricant

APPLICATIONS:

Farm products: insecticides/pesticides (Gafac RS-610)

Industrial applications: metal treatment (Gafac RS-610); textile/leather processing (Gafac RS-610)

Industrial cleaners: drycleaning compositions (Gafac RS-610)

PROPERTIES:

Form:

Hazy, viscous liquid (Gafac RS-610)

Composition:

100% active (Gafac RS-610)

Solubility:

Sol. in butyl Cellosolve (Gafac RS-610)

Sol. in cottonseed oil (Gafac RS-610)

Sol. in ethanol (Gafac RS-610)

Sol. in kerosene (Gafac RS-610)

Sol. in perchloroethylene (Gafac RS-610)

Sol. in Stoddard solvent (Gafac RS-610)

Disp. in water (Gafac RS-610)

Sol. in xylene (Gafac RS-610)

Ionic Nature:

Anionic (Gafac RS-610)

Sp.gr.:
1.04–1.06 (Gafac RS-610)
Density:
8.7 lb/gal (Gafac RS-610)
Pour Pt.:
< 0 C (Gafac RS-610)
Acid No.:
75–85 (Gafac RS-610)
Stability:
Stable to alkaline and neutral media (Gafac RS-610)
pH:
< 2.5 (10% sol'n.) (Gafac RS-610)
TOXICITY/HANDLING:
Protect skin and eyes from contact (Gafac RS-610)

POP (9) methyl diethyl ammonium chloride

SYNONYMS:
Polyoxypropylene (9) methyl diethyl ammonium chloride
PPG-9 diethylmonium chloride (CTFA)
Quaternium-6
STRUCTURE:

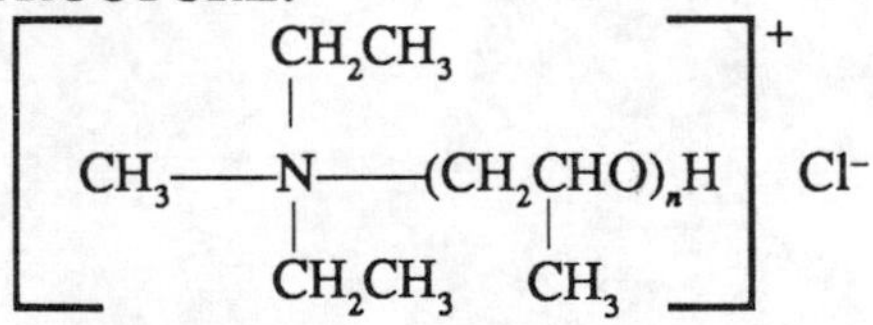

where avg. $n = 9$
CAS No.:
9042-76-6
TRADENAME EQUIVALENTS:
Emcol CC-9 [Witco/Organics]
CATEGORY:
Dispersant, viscosity reducer
APPLICATIONS:
Industrial applications: inhibition of pigment settling in aq. systems (Emcol CC-9); viscosity reduction in industrial slurries (Emcol CC-9)
PROPERTIES:
Form:
Clear liquid (Emcol CC-9)

POP (9) methyl diethyl ammonium chloride (cont'd.)

Color:
Light amber (Emcol CC-9)
Composition:
100% conc. (Emcol CC-9)
Solubility:
Sol. in isopropanol (@ 5%) (Emcol CC-9)
Insol. in kerosene (@ 5%) (Emcol CC-9)
Sol. in water (@ 5%) (Emcol CC-9)
Insol. in xylene (@ 5%) (Emcol CC-9)
Ionic Nature:
Cationic (Emcol CC-9)
Sp.gr.:
1.01 (25/4 C) (Emcol CC-9)
Flash Pt.:
> 93 C (Emcol CC-9)
pH:
6.5 (10% in water) (Emcol CC-9)

POP (25) methyl diethyl ammonium chloride

SYNONYMS:
Polyoxypropylene (25) methyl diethyl ammonium chloride
PPG-25 diethylmonium chloride (CTFA)
Quaternium-20
STRUCTURE:

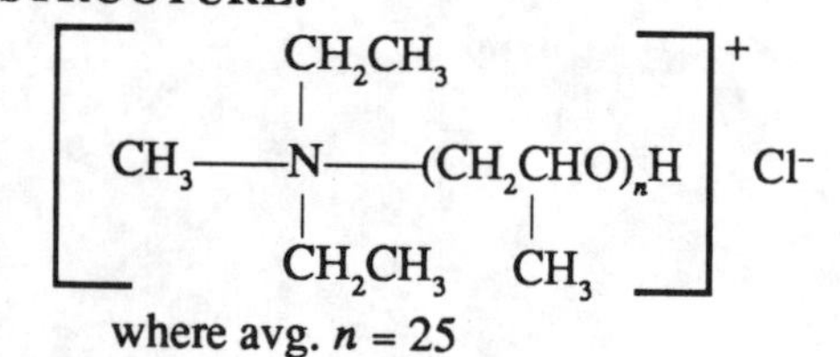

where avg. $n = 25$

CAS No.:
RD No.: 977062-01-3
TRADENAME EQUIVALENTS:
Emcol CC-36 [Witco/Organics]
CATEGORY:
Dispersant
APPLICATIONS:
Industrial applications: hydrocarbon or chlorinated drycleaning solvents (Emcol CC-36); pigments and fillers during grinding into resin systems (Emcol CC-36)

POP (25) methyl diethyl ammonium chloride (cont'd.)

PROPERTIES:

Form:
Clear liquid (Emcol CC-36)

Color:
Light amber (Emcol CC-36)

Composition:
100% conc. (Emcol CC-36)

Solubility:
Sol. in isopropanol (@ 5%) (Emcol CC-36)
Insol. in kerosene (@ 5%) (Emcol CC-36)
Sol. in water (@ 5%) (Emcol CC-36)
Insol. in xylene (@ 5%) (Emcol CC-36)

Ionic Nature:
Cationic (Emcol CC-36)

Sp.gr.:
1.01 (25/4 C) (Emcol CC-36)

Flash Pt.:
> 93 C (Emcol CC-36)

pH:
6.7 (10% in 10:6 isopropanol/water) (Emcol CC-36)

POP (40) methyl diethyl ammonium chloride

SYNONYMS:
Polyoxypropylene (40) methyl diethyl ammonium chloride
PPG-40 diethylmonium chloride (CTFA)
Quaternium-21

STRUCTURE:

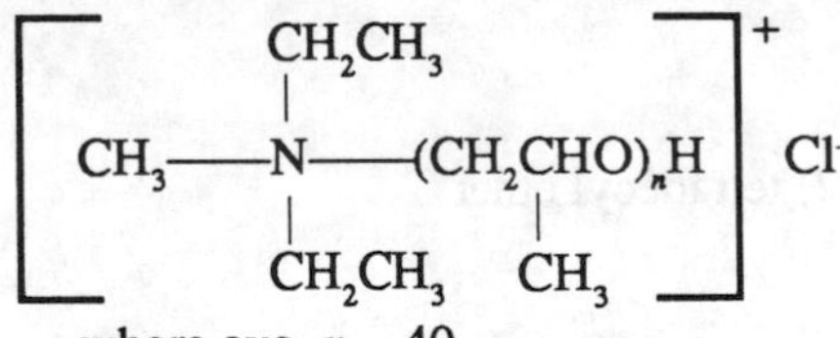

where avg. $n = 40$

CAS No.:
9076-43-1

TRADENAME EQUIVALENTS:
Emcol CC-42 [Witco/Organics]

CATEGORY:
Dispersant, conditioner, emulsifier, solvent, antistat, lubricant, corrosion inhibitor, viscosity reducer, dedusting agent

POP (40) methyl diethyl ammonium chloride (cont'd.)

APPLICATIONS:

Cosmetic industry applications: (Emcol CC-42); germicides (Emcol CC-42); hair rinse toiletries (Emcol CC-42)

Household detergents: surfactant formulations (Emcol CC-42)

Industrial applications: ore flotation (Emcol CC-42); pigments in nonaq. systems (Emcol CC-42); plastics (Emcol CC-42); textile/leather processing (Emcol CC-42); wood oils (Emcol CC-42)

PROPERTIES:

Form:

Oily liquid (Emcol CC-42)

Color:

Light amber (Emcol CC-42)

Solubility:

Sol. in acetone (Emcol CC-42)
Sol. in ethanol (Emcol CC-42)
Sol. in isopropanol (Emcol CC-42)
Sol. in MEK (Emcol CC-42)
Sol. in min. spirits (Emcol CC-42)
Insol. in paraffin oil (Emcol CC-42)
Partly sol. in water (Emcol CC-42)

Ionic Nature:

Cationic (Emcol CC-42)

Sp.gr.:

1.01 (25/4 C) (Emcol CC-42)

Flash Pt.:

> 200 C (Emcol CC-42)

POP (6) POE (12) tetradecyl ether

SYNONYMS:

POE (12) POP (6) tetradecyl ether
Polyoxypropylene (6) polyoxyethylene (12) tetradecyl ether
POE (12) POP (6) decyltetradecyl ether
PPG-6-decyltetradeceth-12 (CTFA)

STRUCTURE:

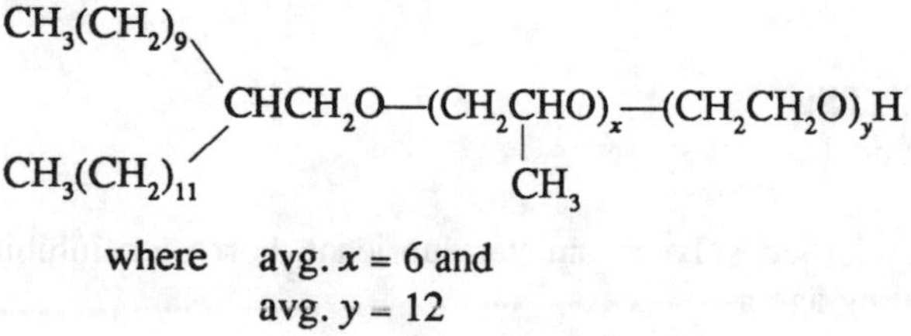

$CH_3(CH_2)_9$ and $CH_3(CH_2)_{11}$ attached to:

$$CHCH_2O—(CH_2\underset{\displaystyle CH_3}{\underset{|}{C}}HO)_x—(CH_2CH_2O)_yH$$

where avg. $x = 6$ and
avg. $y = 12$

TRADENAME EQUIVALENTS:
Nikkol PEN-4612 [Nikko]
CATEGORY:
Solubilizer
APPLICATIONS:
Cosmetic industry preparations: (Nikkol PEN-4612); lotions (Nikkol PEN-4612); toiletries (Nikkol PEN-4612)
PROPERTIES:
Form:
Solid (Nikkol PEN-4612)
Composition:
100% conc. (Nikkol PEN-4612)
Ionic Nature:
Nonionic (Nikkol PEN-4612)
HLB:
8.5 (Nikkol PEN-4612)

POP (6) POE (20) tetradecyl ether

SYNONYMS:
POE (20) POP (6) tetradecyl ether
Polyoxypropylene (6) polyoxyethylene (20) tetradecyl ether
POE (20) POP (6) decyltetradecyl ether
PPG-6-decyltetradeceth-20 (CTFA)
STRUCTURE:

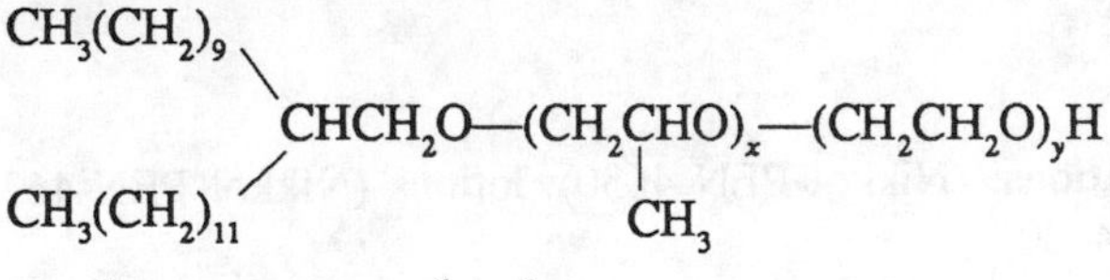

$$\begin{matrix} CH_3(CH_2)_9 \diagdown & & \\ & CHCH_2O{-}(CH_2\underset{\displaystyle CH_3}{\underset{|}{C}}HO)_x{-}(CH_2CH_2O)_yH \\ CH_3(CH_2)_{11} \diagup & & \end{matrix}$$

where avg. $x = 6$ and
avg. $y = 20$
TRADENAME EQUIVALENTS:
Nikkol PEN-4620 [Nikko]
CATEGORY:
Solubilizer
APPLICATIONS:
Cosmetic industry preparations: (Nikkol PEN-4620); lotions (Nikkol PEN-4620); toiletries (Nikkol PEN-4620)

POP (6) POE (20) tetradecyl ether (cont'd.)

PROPERTIES:

Form:

Solid (Nikkol PEN-4620)

Composition:

100% conc. (Nikkol PEN-4620)

Ionic Nature:

Nonionic (Nikkol PEN-4620)

HLB:

11.0 (Nikkol PEN-4620)

POP (6) POE (30) tetradecyl ether

SYNONYMS:

POE (30) POP (6) decyltetradecyl ether
POE (30) POP (6) tetradecyl ether
Polyoxypropylene (6) polyoxyethylene (30) tetradecyl ether
PPG-6-decyltetradeceth-30 (CTFA)

STRUCTURE:

$$\begin{array}{l} CH_3(CH_2)_9 \diagdown \\ \qquad\qquad\quad CHCH_2O\text{—}(CH_2\underset{\displaystyle CH_3}{\underset{|}{C}}HO)_x\text{—}(CH_2CH_2O)_yH \\ CH_3(CH_2)_{11} \diagup \end{array}$$

where avg. $x = 6$ and
avg. $y = 30$

TRADENAME EQUIVALENTS:

Nikkol PEN-4630 [Nikko]

CATEGORY:

Solubilizer

APPLICATIONS:

Cosmetic industry preparations: (Nikkol PEN-4630); lotions (Nikkol PEN-4630); toiletries (Nikkol PEN-4630)

PROPERTIES:

Form:

Solid (Nikkol PEN-4630)

Composition:

100% conc. (Nikkol PEN-4630)

Ionic Nature:

Nonionic (Nikkol PEN-4630)

HLB:

12.0 (Nikkol PEN-4630)

STRUCTURE:

$$R(OCHCH_2)_x(OCH_2CH_2)_yOH$$
(with CH_3 branch on the OCHCH$_2$ unit)

where R represents an alcohol stem with 12 to 18 carbons in the alkyl chain
avg. $x = 6$
avg $y = 11$

TRADENAME EQUIVALENTS:

Plurafac D-25 [BASF Wyandotte]

CATEGORY:

Surfactant, dispersant, wetting agent, emulsifier, demulsifier

APPLICATIONS:

Household detergents: dishwashing (Plurafac D-25); light-duty cleaners (Plurafac D-25); rinse aids (Plurafac D-25)

Industrial cleaners: heavy-duty cleaners (Plurafac D-25); metal cleaners (Plurafac D-25)

PROPERTIES:

Form:

Slightly cloudy liquid (Plurafac D-25)

Composition:

100% active (Plurafac D-25)

Solubility:

Sol. in acetone (Plurafac D-25)
Sol. in benzene (Plurafac D-25)
Sol. in butyl Cellosolve (Plurafac D-25)
Sol. in carbon tetrachloride (Plurafac D-25)
Sol. in chloroform (Plurafac D-25)
Sol. in ethanol (Plurafac D-25)
Sol. in isopropanol (Plurafac D-25)
Sol. in methanol (Plurafac D-25)
Sol. in MEK (Plurafac D-25)
Sol. in perchloroethylene (Plurafac D-25)
Sol. in propylene glycol (Plurafac D-25)
Sol. in water; more sol. in cold water than hot (Plurafac D-25)
Sol. in xylene (Plurafac D-25)

Sp.gr.:

1.007 (Plurafac D-25)

Density:

8.4 lb/gal (Plurafac D-25)

Visc.:

95 cps (Brookfield) (Plurafac D-25)

Pour Pt.:

–18 C (Plurafac D-25)

PPG-6 Pareth-28-11 (cont'd.)

Flash Pt.:
465 F (Plurafac D-25)
Fire Pt.:
515 F (Plurafac D-25)
Cloud Pt.:
59 C (1% aq.) (Plurafac D-25)
HLB:
11 (Plurafac D-25)
Ref. Index:
1.4560 (Plurafac D-25)
pH:
6.0–7.0 (1% aq.) (Plurafac D-25)
Surface Tension:
34.3 dynes/cm (0.1%) (Plurafac D-25)
Biodegradable: (Plurafac D-25)
TOXICITY/HANDLING:
Mild skin irritant (Plurafac D-25)

Propylene glycol dicaprylate/dicaprate (CTFA)

SYNONYMS:
Mixture of propylene glycol diesters of caprylic and capric acids
CAS No.:
RD No.: 977060-55-1
TRADENAME EQUIVALENTS:
Captex 200 [Capital City]
Edenol 302 [Henkel/Canada]
Hodag CC-22, CC-22-S [Hodag]
Lexol PG855, PG865 [Inolex]
Liponate PC [Lipo]
Miglyol 840 [Dynamit-Nobel]
Neobee M-20 [PVO]
Standamul 302 [Henkel]
CATEGORY:
Solubilizer, solvent, cosolvent, emollient, base, carrier, vehicle, moisturizer, thickener, viscosity control agent
APPLICATIONS:
Bath products: (Liponate PC); bath oils (Edenol 302; Lexol PG855, PG865; Miglyol 840; Standamul 302)
Cleansers: skin cleanser (Edenol 302; Standamul 302)
Cosmetic industry preparations: (Edenol 302; Hodag CC-22, CC-22-S; Miglyol 840;

Neobee M-20; Standamul 302); aerosols (Lexol PG855, PG865; Liponate PC); creams and lotions (Captex 200; Edenol 302; Lexol PG855, PG865; Liponate PC; Miglyol 840; Neobee M-20; Standamul 302); makeup (Captex 200; Lexol PG865; Standamul 302); perfumery (Captex 200; Lexol PG855, PG865; Miglyol 840); pigmented cosmetics (Lexol PG855, PG865; Liponate PC; Standamul 302); shaving preparations (Lexol PG855, PG865); skin preparations (Miglyol 840)

Food applications: (Hodag CC-22, CC-22-S); dietetic products (Miglyol 840); flavors (Captex 200; Lexol PG855, PG865; Miglyol 840; Neobee M-20); food additives (Neobee M-20)

Pharmaceutical applications: (Captex 200; Hodag CC-22, CC-22-S; Miglyol 840; Neobee M-20; Standamul 302); antibiotics (Lexol PG865); antiperspirant/deodorant (Edenol 302; Standamul 302); injection products (Miglyol 840); medicinals (Lexol PG865); ointments (Miglyol 840; Neobee M-20); oral products (Miglyol 840); suppositories (Miglyol 840); topical products (Standamul 302); vitamins (Captex 200; Lexol PG865)

PROPERTIES:

Form:

Liquid (Hodag CC-22, CC-22-S; Liponate PC; Miglyol 840)
Low-viscosity liquid (Captex 200; Edenol 302; Lexol PG855, PG865; Neobee M-20)
Clear, low-viscosity oily liquid (Standamul 302)

Color:

Colorless (Liponate PC)
Yellow (Edenol 302)
Gardner 3.0 max. (Miglyol 840)

Odor:

None (Miglyol 840)
Faint, characteristic (Standamul 302)

Taste:

None (Miglyol 840)

Composition:

100% conc. (Hodag CC-22, CC-22-S; Lexol PG855; Standamul 302)

Solubility:

Sol. in acetone (Lexol PG865; Neobee M-20)
Sol. in alcohols (Captex 200; Edenol 302; Lexol PG865; Neobee M-20); sol. in 96% alcohol (20 C) (Miglyol 840)
Sol. in aliphatics (Captex 200)
Sol. in castor oil (Standamul 302); (@ 10%) (Edenol 302)
Sol. in chloroform (20 C) (Miglyol 840)
Sol. in diethyl ether (20 C) (Miglyol 840)
Sol. in ethanol (Liponate PC); sol. in anhyd. ethanol (Standamul 302); sol. in 95% ethanol-SD40 (@ 10%) (Edenol 302)
Sol. in hydroalcohols (Lexol PG855)
Sol. in isopropanol (20 C) (Miglyol 840)

Sol. in isopropyl myristate (Standamul 302); (@ 10%) (Edenol 302)
Sol. in ketones (Edenol 302)
Sol. in min. oils (Edenol 302; Lexol PG865; Liponate PC; Neobee M-20; Standamul 302); (@ 10%) (Edenol 302)
Sol. in oils (Captex 200)
Sol. in oleyl alcohol (Standamul 302); (@ 10%) (Edenol 302)
Sol. in organic solvents (Captex 200)
Sol. in petroleum ether (20 C) (Miglyol 840)
Sol. in silicone fluid (Standamul 302); (@ 10%) (Edenol 302)
Sol. in toluene (20 C) (Miglyol 840)
Sol. in veg. oil (Lexol PG865; Liponate PC)
Insol. in water (Edenol 302; Liponate PC)

Ionic Nature:
Nonionic (Hodag CC-22, CC-22-S; Lexol PG855; Miglyol 840)

Sp.gr.:
0.91–0.92 (20 C) (Miglyol 840)
0.92 g/ml (Edenol 302)
0.920 (Neobee M-20)
0.921 (20 C) (Standamul 302)

Visc.:
9 cps (Neobee M-20)
9–12 cps (20 C) (Miglyol 840)

Setting Pt.:
–20 C (Neobee M-20)

Gel Pt.:
–30 C max. (Standamul 302)

Solidification Pt.:
< 0 C (Edenol 302)

Cloud Pt.:
–30 C (Miglyol 840)
< 0 C (Edenol 302)

HLB:
9.0 (Edenol 302; Standamul 302)

Acid No.:
0.1 max. (Liponate PC; Miglyol 840)
0.5 max. (Standamul 302)
1.0 max. (Edenol 302)

Iodine No.:
0.5 (Neobee M-20)
0.5 max. (Miglyol 840)
2.0 max. (Edenol 302; Standamul 302)

Saponification No.:
315–335 (Edenol 302; Liponate PC; Neobee M-20; Standamul 302)

320–340 (Miglyol 840)

Stability:

Oxidation-stable; excellent low-temp. stability (Standamul 302)

Extremely resistant to oxidation (Edenol 302)

Ref. Index:

1.435 (Edenol 302)

1.4400–1.4420 (Miglyol 840)

Surface Tension:

31.0 dynes/cm (Neobee M-20)

Biodegradable: (Miglyol 840)

STD. PKGS.:

385-lb net closed-head steel drums and bulk (Standamul 302)

Propylene glycol monomethyl ether

SYNONYMS:

Methoxypropanol (CTFA)

1-Methoxy-2-propanol

Monopropylene glycol methyl ether

Monopropylene glycol monomethyl ether

2-Propanol, 1-methoxy-

EMPIRICAL FORMULA:

$C_4H_{10}O_2$

STRUCTURE:

```
CH3OCH2CHOH
          |
          CH3
```

CAS No.:

107-98-2

TRADENAME EQUIVALENTS:

Arcosolv PM [Arco]

Dowanol PM [Dow]

Poly-Solv MPM [Olin]

CATEGORY:

Solvent, solubilizer, plasticizer, coupling agent, dispersant

APPLICATIONS:

Automotive industry: antifreeze (Poly-Solv MPM); brake fluids (Poly-Solv MPM)

Cosmetic industry preparations: (Arcosolv PM; Poly-Solv MPM); nail polish (Poly-Solv MPM)

Farm products: agricultural oils/sprays (Arcosolv PM; Poly-Solv MPM); insect repellents (Poly-Solv MPM)

Household detergents: (Arcosolv PM; Dowanol PM); hard surface cleaner (Poly-Solv MPM)
Industrial applications: adhesives (Poly-Solv MPM); dyes and pigments (Poly-Solv MPM); industrial processing (Arcosolv PM); paint/coatings mfg. (Arcosolv PM; Dowanol PM); plastics (Poly-Solv MPM); polishes and waxes (Poly-Solv MPM); printing inks (Arcosolv PM; Dowanol PM; Poly-Solv MPM); textile/leather processing (Poly-Solv MPM)
Industrial cleaners: wax and paint strippers (Poly-Solv MPM)

PROPERTIES:

Form:
Liquid (Arcosolv PM; Poly-Solv MPM)

Color:
APHA 10 max. (Arcosolv PM; Poly-Solv MPM)

Odor:
Mild, characteristic (Poly-Solv MPM)
Mild, pleasant characteristic (Arcosolv PM)

Composition:
100% conc. (Arcosolv PM; Poly-Solv MPM)

Solubility:
Sol. in many organic liquids (Dowanol PM); miscible with a number of organic solvents (Arcosolv PM; Poly-Solv MPM)
Completely sol. in water (Arcosolv PM); sol. (Dowanol PM); miscible (Poly-Solv MPM)

M.W.:
90.1 (Arcosolv PM)
90.12 (Poly-Solv MPM)

Sp.gr.:
0.918–0.921 (Arcosolv PM)
0.923 (20/20 C) (Poly-Solv MPM)

Density:
7.65 lb/gal (Arcosolv PM)
7.68 lb/gal (20 C) (Poly-Solv MPM)

Visc.:
1.8 cstk (Arcosolv PM)
1.9 cP (20 C) (Poly-Solv MPM)

F.P.:
–96 C (Poly-Solv MPM)
–95 C (Arcosolv PM)

B.P.:
120 C (Dowanol PM)
120.1 C (760 mm Hg) (Arcosolv PM)
121 C (760 mm) (Poly-Solv MPM)

Flash Pt.:
32 C (TCC) (Arcosolv PM)
36 C (TCC) (Poly-Solv MPM)
Ref. Index:
1.4036 (20 C) (Poly-Solv MPM)
1.404 (20 C) (Arcosolv PM)
Surface Tension:
26.5 dynes/cm (Arcosolv PM)
Conductivity:
3.78 K × 10^{4} cal/cm^2-s°C/cm (60 C) (Arcosolv PM)
Specific Heat:
0.57 cal/g/°C (Arcosolv PM)
TOXICITY/HANDLING:
Relatively low acute toxicity; may cause mild adverse effects on skin and eye contact; avoid ingestion, eye contact, and prolonged skin contact and inhalation of vapors (Arcosolv PM)
Mild eye and primary skin irritant; avoid contact with skin and eyes (Poly-Solv MPM)
STORAGE/HANDLING:
Flammable liquid; store in carbon steel vessels; avoid contact with air when storing for long periods of time (Arcosolv PM)
STD. PKGS.:
Tank car, tank truck, drums (Arcosolv PM)
55-gal (420 lb net) drums, tank cars, tank trucks (Poly-Solv MPM)

Sodium borate (CTFA)

SYNONYMS:

Borax
Borax decahydrate
Borax granular
Puffed borax
Sodium tetraborate
Sodium tetraborate decahydrate
Sodium pyroborate
Tincal

Other forms:

Pentahydrate
Anhydrous:
Borax, dehydrated
Sodium borax, anhydrous

EMPIRICAL FORMULA:

$Na_2B_4O_7 \cdot 10H_2O$ (decahydrate)
Other forms:
$Na_2B_4O_7 \cdot 5H_2O$ (pentahydrate)
$Na_2B_4O_7$ (anhydrous)

CAS No.:

1303-96-4 (hydrous)
1330-43-4 (anhydrous)

TRADENAME EQUIVALENTS:

Borax [U.S. Borax] (decahydrate)
Dehybor [U.S. Borax] (anhyd.)
Polybor 3 [U.S. Borax] (more water-sol. form)
Pyrobor [Kerr-McGee] (anhyd.)
Trona [Kerr-McGee] (anhyd.)
Tronabor [Kerr-McGee] (pentahydrate)

CATEGORY:

Dispersant, wetting agent, mold lubricant, herbicide, larvicide, fungicide

APPLICATIONS:

Farm products: fertilizers (generic decahydrate; Tronabor); herbicides (generic decahydrate; generic anhydrous; Polybor 3; Pyrobor); soil sterilant (generic pentahydrate); weed killer (generic pentahydrate)

Food applications: fungus control on citrus fruits (generic pentahydrate)

Household detergents: (generic decahydrate); bleaches (generic decahydrate)
Industrial applications: flux (generic decahydrate); glass and ceramic products (generic decahydrate; generic anhydrous); latex applications (Borax); paint mfg. (generic decahydrate); photography (generic decahydrate); rubber (Borax); textile/leather processing (generic decahydrate)
Pharmaceutical applications (generic decahydrate)

PROPERTIES:

Form:
Free-flowing crystals (generic anhydrous)
Crystalline solid (generic decahydrate)
Free-flowing powder (generic pentahydrate)
Powder (Borax; generic decahydrate)

Color:
White (Borax; generic decahydrate; generic anhydrous)

Odor:
Odorless (Borax; generic decahydrate)

Composition:
14.5% boron (Tronabor)

Solubility:
Insol. in alcohols (Borax; generic decahydrate)
Sol. in glycerin (Borax; generic decahydrate)
Sol. in water (Borax); 5.14 g/100 ml (20 C) (generic decahydrate); slightly sol. cold water (generic anhydrous)

Sp.gr.:
1.73 (Borax;); (20/4 C) (generic decahydrate)
1.815 (generic pentahydrate)
2.367 (generic anhydrous)

M.P.:
741 C (generic anhydrous)

Stability:
Loses water of hydration @ 60 C (Borax)
Loses water of crystallization when heated to melting (75–320 C) (generic decahydrate)
Begins to lose water of hydration @ 122 C (generic pentahydrate)

Storage Stability:
Hygroscopic; forms partial hydrate in damp air (generic anhydrous)

TOXICITY/HANDLING:
Nontoxic (Borax)
Toxic (generic decahydrate; generic anhydrous; generic pentahydrate)
TLV: 1 mg/m^3 (generic anhydrous; generic pentahydrate)

STORAGE/HANDLING:
Noncombustible (generic decahydrate; generic anhydrous; generic pentahydrate)

Sodium cocoyl isethionate (CTFA)

SYNONYMS:
Fatty acids, coconut oil, sulfoethyl esters, sodium salts
STRUCTURE:

$$RC(=O)—OCH_2CH_2SO_3Na$$

where RCO^- represents the coconut acid radical
CAS No.:
61789-32-0
TRADENAME EQUIVALENTS:
Hostapon KA Pdr. Hi Conc. [Amer. Hoechst]
Hostapon KA Special [Hoechst AG]
Igepon AC-78 [GAF]
CATEGORY:
Dispersant, detergent, foaming agent, surfactant
APPLICATIONS:
Bath products: bubble bath (Igepon AC-78)
Cleansers: detergent bars (Hostapon KA Pdr. Hi Conc.; Igepon AC-78)
Cosmetic industry preparations: (Hostapon KA Pdr. Hi Conc.; Igepon AC-78); detergent base (Hostapon KA Special); shampoos (Igepon AC-78)
Pharmaceutical applications: dentifrices (Igepon AC-78)
PROPERTIES:
Form:
Powder (Hostapon KA Pdr. Hi Conc.; Hostapon KA Special; Igepon AC-78)
Composition:
80% conc. (Hostapon KA Special)
≥ 83% active (Igepon AC-78)
Ionic Nature:
Anionic (Hostapon KA Pdr. Hi Conc.; Hostapon KA Special; Igepon AC-78)

Sodium cumenesulfonate (CTFA)

SYNONYMS:
Benzene, (1-methylethyl)-, monosulfo deriv., sodium salt
(1-Methylethyl) benzene, monosulfo deriv., sodium salt
EMPIRICAL FORMULA:
$C_9H_{12}O_3S \cdot Na$

STRUCTURE:

H_3C CH_3
CH

SO_3Na

CAS No.:

32073-22-6

TRADENAME EQUIVALENTS:

Na-Cumene Sulfonate 40, Powder [Chemische Werke Huls AG]
Naxonate 45SC, SC [Nease]
Reworyl NCS40 [Rewo Chem. Werke GmbH]
Stepanate C-S [Stepan]
Ultra SCS Liquid [Witco/Organics]
Witconate SCS [Witco/Organics]

CATEGORY:

Hydrotrope, solubilizer, coupling agent, stabilizer, viscosity control agent, solution aid, antiblocking agent, anticaking agent, cloud pt. depressant

APPLICATIONS:

Cosmetic industry preparations: shampoos (Witconate SCS)

Household detergents: (Reworyl NCS40; Witconate SCS); dishwashing (Naxonate 45SC, SC); heavy-duty cleaner (Naxonate 45SC, SC); liquid detergents (Na-Cumene Sulfonate 40, Powder; Naxonate 45SC, SC; Stepanate C-S; Ultra SCS Liquid); powdered detergents (Na-Cumene Sulfonate 40, Powder; Naxonate 45SC, SC; Ultra SCS Liquid); spray-dried detergents (Stepanate C-S)

Industrial applications: adhesives (Naxonate 45SC, SC; Witconate SCS); aerosols (Witconate SCS); dyes and pigments (Naxonate 45SC, SC); electroplating (Naxonate 45SC, SC); extraction reagent (Witconate SCS); industrial processing (Witconate SCS); lubricating/cutting oils (Naxonate 45SC, SC; Witconate SCS); paper mfg. (Witconate SCS); photography (Witconate SCS); printing inks (Naxonate 45SC, SC; Witconate SCS); textile/leather processing (Naxonate 45SC, SC; Witconate SCS); wood pulping (Witconate SCS)

Industrial cleaners: boiler descaling (Witconate SCS); metal processing surfactants (Witconate SCS); sanitizers/germicides (Naxonate 45SC, SC)

PROPERTIES:

Form:

Liquid (Na-Cumene Sulfonate 40; Naxonate 45SC; Reworyl NCS40; Ultra SCS Liquid; Witconate SCS)

Powder (Na-Cumene Sulfonate Powder; Naxonate SC; Witconate SCS)

Color:

Cream (Naxonate SC)

Sodium cumenesulfonate (*cont'd.*)

Klett 50 (Witconate SCS, liquid)
Klett 70 max. (Stepanate C-S)
Klett 75 max. (Naxonate 45SC)

Composition:

40% active (Na-Cumene Sulfonate 40; Reworyl NCS40; Ultra SCS Liquid)
44–46% active (Stepanate C-S)
45% active (Naxonate 45SC; Witconate SCS, liquid)
93% active (Witconate SCS, powder)
95% active min. (Naxonate SC)
96% active (Na-Cumene Sulfonate Powder)

Ionic Nature:

Anionic (Na-Cumene Sulfonate 40, Powder; Naxonate 45SC, SC; Ultra SCS Liquid)

Density:

0.32 g/cc (apparent) (Witconate SCS, powder)
1.2 g/cc (Witconate SCS, liquid)
9.65 lb/gal (Stepanate C-S)

pH:

7.0–9.0 (Naxonate 45SC)
7.5 (Witconate SCS, liquid)
9.0 (Witconate SCS, 40% sol'n., powder)

TOXICITY/HANDLING:

Avoid prolonged contact with skin; use normal safety precautions (Witconate SCS)
Believed to be nonhazardous; however, avoid contact with eyes and prlonged contact with skin (Stepanate C-S)

STORAGE/HANDLING:

Storage and transfer in stainless steel is recommended; store above 35 C to avoid crystallization; if crystallization occurs, heating will reliquefy (Witconate SCS)

STD. PKGS.:

55-gal lined drums or tankwagons (Naxonate 45SC)
125-lb leverpak (Naxonate SC)

Sodium diamyl sulfosuccinate

SYNONYMS:

Butanedioic acid, sulfo-, 1,4-dipentyl ester, sodium salt
Diamyl sodium sulfosuccinate (CTFA)
1,4-Dipentylsulfobutanedioic acid, sodium salt
Sodium sulfosuccinic acid, diamyl ester

EMPIRICAL FORMULA:

$C_{14}H_{26}O_7S \cdot Na$

STRUCTURE:

```
O=C—OC5H11
  |
H–C–SO3Na
  |
  CH2
  |
O=C—OC5H11
```

CAS No.:
922-80-5

TRADENAME EQUIVALENTS:
Aerosol AY, AY-65, AY-100 [Amer. Cyanamid]
Geropon AY [Geronazzo SpA]

CATEGORY:
Dispersant, emulsifier, wetting agent, antigelling agent, surfactant

APPLICATIONS:
Farm products: (Aerosol AY-65, AY-100)
Household detergents: cleaners of porcelain, tile, brick, cement (Aerosol AY-65, AY-100)
Industrial applications: dyes and pigments (Aerosol AY); electroplating (Aerosol AY, AY-65, AY-100); industrial processing (Aerosol AY); ore leaching (Aerosol AY, AY-65, AY-100); paint mfg. (Aerosol AY); paper mfg. (Aerosol AY); polymers/polymerization (Aerosol AY, AY-65, AY-100); soldering fluxes (Aerosol AY)

PROPERTIES:

Form:
Clear liquid (Aerosol AY-65)
Wax (Geropon AY)
Hard, waxy solid (Aerosol AY, AY-100)

Color:
APHA 100 max. (50% solids) (Aerosol AY)

Composition:
65% active in water/ethanol (Aerosol AY-65)
70% conc. (Geropon AY)
100% active (Aerosol AY-100)
100% solids (Aerosol AY)

Solubility:
Sparingly sol. in nonpolar solvents (Aerosol AY)
Sol. in organic solvents (Aerosol AY-100)
Sol. in polar organic solvents (Aerosol AY; Geropon AY)
Easily sol. in water (Geropon AY); sol. (Aerosol AY-100); sol. 39.2 g/100 ml water (Aerosol AY)

Ionic Nature:
Anionic (Aerosol AY, AY-65, AY-100; Geropon AY)

Sodium diamyl sulfosuccinate (*cont'd.*)

M.W.:
360 (Aerosol AY)
Sp.gr.:
1.081 (Aerosol AY-65)
1.2 (Aerosol AY, AY-100)
Density:
9.0 lb/gal (Aerosol AY-65)
10.0 lb/gal (Aerosol AY-100)
F.P.:
< 18 C (Aerosol AY-65)
M.P.:
222–227 C (Aerosol AY)
Flash Pt.:
None; decomposes @ 235 C (OC) (Aerosol AY)
Acid No.:
2.5 max. (Aerosol AY)
Iodine No.:
0.25 max. (Aerosol AY)
Stability:
Good; good salt tolerance (Aerosol AY-100)
Surface Tension:
29 dynes/cm (Aerosol AY-65, AY-100)
50.2 dynes/cm (0.01% aq.) (Aerosol AY)
Biodegradable: (Aerosol AY-65)
TOXICITY/HANDLING:
Avoid prolonged/repeated skin contact with conc. sol'ns.; prevent entry of product into eye (Aerosol AY)
STD. PKGS.:
50-lb net fiber drums (Aerosol AY)

Sodium dicyclohexyl sulfosuccinate

SYNONYMS:
Dicyclohexyl sodium sulfosuccinate (CTFA)
Succinic acid, sulfo-, 1,4, dicyclohexyl ester, sodium salt
Sulfosuccinic acid, 1,4-dicyclohexyl ester, sodium salt
EMPIRICAL FORMULA:
$C_{16}H_{26}O_7S \cdot Na$

STRUCTURE:

$$
\begin{array}{l}
O{=}C{-}OC_6H_{11} \\
\quad | \\
H{-}C{-}SO_3Na \\
\quad | \\
\quad CH_2 \\
\quad | \\
O{=}C{-}OC_6H_{11}
\end{array}
$$

CAS No.:
23386-52-9

TRADENAME EQUIVALENTS:
Aerosol A-196 [American Cyanamid]
Alconate 2CH [Alcolac]
Protowet 4337 [Procter Chem. Co.]

CATEGORY:
Emulsifier, dispersant, surfactant, stabilizer

APPLICATIONS:
Industrial applications: polymers/polymerization (Aerosol A-196; Alconate 2CH)

PROPERTIES:

Form:
Clear to cloudy liquid (Aerosol A-196)
Paste (Alconate 2CH)
Pellets (Aerosol A-196)

Color:
White (Aerosol A-196 pellets)

Composition:
40% active sol'n. in water (Aerosol A-196 sol'n.)
40% conc. (Protowet 4337)
43% active (Alconate 2CH)
85% active (Aerosol A-196 pellets)

Ionic Nature: Anionic

Stability:
Good mechanical stability (Aerosol A-196)

Surface Tension:
39 dynes/cm (Aerosol A-196)

Biodegradable: (Aerosol A-196)

Sodium dihexyl sulfosuccinate

SYNONYMS:

Bis (1-methylamyl) sodium sulfosuccinate
Butanedioic acid, sulfo-, 1,4-dihexyl ester, sodium salt
Dihexyl sodium sulfosuccinate (CTFA)
Dihexyl sulfosuccinate, sodium salt
Sodium 1,4-dihexyl sulfobutanedioate
Sodium sulfosuccinic acid, dihexyl ester

EMPIRICAL FORMULA:

$C_{16}H_{30}O_7S \cdot Na$

STRUCTURE:

```
          CH3
          |
O=C—OCHCH2CH2CH2CH3
  |
H–C—SO3Na
  |
  CH2
  |
O=C—OCHCH2CH2CH2CH3
          |
          CH3
```

CAS No.:

3006-15-3

TRADENAME EQUIVALENTS:

Aerosol MA-80 [Amer. Cyanamid]
Alcopol OS [Allied Colloids Ltd.]
Astrowet H-80 [Alco]
Empimin MA [Albright & Wilson/Detergents]
Lankropol KMA [Lankro]
Monawet MM-80 [Mona]

CATEGORY:

Dispersant, wetting agent, emulsifier, solubilizer, surfactant, penetrant, stabilizer

APPLICATIONS:

Cosmetic industry preparations: (Monawet MM-80)
Farm products: (Monawet MM-80); insecticides/pesticides (Monawet MM-80)
Food applications: food packaging (Monawet MM-80)
Household detergents: (Monawet MM-80); soaps (Lankropol KMA)
Industrial applications: electroplating (Aerosol MA-80); industrial processing (Empimin MA); ore leaching (Aerosol MA-80); paint mfg. (Monawet MM-80); plastics (Monawet MM-80); polymers/polymerization (Aerosol MA-80; Astrowet H-80; Empimin MA; Lankropol KMA; Monawet MM-80); printing inks (Monawet MM-80); textile/leather processing (Aerosol MA-80; Monawet MM-80)
Industrial cleaners: drycleaning compositions (Monawet MM-80)

PROPERTIES:

Form:

Liquid (Alcopol OS; Astrowet H-80)
Clear liquid (Monawet MM-80); (@ 20 C) (Empimin MA)
Hazy liquid (Lankropol KMA)
Clear, slightly viscous liquid (Aerosol MA-80)

Color:

Pale straw (Lankropol KMA)
Straw (Empimin MA)
APHA 50 max. (Aerosol MA-80)
APHA 75 max. (Monawet MM-80)

Odor:

Ethanolic (Lankropol KMA)

Composition:

60% conc. (Alcopol OS)
60% active; contains ethanol (Lankropol KMA)
63% active min. in water/ethanol sol'n. (10% ethanol) (Empimin MA)
80% active (Aerosol MA-80)
80% conc. (Astrowet H-80)
80 ± 1% active in 15% water/5% isopropanol (Monawet MM-80)

Solubility:

Sol. in alcohols (Aerosol MA-80)
Sol. in nonpolar solvents (Monawet MM-80)
Sol. in organic solvents (Aerosol MA-80)
Sol. in polar solvents (Monawet MM-80)
Sol. in water (Aerosol MA-80; Empimin MA); @ 20 C (Lankropol KMA); 33% sol. @ 20 C (Monawet MM-80)

Ionic Nature:

Anionic (Aerosol MA-80; Astrowet H-80; Lankropol KMA; Monawet MM-80)

M.W.:

388 (Monawet MM-80)

Sp.gr.:

1.082 (20 C) (Lankropol KMA)
1.10 (23 C) (Monawet MM-80)
1.13 (Aerosol MA-80)

Density:

1.1 g/cm³ (20 C) (Empimin MA)
9.2 lb/gal (Monawet MM-80)
9.4 lb/gal (Aerosol MA-80)

Visc.:

31 cs (20 C) (Lankropol KMA)
80 cs (20 C) (Empimin MA)

Sodium dihexyl sulfosuccinate (cont'd.)

F.P.:
-28 C (Aerosol MA-80)

M.P.:
199–292 C (Aerosol MA-80)

Pour Pt.:
< 0 C (Lankropol KMA)

Flash Pt.:
30 C (CC) (Empimin MA)
91 F (Abel CC) (Lankropol KMA)
110 F (PMCC) (Monawet MM-80)

Cloud Pt.:
< 0 C (Monawet MM-80)

Acid No.:
2.5 max. (Monawet MM-80)

Stability:
Indefinitely stable in neutral hot and cold aq. systems; stable in sol'ns. with pH range of 2–10 (Monawet MM-80)
Hydrolyzed by both acids and alkalis but generally stable in the pH range 6–9 (Empimin MA)

pH:
6.0 ± 1 (Monawet MM-80); (5% sol'n.) (Empimin MA)
6.0–7.5 (1% aq.) (Lankropol KMA)

Surface Tension:
28 dynes/cm (Aerosol MA-80)
46 dynes/cm (0.1%) (Monawet MM-80)

Biodegradable: Slowly biodegradable (Aerosol MA-80)

TOXICITY/HANDLING:
Irritating to eyes, skin (Aerosol MA-80)
Low toxicity; exerts a pronounced defatting effect on skin; strongly irritating to eyes (Monawet MM-80)
Skin irritant on prolonged contact with conc. form; protective gloves and goggles should be worn; in a fire, may produce toxic fumes of sulfur dioxide; spillages may be slippery (Lankropol KMA)

STORAGE/HANDLING:
Contains ethanol—keep away from heat, flames, and sparks; stainless steel, glass, resin, or polythene-lined equipment recommended (Lankropol KMA)

STD. PKGS.:
Polyethylene-lined fiber drums (Monawet MM-80)

Sodium diisobutyl sulfosuccinate

SYNONYMS:

1,4 Bis (2-methylpropyl) sulfobutanedioate, sodium salt
Butanedioic acid, sulfo-, 1,4-bis (2-methylpropyl) ester, sodium salt
Diisobutyl sodium sulfosuccinate (CTFA)
Sodium sulfosuccinic acid, diisobutyl ester

EMPIRICAL FORMULA:

$C_{12}H_{22}O_7S \cdot Na$

STRUCTURE:

```
O=C—OCH2CH(CH3)2
  |
H–C—SO3Na
  |
  CH2
  |
O=C—OCH2CH(CH3)2
```

CAS No.:

127-39-9

TRADENAME EQUIVALENTS:

Aerosol IB-45 [Amer. Cyanamid]
Alcopol OB [Allied Colloids Ltd.]
Astrowet B-45 [Alco]
Geropon CYA/45 [Geronazzo SpA]
Monawet MB-45 [Mona]

CATEGORY:

Emulsifier, dispersant, wetting agent, surfactant, penetrant, solubilizer

APPLICATIONS:

Cosmetic industry preparations: (Monawet MB-45)
Farm products: (Monawet MB-45)
Food applications: food contact applications (Monawet MB-45)
Household detergents: (Monawet MB-45)
Industrial applications: carpet backing (Monawet MB-45); dyes and pigments (Aerosol IB-45); paint mfg. (Monawet MB-45); polymers/polymerization (Aerosol IB-45; Astrowet B-45; Geropon CYA/45; Monawet MB-45); printing inks (Monawet MB-45); textile/leather processing (Monawet MB-45)
Industrial cleaners: drycleaning compositions (Monawet MB-45); sanitation (Monawet MB-45)

PROPERTIES:

Form:

Liquid (Alcopol OB; Astrowet B-45; Geropon CYA/45)
Clear liquid (Aerosol IB-45; Monawet MB-45)

Color:

APHA 75 max. (Monawet MB-45)

Sodium diisobutyl sulfosuccinate (cont'd.)

Composition:

45% active (Aerosol IB-45; Alcopol OB; Astrowet B-45; Gerapon CYA/45)
45 ± 1% active in water (Monawet MB-45)

Solubility:

Fair sol. in nonpolar solvents (Monawet MB-45)
Fair sol. in polar solvents (Monawet MB-45)
Very sol. in water (Aerosol IB-45; Monawet MB-45)

Ionic Nature:

Anionic (Aerosol IB-45; Astrowet B-45; Geropon CYA/45; Monawet MB-45)

M.W.:

332 (Monawet MB-45)

Sp.gr.:

1.12 (23 C) (Monawet MB-45)

Density:

9.3 lb/gal (Monawet MB-45)

F.P.:

20–21 C (Aerosol IB-45)

Flash Pt.:

215 F (PMCC) (Monawet MB-45)

Cloud Pt.:

13 C (Monawet MB-45)

Acid No.:

2.5 max. (Monawet MB-45)

Stability:

Excellent in conc. salt sol'ns. (Aerosol IB-45)
Indefinitely stable in neutral hot and cold aq. systems; stable in sol'ns. with pH range of 2–10 (Monawet MB-45)

pH:

6.0 ± 1 (Monawet MB-45)

Surface Tension:

49 dynes/cm (Aerosol IB-45)
54 dynes/cm (0.1%) (Monawet MB-45)

Biodegradable: (Aerosol IB-45)

TOXICITY/HANDLING:

May be irritating to eyes (Aerosol IB-45)
Low toxicity; exerts a pronounced defatting effect on skin; strongly irritating to eyes (Monawet MB-45)

STD. PKGS.:

Polyethylene-lined fiber drums (Monawet MB-45)

SYNONYMS:

1,4-Bis (2-ethylhexyl) sulfobutanedioate, sodium salt
Butanedioic acid, sulfo-, 1,4-bis (2-ethylhexyl) ester, sodium salt
Di-(2-ethylhexyl) sodium sulfosuccinate
Dioctyl ester of sodium sulfosuccinic acid
Dioctyl sodium sulfosuccinate (CTFA)
Docusate sodium
Sodium di-(2-ethylhexyl) sulfosuccinate

EMPIRICAL FORMULA:

$C_{20}H_{38}O_7S{\bullet}Na$

STRUCTURE:

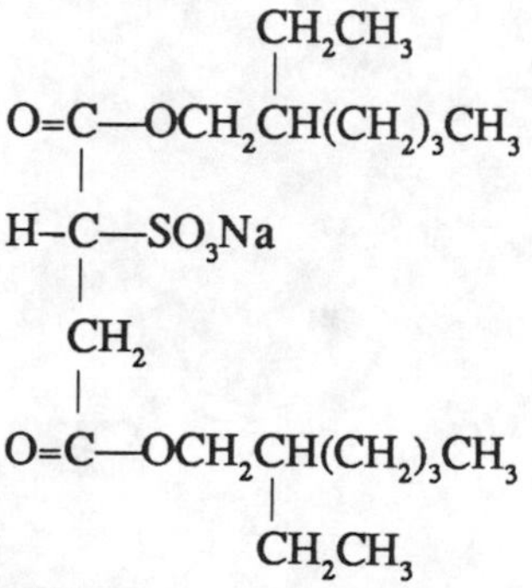

```
              CH2CH3
              |
O=C—OCH2CH(CH2)3CH3
  |
H–C—SO3Na
  |
  CH2
  |
O=C—OCH2CH(CH2)3CH3
              |
              CH2CH3
```

CAS No.

577-11-7

TRADENAME EQUIVALENTS:

Aerosol GPG, OT-70PG, OT-75, OT-100, OT-B, OTS [Amer. Cyanamid]
Alcopol O-60%, O-70PG, O-100% [Allied Colloids Ltd.]
Alkasurf SS-O-40, SS-O-60, SS-O-75 [Alkaril]
Alrowet D-65 [Ciba-Geigy]
Arylene M40, M60 [Hart]
Astrowet O-70-PG, O-75 [Alco]
Atcowet W [Bostik South]
Avirol SO-70P [Henkel]
Chemax DOSS-70, DOSS-75E [Chemax]
Complemix 100 [Amer. Cyanamid]
Condanol SBDO60 [Dutton & Reinisch]
Cyclopol SBDO [Cyclo]
Denwet CM [Graden]
Drewfax 528 [Drew Produtos]
Emcol 4500, 4560 [Witco/Organics]
Empimin OP-45, O-70, OT [Albright & Wilson]
Gemtex PA-70, PA-75, PA-85P, PAX-60, SC-40, SC-70, SC-75 [Finetex]
Hipochem EK-18 [High Point]

Sodium dioctyl sulfosuccinate (cont'd.)

TRADENAME EQUIVALENTS *(cont'd.):*

Lankropol KO2, KO Special [Lankro]
Mackanate DOS-70M5, DOS-75 [McIntyre]
Manoxol OT, OT60%, OT/B [Manchem Ltd.]
Manoxol OT/P [Manchem Ltd.] (pharmaceutical grade)
Marlinat DF8 [Chemische Werke Huls AG]
Merpasol DIO60, DIO75 [Kempen]
Monawet MO-65-150, MO-65PEG, MO-70, MO-70-150, MO-70E, MO-70R, MO-70RP, MO-70S, MO-75E, MO-84R2W, MO85P [Mona]
Nekal WT-27 [GAF]
Nissan Rapisol B-30, B-80 [Nippon Oil & Fats]
Penetron Conc. [Hart Prod. Corp.]
Pentex 99 [Colloids Inc.]
Rewopol NEHS40, SBDO70 [Rewo]
Sanmorin OT70 [Sayno Chem. Ind. Ltd.]
Schercopol DOS-70, DOS-PG-85 [Scher]
Secosol DOS/70 [Stepan Europe]
Solusol 75%, 84%, 85%, 100% [Amer. Cyanamid]
Stantex T-14 [Henkel]
Tex-Wet 1001 [Intex]
Thorowet G-40, G-75 [Clough]
Triton GR-5M, GR-7M [Rohm & Haas]
Varsulf OT [Sherex]

CATEGORY:

Wetting agent, surfactant, antistat, detergent, dispersant, emulsifier, penetrating agent, release agent, rust preventative, solubilizing agent, stabilizer, surface tension depressant, water carrier

APPLICATIONS:

Bath products: bubble bath (Gemtex PA-70, PA-75, PA-85P, PAX-60, SC-40, SC-70, SC-75); bath oils (Gemtex PA-70, PA-75, PA-85P, PAX-60, SC-40, SC-70, SC-75; Monawet MO-70, MO-70E, MO-70R, MO-75E, MO-84R2W)

Cosmetic industry preparations: (Aerosol GPG, OT-70PG, OT-75, OT-100, OT-B, OT-S; Alkasurf SS-O-60; Emcol 4500; Marlinat DF8; Monawet MO-70, MO-70E, MO-70R, MO-75E, MO-84R2W; Pentex 99); shampoos (Monawet MO-70, MO-70E, MO-70R, MO-75E, MO-84R2W)

Farm products: (Aerosol GPG, OT-70PG, OT-75, OT-100, OT-B, OT-S; Alkasurf SS-O-60; Mackanate DOS-70M5; Manoxol OT, OT60%, OT/B; Triton GR-7M); fertilizers (Monawet MO-70, MO-70E, MO-70R, MO-75E, MO-84R2W); insecticides/pesticides (Alkasurf SS-O-75; Avirol SO-70P; Monawet MO-70, MO-70E, MO-70R, MO-75E, MO-84R2W)

Food applications: (Complemix 100; Monawet MO-70, MO-70E, MO-70R, MO-75E, MO-84R2W); food packaging (Avirol SO-70P); food plant cleaning (Monawet MO-70E)

Household detergents: (Alkasurf SS-O-60; Avirol SO-70P; Monawet MO-70E; Nekal WT-27; Penetron Conc.; Pentex 99); dishwashing (Marlinat DF8; Pentex 99); glass cleaners (Marlinat DF8); heavy-duty cleaner (incl. lime soap disp.) laundry detergent (Manoxol OT, OT60%, OT/B; Penetron Conc.)

Industrial applications: (Alkasurf SS-O-40; Sanmorin OT70; Thorowet G-40, G-75); dyes and pigments (Alkasurf SS-O-75; Avirol SO-70P; Monawet MO-70, MO-70E, MO-70R, MO-75E, MO-84R2W; Nissan Rapisol B-30, B-80; Penetron Conc.; Pentex 99); electroplating (Aerosol OT-100; Alkasurf SS-O-60; Avirol SO-70P; Nekal WT-27); fire fighting (Aerosol OT-100; Avirol SO-70P; Nekal WT-27); industrial processing (Aerosol OT-100; Manoxol OT, OT60%, OT/B); metal-working (Aerosol GPG, OT-70PG, OT-75, OT-100, OT-B, OT-S; Manoxol OT, OT60%, OT/B); paint mfg. (Aerosol GPG, OT-70PG, OT-75, OT-100, OT-B, OT-S; Alkasurf SS-O-75; Avirol SO-70P; Manoxol OT, OT60%, OT/B; Marlinat DF8; Monawet MO-70E; Penetron Conc.; Pentex 99); paper mfg. (Aerosol GPG, OT-70PG, OT-75, OT-100, OT-B, OT-S; Marlinat DF8; Merpasol DIO60, DIO75; Nekal WT-27; Penetron Conc.; Pentex 99; Varsulf OT); petroleum industry (Aerosol GPG, OT-70PG, OT-75, OT-100, OT-B, OT-S; Avirol SO-70P; Monawet MO65-150); photography (Aerosol OT-100; Manoxol OT, OT60%, OT/B); plastics (Aerosol GPG, OT-70PG, OT-75, OT-100, OT-B, OT-S; Avirol SO-70P; Nissan Rapisol B-30, B-80); polishes and waxes (Aerosol GPG, OT-70PG, OT-75, OT-100, OT-B, OT-S); printing inks (Aerosol OT-100; Avirol SO-70P; Manoxol OT, OT60%, OT/B; Penetron Conc.); rubber (Aerosol GPG, OT-70PG, OT-75, OT-100, OT-B, OT-S); textile/leather processing (Aerosol GPG, OT-70PG, OT-75, OT-100, OT-B, OT-S; Alkasurf SS-O-60, SS-O-75; Alrowet D-65; Atcowet W; Avirol SO-70P; Empimin OP-45, OP70; Gemtex PA-70, PA-75, PA-85P, PAX-60, SC-40, SC-70, SC-75; Hipochem EK-18; Manoxol OT, OT60%, OT/B; Marlinat DF8; Merpasol DIO60, DIO75; Monawet MO-70, MO-70E, MO-70R, MO-75E, MO-84R2W; Nekal WT-27; Penetron Conc.; Secosol DOS/70; Tex-Wet 1001; Varsulf OT); wettable powders (Manoxol OT/B)

Industrial cleaners: (Aerosol GPG, OT-70PG, OT-75, OT-100, OT-B, OT-S; Monawet MO65-150; Penetron Conc.); drycleaning compositions (Aerosol GPG, OT-70PG, OT-75, OT-100, OT-B, OT-S; Alkasurf SS-O-60, SS-O-75; Avirol SO-70P; Lankropol KO Special; Mackanate DOS-70M5; Manoxol OT, OT60%, OT/B; Monawet MO65-150, MO-70, MO-70R; Nekal WT-27; Pentex 99; Triton GR-7M; Varsulf OT); metal processing surfactants (Marlinat DF8); textile cleaning (Arylene M40, M60; Hipochem EK-18; Pentex 99)

Pharmaceutical applications: (Manoxol OT/P)

PROPERTIES:

Form:

Liquid (Aerosol OT-70PG; Alcopol O70PG; Alkasurf SS-O-40; Astrowet O-75; Atcowet W; Condanol SBDO60; Cyclopol SBDO; Gemtex PA-70, PA-75, PA-85P, PAX-60, SC-40, SC-70, SC-75; Hipochem EK-18; Mackanate DOS-70M5, DOS-75; Manoxol OT60%; Marlinat DF8; Rewopol SBDO70; Sanmorin OT70;

Sodium dioctyl sulfosuccinate (cont'd.)

Secosol DOS/70; Solusol 75%, 84%; Stantex T-14; Thorowet G-40, G-75; Triton GR-5M, GR-7M; Varsulf OT)

Clear liquid (Aerosol OT-S; Alkasurf SS-O-60, SS-O-75; Arylene M40, M60; Avirol SO-70P; Denwet CM; Empimin OP-45, OP-70, OT; Lankropol KO2, KO Special; Monawet MO65-150, MO-70, MO-70E, MO-70R, MO-75E; Nekal WT-27; Pentex 99; Tex-Wet 1001)

Viscous liquid (Aerosol GPG, OT-75; Merpasol DIO60, DIO75; Monawet MO-84R2W)

Paste (Drewfax 528)

Solid (Complemix 100; Solusol 85%, 100%)

Powder (Manoxol OT/B)

Fine powder (Monawet MO-85P)

Free-flowing powder (Aerosol OT-B)

Waxy lumps (Manoxol OT)

Waxy solid (Aerosol OT-100)

Wax (Alcopol O 60%)

Color:

Almost colorless (Rewopol SBDO70; Solusol 75%, 84%)

Colorless (Aerosol OT-75; Alkasurf SS-O-60, SS-O-75; Denwet CM; Monawet MO65-150, MO-70, MO-70E, MO-70R, MO-75E; Penetron Conc.; Pentex 99; Tex-Wet 1001)

Water-white (Manoxol OT60%)

White (Aerosol OT-B; Complemix 100; Monawet MO-85P; Nekal WT-27; Solusol 85%, 100%)

Off-white (Manoxol OT)

Light amber (Aerosol OT-S; Hipochem EK-18)

Pale straw (Lankropol KO2, KO Special)

Straw (Empimin OP45, OP-70, OT)

Clear to slightly yellow (Aerosol OT-70PG)

Light/pale yellow (Condanol SBDO60; Monawet MO-84R2W)

APHA 100 (Aerosol OT-100; Avirol SO-70P)

APHA 200 (Triton GR-5M)

VCS 2 (Triton GR-7M)

Odor:

Alcoholic (Penetron Conc.)

Ethanolic (Lankropol KO2)

Typical (Alkasurf SS-O-75)

Mild (Lankropol KO Special)

Strong typical (Condanol SBDO60)

Composition:

30% conc. (Nissan Rapisol B-30)

40% conc. (Alkasurf SS-O-40; Arylene M40; Gemtex SC-40; Rewopol NEHS40)

42–45% conc. (Thorowet G-40)

44 ± 2% active (Empimin OP-45)
60 ± 1% solids (Tex-Wet 1001)
60% active (Alkasurf SS-O-60; Merpasol DIO60)
60% active min. (Empimin OT)
60% active (contains ethanol) (Lankropol KO2)
60% active (contains mineral oil) (Lankropol KO Special)
60% active in aq. sol'n. with 20% 2-propanol (Triton GR-5M)
60% active in water and isopropanol (Condanol SBDO60)
60% active in 15% methylated spirits and 25% water (Manoxol OT60%)
60% conc. (Arylene M60; Gemtex PAX-60)
64% active in light petroleum distillate (Triton GR-7M)
64% active min. (Rewopol SBDO70)
65% active (Monawet MO-65-150, MO-65PEG; Stantex T-14)
65% conc. (Alrowet D-65)
68–70% active (Cyclopol SBDO)
70% active (Aerosol GPG, OT-70PG, OTS; Drewfax 528; Marlinat DF8; Monawet MO-70, MO-70-150, MO-70E, MO-70R, MO-70RP, MO-70S; Nekal WT-27)
70% conc. (Gemtex PA-70, SC-70; Mackanate DOS-70M5; Sanmorin OT70; Secosol DOS/70)
70% conc. in propylene glycol/water (Astrowet O-70-PG)
70% solids (Avirol SO-70P; Schercopol DOS-70)
72 ± 2% active (Empimin O-70)
73–76% conc. (Thorowet G-75)
75% active (Aerosol OT-75; Alkasurf SS-O-75; Hipochem EK-18; Merpasol DIO75; Monawet MO-75E)
75% active in lower alcohol and water (Solusol 75%)
75% conc. (Astrowet O-75; Atcowet W; Chemax DOSS-75E; Gemtex PA-75, SC-75; Mackanate DOS-75; Pentex 99; Varsulf OT)
75 ± 1% active (Denwet CM)
80% conc. (Nissan Rapisol B-80)
84% active (Solusol 84%; Monawet MO-84R2W)
85% active (Aerosol OT-B, OTS; Monawet MO-85P; Solusol 85%)
85% active with 15% sodium benzoate (Manoxol OT/B)
85% conc. (Gemtex PA-85P)
85% solids (Schercopol DOS-PG-85)
100% active (Aerosol OT-100; Complemix 100; Manoxol OT; Solusol 100%

Solubility:

Sol. in nonpolar solvents (Aerosol OT-100; Monawet MO-70, MO-70E, MO-70R, MO-75E, MO-84R2W; Nekal WT-27; Penetron Conc.); insol. (Aerosol OT-B)
Sol. in oils (Manoxol OT)
Sol. in oils, fats, and waxes by heating to 75 C (Aerosol OT-100)
Sol. in organic solvents (Aerosol GPG, OT-70PG, OT-75, OT-S; Alkasurf SS-O-60, SS-O-75; Manoxol OT)

Sol. in polar solvents (Aerosol OT-100; Monawet MO-70, MO-70E, MO-70R, MO-75E, MO-84R2W; Nekal WT-27; Penetron Conc.); insol. (Aerosol OT-B)
Very sol. in most solvents (Manoxol OT)
Completely water-sol. (Emcol 4500); sol. in water @ 25 C (10% sol'n.) (Hipochem EK-18); 1.5% sol. (Manoxol OT); sol. @ < 1.3% (Avirol SO-70P); sol. up to 1.2% solids @ 20 C and 5.5% @ 70 C (Nekal WT-27); 0.85% sol. @ R.T. (Monawet MO-85P); sol. at concs. up to 0.5% (Empimin OP45, OP70; Lankropol KO2); forms clear sol'ns. in deionized water @ 20 C at conc. ≤ 0.5% (Empimin OT); 1.5 g in 100 ml water @ 25 C (Complemix 100; Solusol 100%); limited sol. in water (Aerosol GPG, OT-70PG, OT-75, OT-S; Alkasurf SS-O-60, SS-O-75); disp. (Aerosol OT-100, OT-B; Empimin OT); readily disp. (Manoxol OT60%); readily disp. at concs. up to 10% (Empimin OP45, OP70); disp. @ 20 C (Lankropol KO Special); gelation occurs at conc. range of 10–60% but above 60% clear miscible liquids are produced (Empimin OP45, OP70)

Ionic Nature:
Anionic

M.W.:
444 (Manoxol OT; Monawet MO-70, MO-70E, MO-70R, MO-75E, MO-84R2W)

Sp.gr.:
1.0 (Aerosol OT-S; Condanol SBDO60)
1.03 (Arylene M40)
1.05 (Manoxol OT60%; Monawet MO65-150; Penetron Conc.); (20 C) (Lankropol KO2, KO Special)
1.06 (Arylene M60; Avirol SO-70P; Monawet MO-70R)
1.07 (Alkasurf SS-O-60; Hipochem EK-18)
1.08 (Monawet MO-70, MO-70E, MO-75E)
1.08–1.13 (70 F) (Pentex 99)
1.09 (Aerosol OT-70PG, OT-75; Alkasurf SS-O-75; Solusol 75%)
1.10 (Aerosol OT-100, OT-B; Complemix 100; Monawet MO-84R2W; Solusol 75%, 85%, 100%)

Density:
1.0 g/cm^3 (20 C) (Empimin OT)
1.05 g/cm^3 (20 C) (Empimin OP45)
1.1 g/cm^3 (Alkasurf SS-O-75; Empimin OP70)
8.4 lb/gal (Triton GR-5M, GR-7M)
8.75 lb/gal (Monawet MO65-150)
8.8 lb/gal (Avirol SO-70P; Monawet MO-70R)
8.91 lb/gal (Hipochem EK-18)
9.0 lb/gal (Monawet MO-70, MO-70E, MO-75E)
9.2 lb/gal (Monawet MO-84R2W)

Visc.:
35 cs (20 C) (Lankropol KO2)
55 cs (20 C) (Empimin OT)

100 cs (20 C) (Empimin OP45)
450 cs (20 C) (Empimin OP70)
40 cps (Triton GR-5M)
50 cps (Arylene M40, M60)
110 cps (Triton GR-7M)
200 cps (Aerosol GPG, OT-75)
200–300 cps (Aerosol OT-S)
200–400 cps (Aerosol OT-70PG)
500–1000 cps (70 F) (Pentex 99)
1500 cs (20 C) (Lankropol KO Special)

F.P.:
< –20 C (Alkasurf SS-O-75)
–40 C (Aerosol GPG, OT-75)

B.P.:
350 F (Soluso 84%)

M.P.:
153–157 C (Aerosol OT-100; Solusol 100%)
< 300 C (Aerosol OT-B; Solusol 85%)

Pour Pt.:
< 0 C (Lankropol KO2, KO Special)
–70 F (Triton GR-7M)
–60 F (Triton GR-5M)
40 F (Pentex 99)

Flash Pt.:
25 C (Empimin OT)
42–43 C (COC) (Monawet MO65-150)
85 C (Aerosol OT-75)
> 100 C (Empimin OP45, OP70)
75 F (Setaflash CC) (Triton GR-5M)
80 F (Manoxol OT60%); (PMCC) (Monawet MO-75E)
81 F (Abel CC) (Lankropol KO2)
82 F (PMCC) (Monawet MO-70E)
130 F (TOC) (Triton GR-7M)
> 200 F (COC) (Lankropol KO Special)
223 F (PMCC) (Monawet MO-84R2W)
280 F (PMCC) (Avirol SO-70P; Monawet MO-70R)
325 F (PMCC) (Monawet MO-70)
High (Aerosol OT-70PG)
None; decomposes at 235 C (Aerosol OT-100)
Chars before flash pt. is reached (Aerosol OT-B)

Cloud Pt.:
< –5 C (Monawet MO-70, MO-70E, MO-70R, MO-75E)
–5 C (Avirol SO-70P; Monawet MO-70E)

Sodium dioctyl sulfosuccinate (cont'd.)

< –10 C (Monawet MO-84R2W)

Acid No.:

2.5 max. (Aerosol OT-75, OT-100, OT-B; Monawet MO65-150, MO-70, MO-70E, MO-70R, MO-75E, MO-84R2W)

4 max. (Avirol SO-70P)

Iodine No.:

0.25 max. (Aerosol OT-75, OT-100, OT-B)

Stability:

Excellent (Hipochem EK-18)

Good at pH 2–9 (Nekal WT-27)

Stable at pH 2–10 (Monawet MO-70, MO-70E, MO-70R, MO-75E)

Stable at pH 3–9 (Rewopol SBDO70)

Stable at pH 7–10; effective in hard and soft waters up to boiling temp. (Pentex 99)

Hydrolyzed by acids and alkalies; generally stable at pH 6–9 (Empimin OP45, OP70, OT)

Hydrolyzes in acid and alkaline sol'ns.; generally used at pH 1–10 (Manoxol OT, OT60%)

Stable in hard water, mild acid, and mild alkaline liquors; efficient in hot and cold sol'ns. (Tex-Wet 1001)

Storage Stability:

Stable under normal storage conditions (Avirol SO-70P)

Ref. Index:

1.3932 (Arylene M40, M60)

pH:

5.0–6.0 (10% DW) (Alkasurf SS-O-75); (1% sol'n.) (Nekal WT-27)

5–7 (Avirol SO-70P)

5.5 ± 1.0 (10%) (Monawet MO65-150, MO-84R2W)

6 0 ± 0.5 (Tex-Wet 1001)

6.0 ± 1 (Empimin OP45, OP70, OT; Monawet MO-70, MO-70E, MO-70R, MO-75E)

6.0–7.0 (1% aq.) (Lankropol KO2, KO Special)

6.0–8.0 (Schercpol DOS-70, DOS-PG-85)

6.5–7.5 (5% solids) (Rewopol SBDO70)

7.0 ± 0.5 (2%) (Denwet CM)

Surface Tension:

23 dynes/cm (Triton GR-5M)

26 dynes/cm (Aerosol GPG, OT-70PG, OT-75, OT-S; Monawet MO-85P)

28.7 dynes/cm (Aerosol OT-100, OT-B), (0.1% solids) (Solusol 100%)

29 dynes/cm (Alkasurf SS-O-60, SS-O-75; Monawet MO-70, MO-70E, MO-70R, MO-75E, MO-84R2W)

< 30 dynes/cm (0.1% sol'n. @ 20 C) (Rewopol SBDO 70)

30 dynes/cm (0.1%, 20 C) (Manoxol OT)

Wetting (Draves):

1 s (0.05 %) (Avirol SO-70P)

Biodegradable: (Aerosol GPG, OT-70PG, OT-S; Rewopol SBDO70; Tex-Wet 1001)

TOXICITY/HANDLING:

Strongly irritating to the eyes; repeated/prolonged exposure will cause skin dryness (Aerosol OT-75, OT-100; Alkasurf SS-O-60, SS-O-75; Avirol SO-70P; Monawet MO-70, MO-70E, MO-70R, MO-75E, MO-84R2W)

Dusts may cause sneezing during transfer operations (Aerosol OT-B)

Skin and eye irritant (Arylene M40, M60)

Eye irritant (Solusol 75%, 84%, 85%, 100%)

Eye irritant; avoid eye contact (Complemix 100)

Eye irritant; degreasing to skin; avoid eye and skin contact, inhalation, and ingestion (Manoxol OT, OT60%)

Burning produces toxic fumes (Manoxol OT)

Skin irritant on prolonged contact with conc. form; wear protective gloves and goggles; spillages may be slippery; in a fire may produce toxic fumes of sulfur dioxide (Lankropol KO2, KO Special)

STORAGE/HANDLING:

Keep containers closed when not in use (Avirol SO-70P)

Combustible (Manoxol OT, OT60%)

Contains ethanol—keep away from heat, flames, and sparks (Lankropol KO2)

Store in stainless steel, glass, resin, or polythene-lined equipment (Lankropol KO Special)

STD. PKGS.:

Drums (Penetron Conc.)

Multiply paper sacks with polythene liners (Manoxol OT)

Bulk or drums (Arylene M40, M60; Hipochem EK-18)

Polyethylene-lined fiber containers (Monawet MO-70, MO-70E, MO-70R, MO-75E, MO-84R2W)

Lacquered drums (Condanol SBDO60)

Liquipak drums (Solusol 84%, 85%)

Steel or Liquipak drums (Solusol 75%)

Fiber and Leverpak drums (Solusol 100%)

125-kg drums (Merpasol DIO60, DIO75)

10-gal drums with polythene inner containers (Manoxol OT60%)

55-gal (450 lb net) lined steel drums; 15-gal (125 lb net) Liquipak fiber drums; tank trunks (Aerosol OT-75)

55-gal liquipak drums (Denwet CM)

50-lb net fiber drums (Aerosol OT-100)

100-lb net fiber drums (Aerosol OT-B)

480 lb fiber drums, bulk, tank wagons, rail cars (Avirol SO-70P)

Sodium ditridecyl sulfosuccinate

SYNONYMS:

Butanedioic acid, sulfo-, 1,4-ditridecyl ester, sodium salt
Ditridecyl sodium sulfosuccinate (CTFA)
Sodium bistridecyl sulfosuccinate
Sodium sulfosuccinic acid, bistridecyl ester

EMPIRICAL FORMULA:

$C_{30}H_{58}O_7S \cdot Na$

STRUCTURE:

$O{=}C{-}OCH_2(CH_2)_{11}CH_3$
|
$H{-}C{-}SO_3Na$
|
CH_2
|
$O{=}C{-}OCH_2(CH_2)_{11}CH_3$

CAS No.:

2673-22-5

TRADENAME EQUIVALENTS:

Aerosol TR-70 [American Cyanamid]
Alcopol OD [Allied Colloids]
Emcol 4600 [Witco/Organics]
Monawet MT-70, MT-70E, MT-80H2W [Mona]

CATEGORY:

Dispersant, emulsifier, surfactant, wetting agent, penetrant, solubilizer, rust preventative, foam modifier, detergent

APPLICATIONS:

Food applications: indirect food additives (Monawet MT-70, MT-70E, MT-80H2W)
Industrial applications: aerosols (Emcol 4600); dyes and pigments (Aerosol TR-70; Emcol 4600; Monawet MT-80H2W); plastics (Aerosol TR-70); polymers/polymerization (Aerosol TR-70; Monawet MT-70, MT-70E); printing inks (Aerosol TR-70; Monawet MT-80H2W)
Industrial cleaners: drycleaning compositions (Emcol 4600)

PROPERTIES:

Form:

Liquid (Alcopol OD)
Clear liquid (Aerosol TR-70; Monawet MT-70, MT-70E)
Viscous liquid (Monawet MT-80H2W)

Color:

Light straw (Monawet MT-70. MT-70E)
Light yellow (Monawet MT-80H2W)

Composition:

60% conc. (Alcopol OD)

70% active (Monawet MT-70, MT-70E)
70% active in water/alcohol (Aerosol TR-70)
80% active (Monawet MT-80H2W)

Solubility:
Excellent sol. in nonpolar solvents (Monawet MT-70, MT-70E, MT-80H2W)
High sol. in organic media (Aerosol TR-70)
Excellent sol. in polar solvents (Monawet MT-70, MT-70E, MT-80H2W)
Limited sol. in water (Aerosol TR-70)

Ionic Nature:
Anionic (Aerosol TR-70; Alcopol OD; Monawet MT-70, MT-70E, MT-80H2W)

M.W.:
584 (Monawet MT-70, MT-70E, MT-80H2W)

Sp.gr.:
0.995 (Aerosol TR-70)
1.01 (Monawet MT-70E)
1.02 (Monawet MT-70, MT-80H2W)

Density:
8.3 lb/gal (Aerosol TR-70)
8.4 lb/gal (Monawet MT-70E)
8.5 lb/gal (Monawet MT-70, MT-80H2W)

Visc.:
110 cps (Aerosol TR-70)

F.P.:
–40 C (Aerosol TR-70)

Flash Pt.:
86 F (PMCC) (Monawet MT-70E)
225 F (PMCC) (Monawet MT-80H2W)
230 F (PMCC) (Monawet MT-70)

Cloud Pt.:
–15 C (Monawet MT-70E)
–2 C (Monawet MT-70)
< 0 C (Monawet MT-80H2W)

Acid No.:
2.5 max. (Monawet MT-70, MT-70E, MT-80H2W)

Stability:
Good (Aerosol TR-70)
Stable @ pH 2–10 (Monawet MT-70, MT-70E, MT-80H2W)

pH:
5.5 ± 1 (10%) (Monawet MT-80H2W)
6 ± 1 (Monawet MT-70, MT-70E)

Surface Tension:
26 dynes/cm (Aerosol TR-70)
29 dynes/cm (0.1%) (Monawet MT-70, MT-70E, MT-80H2W)

Sodium ditridecyl sulfosuccinate (cont'd.)

Biodegradable: (Aerosol TR-70)
TOXICITY/HANDLING:
Eye irritant; defatting to skin (Monawet MT-70, MT-70E, MT-80H2W)
STD. PKGS.:
Polyethylene-lined fiber containers (Monawet MT-70, MT-70E, MT-80H2W)

Sodium hexametaphosphate (CTFA)

SYNONYMS:
Graham's salt
Metaphosphoric acid, hexasodium salt
EMPIRICAL FORMULA:
$H_6O_{18}P_6 \cdot 6Na$
STRUCTURE:
$(NaPO_3)_6$
CAS No.:
10124-56-8
TRADENAME EQUIVALENTS:
Calgolac [Merck]
Calgon [Merck]
Fosfodril [FMC]
Hexaphos [FMC]
Polyphos [Olin]
Generically sold by: Monsanto (glassy, amorphous)
CATEGORY:
Dispersant, deflocculant, antiprecipitant, corrosion inhibitor, sequestering agent, thickener
APPLICATIONS:
Food applications: (Calgon)
Household detergents: (Polyphos); detergent bars (Calgolac); dishwashing (Hexaphos); laundry detergent (Calgon; Hexaphos)
Industrial applications: cement slurries (generic); clays (generic; Calgon; Polyphos); corrosion inhibition (Calgon; Polyphos); dyes and pigments (Calgon; Polyphos); ore flotation (Polyphos); paint mfg. (generic); paper mfg. (generic; Polyphos); petroleum industry (generic; Fosfodril); photography (Polyphos); pulp industry (Hexaphos); textile/leather processing (Calgon; Hexaphos; Polyphos); water treatment (Calgon; Fosfodril; Hexaphos; Polyphos)
Industrial cleaners: glass cleaners (Calgolac; Calgon); laboratory cleaners (Calgolac); lime soap dispersant (Hexaphos)

PROPERTIES:

Form:

Granular (generic)
Lump (walnut and pea-sized) (Polyphos)
Ground (Polyphos)
Powder (generic; Calgolac; Calgon; Polyphos)
Particulate (Calgon)
Plate (Calgon; Polyphos)
Coarsely milled (generic)

Fineness:

15% thru 100 mesh (Polyphos, ground)
49% thru 100 mesh (Polyphos, powdered)

Color:

White (generic)

Composition:

67% P_2O_5 (Calgon)
99% sodium polymetaphosphate; 67.1% phosphorus pentoxide (Polyphos)

Solubility:

Insol. in organic solvents (Calgon)
Sol. in water (> 150 g/100 g @ 24 C) (Polyphos); miscible with water (Calgon)

M.W.:

612.1 (generic)

Sp.gr.:

2.181 (generic)

Bulk Density:

70 lb/ft^3 (Polyphos, powdered)
84 lb/ft^3 (Polyphos, ground)

M.P.:

640 C (generic)

pH:

6.8 (1% sol'n.) (Polyphos)
Alkaline (Calgolac)

TOXICITY/HANDLING:

Toxic by oral ingestion; may produce eye, skin, and mucous membrane irritation; wear protective goggles and clothing; use adequate ventilation to control airborne dust (Polyphos)

STD. PKGS.:

100 lb net polyethylene-lined paper bags, 100-, 300-, 320-, and 350-lb net steel drums (Polyphos)

Sodium lignosulfonate (CTFA)

SYNONYMS:

Lignosulfonic acid, sodium salt
Sodium polignate
Sodium sulfonated lignin

CAS No.:

8061-51-6

TRADENAME EQUIVALENTS:

Amasperse N [AC&C]
Darvan #2 [Vanderbilt]
Dynasperse A, B [Reed Lignin] (modified)
Dyqex [Georgia-Pacific]
Lignosite 431, 458, 823, 854 [Georgia-Pacific/Bellingham Div.]
Lignosol AXD [Reed] (acidic desugarized, partially desulfonated)
Lignosol D-10, D-30 [Reed]
Lignosol DXD [Reed] (alkaline desugarized, partially desulfonated)
Lignosol HCX [Reed] (alkaline sugar-free)
Lignosol NSX-110, SFX-65, XD, X [Reed Ltd.]
Lignosol FTA [Reed] (desugarized, desulfonated)
Lignosol NSX-120 [Reed] (modified)
Lignosol SFS [Reed] (sugar-free)
Lignosol SFX [Reed] (desugarized)
Lignosol WT [Reed]
Maracell E [American Can] (partially desulfonated)
Marasperse CB [American Can]
Marasperse CBOS-3 [Reed] (partially desulfonated)
Marasperse CBX-2 [Reed]
Marasperse N-22 [American Can]
Norlig 12 [American Can]
Orzan LS, LS-50, SL-50 [ITT Rayonier]
Orzan S [Crown Zellerbach]
Polyfon F, H, O, T [Westvaco]
Raylig [ITT Rayonier]
Raymix [ITT Rayonier] (desugared)
Reax 15B [Westvaco] (modified, Kraft)
Reax 45A, 45L, 80C, 81A, 82, 83A, 85A [Westvaco] (modified)
Reax 88B [Westvaco] (sugar-free, modified Kraft)
Reax SR-1, SR-7 [Westvaco] (modified Kraft)
Temsperse S002, S003 [Temfibre] (modified)

CATEGORY:

Dispersant, suspending agent, emulsifier, emulsion stabilizer, stabilizer, sequestrant, chelating agent, extender, diluent, wetting agent, conditioner, viscosity reducer, binder, scale inhibitor, antifoam synergist, deflocculant

APPLICATIONS:

Farm products: agricultural chemicals (Marasperse CB, N-22); animal feed pellets (Norlig 12); fertilizers (Reax 88B); herbicides/fungicides (Lignosol AXD, DXD, HCX; Orzan S); insecticides/pesticides (Lignosol AXD, DXD, HCX, X, XD; Marasperse CBOS-3, N-22; Orzan LS, LS-50, S; Reax 15B, 88B)

Industrial applications: adhesives (Norlig 12; Orzan SL-50); asphalt/bitumen emulsions (Amasperse N; Lignosol DXD; Orzan S); briquetting (Norlig 12; Orzan SL-50); carbon black pellets (Orzan SL-50); carbon black slurries (Marasperse CB, CBOS-3); cement/concrete (Lignosite 458, 854; Lignosol SFX; Orzan S, SL-50; Raymix); clay/refractory brick (Norlig 12); dyes and pigments (Amasperse N; Dynasperse A, B; Dyqex; Lignosite 431, 458, 823, 854; Lignosol AXD, D-10, D-30, DXD, FTA, Lignosol NSX-110, NSX-120, SFS, SFX, SFX-65, XD; Marasperse CB, CBOS-3, CBX-2, N-22; Orzan S; Reax 15B, 88B, SR-1, SR-7); gypsum board (Marasperse CB, N-22; Orzan SL-50); industrial processing (Lignosite 823; Raylig; Reax 88B); latexes (Darvan #2); limestone slurries (Orzan SL-50); metallurgy (Amasperse N; Orzan S); mines and minerals (Amasperse N); ore flotation (Orzan LS, LS-50); plywood/particle board (Orzan SL-50); printing (Amasperse N); resins (Lignosol AXD, XD; Orzan S, SL-50); road surface treatment (Norlig 12); rubber (Darvan #2); sizing (Amasperse N); storage batteries (Maracell E; Reax 15B); textile/leather processing (Amasperse N; Lignosol X, XD; Reax SR-1, SR-7); water treatment (Lignosite 431, 458, 854; Lignosol AXD, SFX, WT, XD; Maracell E; Marasperse N-22; Orzan LS, LS-50; Reax 15B); wax emulsions (Lignosite 431, 458, 823, 854; Lignosol XD; Orzan LS, LS-50); wettable powders (Lignosite 431, 458, 823, 854; Lignosol X, XD; Marasperse N-22; Reax 15B)

Industrial cleaners: (Lignosite 431, 458, 854; Lignosol DXD; Maracell E; Marasperse CB, N-22; Orzan LS, LS-50); metal processing surfactants (Lignosol XD)

PROPERTIES:

Form:

Liquid (Lignosite 431 sol'n., 458 sol'n., 854 sol'n.; Lignosol X; Orzan LS-50, SL-50; Raylig; Reax 88B sol'n.; Temsperse S002, S003)

Clear liquid (Amasperse N)

Powder (Darvan #2; Dynasperse A, B; Dyqex; Lignosol D-10, D-30, HCX, NSX-110, NSX-120, SFS, SFX-65, WT; Maracell E; Marasperse CB, CBOS-3, CBX-2, N-22; Orzan S; Temsperse S002, S003)

Fine powder (Lignosite 431, 458, 854)

Spray-dried powder (Lignosol AXD, DXD, FTA, SFX, XD)

Free-flowing powder (Reax 15B, 88B, SR-1, SR-7)

Free-flowing, spray-dried powder (Orzan LS)

Color:

Light (Dyqex)

Yellow (Lignosol WT, XD; Orzan LS)

Brown (Lignosite 431, 458, 854; Lignosol AXD, D-10, D-30, FTA, HCX, SFX; Marasperse N-22; Orzan S; Raylig; Reax SR-1, SR-7)

Dark brown (Darvan #2)
Black (Lignosol DXD; Marasperse CB)

Odor:

Pleasant, vanilla-like (Raylig)

Composition:

35% solids in water (Reax 88B sol'n.)
40% active (Amasperse N)
46–48% solids (Lignosite 458 sol'n.)
47% active in water (Orzan LS-50)
48% active (Temsperse S002, S003 Liquids)
50% active (Lignosite 431 sol'n., 854 sol'n.; Lignosol X; Raylig)
50% active in water (Orzan SL-50)
59% active (Orzan LS)
70% active (Lignosite 823)
80% active (Lignosite 431, 458, 854)
84.5% active min. (Darvan #2)
95% active (Lignosol AXD, D-10, D-30, DXD, FTA, HCX, SFX, WT, XD; Orzan S; Temsperse S002, S003 Powders)
100% active (Dynasperse A, B; Lignosol NSX-110, SFS, SFX-65; Marasperse CBOS-3, CBX-2, N-22; Reax 15B, 88B, SR-1, SR-7)

Solubility:

Sol. in acidic sol'ns. to varying degrees (Polyfon F, H, O, T; Reax 45A, 45L, 80C, 81A, 82, 83A, 85A, 88B)
Sol. in alkaline sol'ns. (Polyfon F, H, O, T; Reax 45A, 45L, 80C, 81A, 82, 83A, 85A, 88B)
Virtually no sol. in most organic solvents; increased sol. with the addition of water in water-miscible solvents, e.g., DMF, ethylene glycol, methanol (Reax SR-1)
Sol. in water (Lignosol AXD, D-10, D-30, HCX, WT, XD; Maracell E; Polyfon F, H, O, T; Raylig; Reax 45A, 45L, 80C, 81A, 82, 83A, 85A, 88B); rapidly and completely sol. (Orzan LS); completely sol. (Amasperse N; Lignosol DXD, FTA; Marasperse CB, N-22); readily sol. (Reax 15B, 88B); very sol. in warm water; max. recommended conc. 25% in water at 21 C (Darvan #2); dissolves in hot or cold water (Lignosite 431, 458, 823, 854); 60% sol. (Orzan S)

Ionic Nature:

Anionic (Amasperse N; Dynasperse A, B; Lignosite 431, 458, 823, 854; Lignosol AXD, D-10, D-30, DXD, FTA, HCX, NSX-110, SFS, SFX, SFX-65, WT, X, XD; Marasperse CB, CBOS-3, CBX-2; Orzan S; Raylig; Reax 15B, 88B, SR-1, SR-7; Temsperse S002, S003)

Sp.gr.:

1.23 (Raylig)
1.235–1.245 (Orzan SL-50)
1.255–1.265 (Orzan LS-50)
1.264 (50% aq.) (Orzan S)

Density:
10.3 lb/gal (Lignosite 458 sol'n.)
10.5 lb/gal (Lignosite 431 sol'n., 854 sol'n.)
Bulk Density:
23 lb/ft^3 (Lignosite 823, 854)
24 lb/ft^3 (loose) (Reax SR-1)
28–32 lb/ft^3 (Lignosol AXD, SFX, XD)
30 lb/ft^3 (Orzan LS)
30–32 lb/ft^3 (Lignosol DXD, FTA)
35–40 lb/ft^3 (Marasperse N-22)
43–47 lb/ft^3 (Marasperse CB)
Visc.:
100 cps (20 C) (Orzan LS-50)
200 cps (20 C) (Orzan SL-50)
250 cps (50% aq.) (Orzan S)
500 cps (Raylig)
F.P.:
0 C (Raylig)
B.P.:
100 C (Raylig)
Stability:
Stable (Orzan S; Raylig)
Stable to 250 C (Lignosol AXD, D-10, D-30, DXD, FTA, HCX, XD)
Stable to 300 C (Lignosol WT)
Storage Stability:
Good under proper storage conditions (Orzan LS)
pH:
4.0 (10% sol'n.) (Lignosite 431)
4.9 (27% sol'n.) (Lignosol AXD)
5.0 (10% aq.) (Lignosite 854)
5.5 (Raylig)
6.5 (25% sol'n.) (Orzan LS, SL-50)
6.7 (2% aq.) (Reax SR-7)
6.8 (27% sol'n.) (Lignosol SFX)
7.0 (25% sol'n.) (Orzan LS-50)
7.0–8.0 (10% aq.) (Lignosite 823)
7.0–8.5 (1% sol'n.) (Darvan #2)
7.5 (27% sol'n.) (Lignosol XD)
7.5–8.5 (3% sol'n.) (Marasperse N-22)
9.0 (27% sol'n.) (Lignosol DXD)
9.5 (27% sol'n.) (Lignosol FTA)
9.9 (2% aq.) (Reax SR-1)
10.5 (Reax 15B)

Sodium lignosulfonate (cont'd.)

10.8 (3% sol'n.) (Maracell E)
11.5 (aq. sol'n.) (Reax 88B)
Neutral (Orzan S)
Alkaline (Lignosol WT)

Surface Tension:
51.4 dynes/cm (1% sol'n.) (Marasperse CB)
55.8 dynes/cm (1% aq.) (Reax SR-1)
58.8 dynes/cm (1% aq.) (Reax SR-7)

Biodegradable: (Orzan LS, LS-50, SL-50)

TOXICITY/HANDLING:
Nontoxic (Orzan LS, LS-50, SL-50)
Mild eye irritant (Raylig)
Slight eye and skin irritant (Orzan S)

STORAGE/HANDLING:
Organic powder—avoid contact with oxidizing agents (Lignosol AXD)
Avoid excessively high humidities or temps. for prolonged periods (Orzan LS)

STD. PKGS.:
Bags, drums, and bulk (Orzan S)
Tank trucks, cars (Raylig)
50-lb net multiwall paper bags (Maracell E; Marasperse CB, N-22; Reax SR-1)
50-lb paper bags (Lignosol AXD)
450-lb net drums (Amasperse N)

Sodium metasilicate (CTFA)

SYNONYMS:
Silicic acid, disodium salt

EMPIRICAL FORMULA:
$H_2O_3Si \cdot 2Na$

STRUCTURE:
Na_2SiO_3

CAS No.:
6834-92-0

TRADENAME EQUIVALENTS:
Crystamet [Stauffer] (pentahydrate)
Drymet [Stauffer] (anhydrous)
Metso 20 [PQ Corp.]
Metso Beads 2048 [PQ Corp.] (anhydrous)
Metso Pentabead 20 [PQ Corp.] (pentahydrate)

CATEGORY:
Dispersant, corrosion inhibitor

APPLICATIONS:

Household detergents: detergent base (generic); floor cleaning (generic); laundry detergent (generic)
Industrial applications: ceramics (Metso); glass (Metso); paper deinking (generic)
Industrial cleaners: dairy cleaning (generic); metal processing surfactants (generic; Metso)

PROPERTIES:

Form:
Granules (generic anhydrous)
Beads (Metso Beads 2048, Pentabead 20)

Color:
White (generic anhydrous)

Composition:
29.3% Na_2O (generic pentahydrate)
51.5% Na_2O (generic anhydrous)

Solubility:
Sol. in water (generic; Drymet; Metso Beads 2048, Pentabead 20)

M.W.:
122.05 (generic)

Sp.gr.:
1.75 (generic pentahydrate)
2.61 (generic anhydrous)

Bulk Density:
55 lb/ft^3 (generic pentahydrate)
75 lb/ft^3 (generic anhydrous)

M.P.:
72.2 C (generic pentahydrate)
1089 C (generic anhydrous)

Stability:
Precipitated by acids and alkaline earths and by heavy metal ions (generic anhydrous)

pH:
12.6 (1% sol'n.) (generic anhydrous)

Sodium methyl cocoyl taurate (CTFA)

SYNONYMS:

Amides, coconut oil, with N-methyltaurine, sodium salts
Sodium N-coconut acid-N-methyl taurate
Sodium N-cocoyl-N-methyl taurate
Sodium N-methyl-N-cocoyl taurate

Sodium methyl cocoyl taurate (cont'd.)

STRUCTURE:

$$RC(=O)—N(CH_3)—CH_2CH_2SO_3Na$$

where RCO^- represents the coconut acid radical

CAS No.:

12765-39-8; 61791-42-2

TRADENAME EQUIVALENTS:

Adinol CT95 [Croda]
Hostapon CT Paste [Hoechst AG]
Igepon TC-42 [GAF]
Nikkol CMT-30 [Nikko]
Tauranol WS, WS Conc., WS H.P., WSP [Finetex]

CATEGORY:

Dispersant, detergent, emulsifier, surfactant, foaming agent

APPLICATIONS:

Bath products: bubble bath (Igepon TC-42)
Cosmetic industry preparations: (Adinol CT95; Igepon TC-42; Tauranol WS, WS Conc., WS H.P., WSP); shampoos (Hostapon CT Paste; Igepon TC-42; Nikkol CMT-30)
Pharmaceutical applications: (Adinol CT95; Tauranol WS H.P.); dental preparations (Tauranol WS H.P.)

PROPERTIES:

Form:

Liquid (Nikkol CMT-30)
Slurry (Tauranol WS, WS Conc.)
Paste (Hostapon CT Paste; Igepon TC-42)
Powder (Adinol CT95; Tauranol WS H.P., WSP)

Color:

White (Adinol CT95; Igepon TC-42)

Composition:

24% conc. (Tauranol WS)
$\geq$ 24% active (Igepon TC-42)
30% conc. (Hostapon CT Paste; Nikkol CMT-30)
31% conc. (Tauranol WS Conc.)
70% conc. (Tauranol WSP)
95% min. conc. (Adinol CT95; Tauranol WS H.P.)

Ionic Nature:

Anionic (Adinol CT95; Hostapon CT Paste; Nikkol CMT-30; Tauranol WS, WS Conc., WS H.P., WSP)

Sodium methyl cocoyl taurate (cont'd.)

M.W.:
363 (Igepon TC-42)
Stability:
Chemically stable (Igepon TC-42)
Tolerant to electrolytes incl. hard water (Tauranol WS H.P.)
pH:
7.0–8.5 (10% sol'n.) (Igepon TC-42)

Sodium methyl oleoyl taurate (CTFA)

SYNONYMS:
Ethanesulfonic acid, 2-[methyl (1-oxo-9-octadecenyl) amino]-, sodium salt
2-[Methyl (1-oxo-9-octadecenyl) amino] ethanesulfonic acid, sodium salt
Sodium N-methyl-N-oleoyl taurate
Sodium N-oleoyl-N-methyl taurate
EMPIRICAL FORMULA:
$C_{21}H_{41}NO_4S \cdot Na$
STRUCTURE:

```
(CH2)7CH3
|
CH
||
CH     O
|      ||
(CH2)7C—N—CH2CH2SO3Na
        |
        CH3
```

CAS No.:
137-20-2
TRADENAME EQUIVALENTS:
Adinol T, T35 [Croda]
Hostapon T [Hoechst AG]
Hostapon T Paste 33, T Powder H.C. [Amer. Hoechst]
Igepon T-33, T-43, T-51, T-77 [GAF]
Igepon TM-43 [GAF]
Tauranol M, ML, MS [Finetex]
Tergenol G, S Liquid, Slurry [Hart Prod.]
CATEGORY:
Surfactant, dispersant, detergent, wetting agent, emulsifier, foaming agent, release agent, stabilizer, corrosion inihibitor

Sodium methyl oleoyl taurate (cont'd.)

APPLICATIONS:

Bath products: bubble bath (Adinol T; Igepon TM-43; Tauranol M, ML, MS); foaming bath oils (Tauranol M, ML)

Cleansers: skin cleansers (Adinol T); sudsing creams and lotions (Igepon TM-43); syndet/soap bars (Igepon TM-43)

Cosmetic industry preparations: (Adinol T35; Hostapon T, T Paste 33, T Powder H.C.; Tergenol G, S Liquid, Slurry); shampoos (Adinol T; Igepon TM-43; Tauranol M, ML, MS)

Farm products: insecticides/herbides (Igepon T-77); wettable powders (Igepon T-77)

Household detergents: (Hostapon T Powder H.C.; Tergenol G, S Liquid, Slurry); carpet & upholstery shampoos (Igepon T-33, T-43, T-51, T-77); laundry detergent (Igepon T-33, T-43, T-51, T-77; Tergenol G, S Liquid, Slurry)

Industrial applications: dyes and pigments (Igepon T-33, T-43, T-51, T-77; Tergenol G, S Liquid, Slurry); latex emulsions (Igepon T-51); paper mfg. (Igepon T-33, T-43, T-51, T-77); printing inks (Tergenol G, S Liquid, Slurry); resins (Tergenol G, S Liquid, Slurry); rubber (Igepon T-33, T-43, T-51, T-77); textile/leather processing (Adinol T35; Igepon T-33, T-43, T-51, T-77; Tergenol G, S Liquid, Slurry)

Industrial cleaners: (Igepon T-33, T-43, T-51, T-77; Tergenol G, S Liquid, Slurry); bottle washing (Igepon T-33, T-43, T-51, T-77); metal processing surfactants (Igepon T-33, T-43, T-51, T-77); textile/leather cleaners (Hostapon T Powder H.C.); textile scouring (Igepon T-33, T-43, T-51, T-77; Tauranol MS)

PROPERTIES:

Form:

Liquid (Adinol T, T35; Tauranol ML; Tergenol S Liquid)
Clear liquid (Igepon T-33)
Viscous heterogeneous liquid slurry (Igepon T-43)
Slurry (Igepon TM-43)
Gel (Adinol T; Igepon T-51; Tergenol G)
Paste (Adinol T; Hostapon T Paste 33; Tauranol MS; Tergenol Slurry)
Powder (Adinol T; Hostapon T, T Powder H.C.; Tauranol M)
Soft, nondusting flakes (Igepon T-77)

Color:

Light (Tergenol G)
White (Tergenol Slurry)
Milky white (Igepon T-43)
Cream (Igepon T-77)
Light amber (Igepon T-51)
Amber (Tergenol S Liquid)
Pale yellow (Igepon T-33)

Odor:

Perfumed (Igepon T-51)
Perfumed, alcoholic (Igepon T-33)
Fatty (Igepon T-43, T-77)

Sodium methyl oleoyl taurate (cont'd.)

Composition:
13.9% active min. (Igepon T-51)
15% active (Tergenol G)
16% conc. min. (Adinol T35)
32% active min. (Igepon T-33)
33% active (Hostapon T Paste 33; Igepon TM-43; Tauranol ML, MS)
33% active min. (Igepon T-43)
38% active (Tergenol S Liquid)
60% active (Tergenol Slurry)
63% conc. (Hostapon T)
64% conc. (Hostapon T Powder H.C.)
67% active min. (Igepon T-77)
75% conc. (Tauranol M)

Solubility:
Sol. in water (Igepon T-33; Tergenol G, Slurry); good in hard water (Adinol T); readily sol. @ 54 C (Igepon T-51); readily sol. @ > 54 C, slowly sol. @ 25 C (Igepon T-43); slowly sol. @ 54 C (Igepon T-77); sol. 20 g/100 ml (Igepon TM-43)

Ionic Nature:
Anionic (Adinol T35; Hostapon T, T Paste 33, T Powder H.C.; Igepon T-33, T-43, T-51, T-77, TM-43; Tauranol M, ML, MS; Tergenol G, S Liquid, Slurry)

M.W.:
425 (Igepon T-33, T-43, T-51, T-77)

Stability:
Good resistance to hydrolysis over wide pH range (Igepon TM-43)

pH:
Neutral (Tergenol S Liquid, Slurry)
6.5–8.0 (10% sol'n.) (Igepon T-33, T-51); (5% sol'n.) (Igepon T-43); (2% sol'n.) (Igepon T-77)

Surface Tension:
36.8 dynes/cm (0.05%) (Igepon TM-43)

Biodegradable: (Adinol T35)

STD. PKGS.:
Drums (Tergenol G, S Liquid, Slurry)

Sodium nonoxynol-9 phosphate (CTFA)

CAS No.:
RD No.: 977058-21-1

TRADENAME EQUIVALENTS:
Emphos CS-1361 [Witco/Organics]

Sodium nonoxynol-9 phosphate (cont'd.)

CATEGORY:

Dispersant, solubilizer, antistat, emulsifier, detergent

APPLICATIONS:

Household detergents: hard surface cleaner (Emphos CS-1361); nonionic surfactant systems (Emphos CS-1361)

Industrial cleaners: drycleaning compositions (Emphos CS-1361); metal processing surfactants (Emphos CS-1361)

PROPERTIES:

Form:

Clear liquid (Emphos CS-1361)

Solubility:

Sol. in carbon tetrachloride (@ 25%) (Emphos CS-1361)
Sol. in ethanol (@ 25%) (Emphos CS-1361)
Sol. in kerosene (@ 25%) (Emphos CS-1361)
Insol. in min. oil (@ 25%) (Emphos CS-1361)
Sol. in heavy aromatic naphtha (@ 25%) (Emphos CS-1361)
Sol. in perchloroethylene (@ 25%) (Emphos CS-1361)
Disp. in 10% sodium hydroxide (@ 25%) (Emphos CS-1361)Sol. in Stoddard solvent
Sol. in water (@ 25%) (Emphos CS-1361)

Ionic Nature:

Anionic (Emphos CS-1361)

Sp.gr.:

1.10 (Emphos CS-1361)

Pour Pt.:

4 C (Emphos CS-1361)

Acid No.:

28 (to pH 9.5) (Emphos CS-1361)

pH:

5.0 (3% aq.) (Emphos CS-1361)

Surface Tension:

31.9 dynes/cm (0.05% in distilled water) (Emphos CS-1361)

Sodium nonoxynol-4 sulfate (CTFA)

SYNONYMS:

Sodium nonyl phenol ether sulfate (4 EO)
Sodium nonyl phenol polyglycol ether sulfate (4 EO)
Sodium salt of sulfated nonoxynol-4

CAS No.:

9014-90-8 (generic)
RD No.: 977069-67-2

TRADENAME EQUIVALENTS:
Alipal CO-433 [GAF]
Serdet DNK30 [Servo B.V.]
Sermul EA54 [Servo B.V.]
CATEGORY:
Dispersant, emulsifier, detergent, wetting agent, foaming agent
APPLICATIONS:
Cosmetic industry preparations: (Alipal CO-433)
Farm products: insecticides/pesticides (Alipal CO-433)
Household detergents: all-purpose cleaner (Alipal CO-433); dishwashing (Alipal CO-433)
Industrial applications: polymers/polymerization (Alipal CO-433; Serdet DNK30; Sermul EA54); textile/leather processing (Alipal CO-433)
PROPERTIES:
Form:
Liquid (Serdet DNK30; Sermul EA54)
Clear liquid (Alipal CO-433)
Color:
Varnish 4 max. (Alipal CO-433)
Odor:
Mild aromatic (Alipal CO-433)
Composition:
28% active (Alipal CO-433)
31% conc. (Serdet DNK30; Sermul EA54)
Solubility:
Sol. in water (Alipal CO-433)
Ionic Nature:
Anionic (Alipal CO-433; Serdet DNK30; Sermul EA54)
Sp.gr.:
1.065 (Alipal CO-433)
Density:
8.9 lb/gal (Alipal CO-433)
Visc.:
2500 cps (Alipal CO-433)
Stability:
Good (Alipal CO-433)
Surface Tension:
32 dynes/cm (1% sol'n.) (Alipal CO-433)

Sodium toluenesulfonate (CTFA)

SYNONYMS:

Benzenesulfonic acid, methyl-, sodium salt
Methylbenzenesulfonic acid, sodium salt
Toluene sulfonic acid, sodium salt

EMPIRICAL FORMULA:

$C_7H_8O_3S \cdot Na$

STRUCTURE:

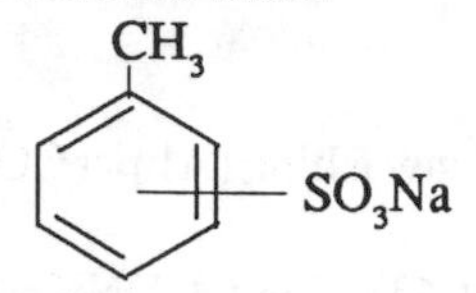

CAS No.:

12068-03-0

TRADENAME EQUIVALENTS:

Eltesol ST34, ST40, ST90 [Albright & Wilson/Marchon]
Eltesol ST Pellets [Albright & Wilson/Detergents]
Hartotrope STS40, STS Powder [Hart Chem.]
Manro STS40 [Manro]
Na-Toluene Sulfonate 30, 40 [Chemische Werke Huls AG]
Naxonate 4ST, ST [Ruetgers Nease]
Reworyl NTS40 [Rewo Chemische]
Stepanate T [Stepan]
Witconate STS [Witco/Organics]

CATEGORY:

Hydrotrope, solubilizer, coupling agent, processing aid, viscosity modifier, cloud point depressant, antiblocking agent, anticaking agent, solvent, fluidizer, freeze-thaw stabilizer, detergent

APPLICATIONS:

Cosmetic industry preparations: shampoos (Witconate STS)

Household detergents: (Reworyl NTS40; Stepanate T; Witconate STS); built detergents (Hartotrope STS40, STS Powder); dishwashing (Naxonate 4ST, ST; Stepanate T); heavy-duty cleaner (Eltesol ST34, ST40, ST90, ST Pellets; Naxonate 4ST, ST; Stepanate T); light-duty cleaners (Eltesol ST34, ST40, ST90, ST Pellets; Hartotrope STS40, STS Powder); liquid detergents (Eltesol ST34, ST40, ST90, ST Pellets; Hartotrope STS40, STS Powder; Manro STS40; Na-Toluene Sulfonate 30, 40; Naxonate 4ST, ST); powdered detergents (Eltesol ST34, ST40, ST90, ST Pellets; Hartotrope STS Powder; Naxonate 4ST, ST; Na-Toluene Sulfonate 30, 40; Witconate STS)

Industrial applications: adhesives (Naxonate 4ST, ST; Witconate STS); aerosols (Witconate STS); dyes and pigments (Naxonate 4ST, ST; Witconate STS); electroplating (Naxonate 4ST, ST); extraction processes (Witconate STS); industrial processing (Witconate STS); lubricating/cutting oils (Naxonate 4ST, ST; Witconate STS); paper mfg. (Witconate STS); photography (Witconate STS); printing inks (Naxonate 4ST, ST; Witconate STS); textile/leather processing (Naxonate 4ST, ST; Witconate STS); wood pulping (Witconate STS)

Industrial cleaners: (Witconate STS); germicides (Naxonate 4ST, ST); metal processing surfactants (Witconate STS); wax strippers (Stepanate T)

PROPERTIES:

Form:

Liquid (Eltesol ST34, ST40; Na-Toluene Sulfonate 30, 40; Naxonate 4ST; Reworyl NTS40; Witconate STS)

Clear liquid (Hartotrope STS40; Manro STS40)

Pellets (Eltesol ST Pellets)

Powder (Eltesol ST90; Hartotrope STS Powder; Naxonate ST; Witconate STS)

Color:

White (Eltesol ST Pellets; Hartotrope STS Powder)

Off-white (Eltesol ST90)

Cream (Naxonate ST)

Straw (Eltesol ST40)

Pale yellow (Eltesol ST34; Manro STS40)

Klett 50 max. (Naxonate 4ST)

Klett 150 (Witconate STS)

Odor:

Negligible (Manro STS40)

Composition:

30% active (Na-Toluene Sulfonate 30)

34% active in water (Eltesol ST34)

39.5% ative min. in water (Manro STS40)

40% active (Hartotrope STS40; Na-Toluene Sulfonate 40; Stepanate T; Witconate STS Liquid)

40% active min. (Naxonate 4ST)

40% conc. (Reworyl NTS40)

40 ± 1% active in water (Eltesol ST40)

85% active min.; 8% water max. (Eltesol ST Pellets)

88% active (Witconate STS Powder)

89% active in water (Eltesol ST90)

93% active (Hartotrope STS Powder)

95% active min. (Naxonate ST)

Ionic Nature:

Anionic (Hartotrope STS40; Naxonate 4ST, ST; Stepanate T)

Sp.gr.:

1.19 (20 C) (Manro STS40)

Density:

0.32 g/cm^3 (Witconate STS Powder)

0.55 g/cm^3 (20 C) (Eltesol ST Pellets)

Sodium toluenesulfonate (cont'd.)

1.10 g/cm^3 (20 C) (Eltesol ST40)
1.2 g/cm^3 (Witconate STS Liquid)
10.1 lb/gal (Stepanate T)

pH:

7.0–9.0 (Naxonate 4ST); (1% aq.) (Stepanate T)
7.0–10.0 (Eltesol ST34); (10% aq.) (Manro STS40)
7.0–10.5 (10% aq.) (Eltesol ST40)
9.0 (Witconate STS Liquid)
9.0–10.5 (Eltesol ST90); (3% aq. sol'n.) (Eltesol ST Pellets)
10.0 (40% sol'n.) (Witconate STS Powder)

TOXICITY/HANDLING:

Avoid prolonged contact with skin; use normal safety precautions (Witconate STS)
Avoid contact with skin and eyes (Hartotrope STS40, STS Powder)

STORAGE/HANDLING:

Storage and transfer in stainless steel is recommended; store above 30 C to avoid crystallization; if crystallization occurs, heating will reliquefy (Witconate STS)

STD. PKGS.:

45-gal drums or road tankers (Manro STS40)
55-gal lined drums or tankwagons (Naxonate 4ST)
175 leverpak (Naxonate ST)
Drums, T/T (Hartotrope STS40)
Bags (Hartotrope STS Powder)

Sodium xylenesulfonate (CTFA)

SYNONYMS:

Benzenesulfonic acid, dimethyl-, sodium salt
Dimethylbenzenesulfonic acid, sodium salt

EMPIRICAL FORMULA:

$C_8H_{10}O_3S \cdot Na$

STRUCTURE:

$(CH_3)_2C_6H_3SO_3Na$

CAS No.:

1300-72-7

TRADENAME EQUIVALENTS:

Alkatrope SX-40 [Alkaril]
Carsosulf SXS Liquid [Carson]
Conco SXS [Continental]
Eltesol SX30, SX93, SX Pellets [Albright & Wilson/Marchon]
ESI-Terge SXS [Emulsion Systems]
Hartotrope SXS40, SXS Powder [Hart]
Lakeway SXS [Bofors Lakeway]

TRADENAME EQUIVALENTS *(cont'd.)*:

Lankrosol SXS-30 [Lankro]
Manro SXS30, SXS40, SXS93 [Manro]
Naxonate 4L, 5L, G [Nease]
Ninex 303 [Stepan/Europe]
Norfox SXS-40, SXS-96 [Norman, Fox & Co.]
Pilot SXS-40, SXS-96 [Pilot]
Polystep A-2 [Stepan]
Reworyl NXS40 [Rewo]
Richonate SXS [Richardson]
Siponate SXS [Alcolac]
Stepanate X [Stepan]
Sterling 2XS [Canada Packers]
Sulfotex SXS [Henkel]
Surco SXS [Onyx/Millmaster]
SXS [Continental]
Ultra SXS Liquid, SXS Powder [Witco/Organics]
Ultrawet 40SX [Arco]
Witconate SXS Liquid, SXS Powder [Witco/Organics]

CATEGORY:

Coupling agent, solubilizer, cloud point depressant, detergency aid, dispersant, emulsifier, hydrotrope, leveling agent, stabilizer, viscosity modifier

APPLICATIONS:

Cosmetic industry preparations: shampoos (Lakeway SXS; Witconate SXS)

Household detergents: (Conco SXS; Reworyl NXS40; Richonate SXS; Stepanate X; SXS); dishwashing (Eltesol SX30, SX93, SX Pellets; Lakeway SXS; Naxonate 4L, G; Richonate SXS; Stepanate X); hard surface cleaner (Lankrosol SXS-30); heavy-duty detergent (Eltesol SX30, SX93, SX Pellets; Naxonate 4L, 5L, G; Stepanate X); light-duty cleaners (Hartrotrope SXS40); liquid detergents (Alkatrope SX-40; Carsosulf SXS Liquid; Conco SXS; Eltesol SX30, SX93, SX Pellets; Hartotrope SXS40; Manro SXS30, SXS40; Naxonate 4L, G; Pilot SXS-40, SXS-96; Sterling 2XS; Sulfotex SXS; Ultra SXS Liquid, SXS Powder; Witconate SXS); powdered detergents (Eltesol SX30, SX93, SX Pellets; Naxonate 4L, G; Ultra SXS Liquid, SXS Powder; Witconate SXS)

Industrial applications: adhesives (Naxonate 4L, 5L, G; Witconate SXS); dyes and pigments (Conco SXS; Lankrosol SXS-30; Naxonate 4L, 5L, G); electroplating (Naxonate 4L, 5L, G); lubricating/cutting oils (Naxonate 4L, 5L, G; Witconate SXS); photography (Witconate SXS); printing inks (Naxonate 4L, 5L, G; Sulfotex SXS; Witconate SXS); textile/leather processing (Naxonate 4L, 5L, G; Witconate SXS); wood pulping (Conco SXS)

Industrial cleaners: (Richonate SXS; Witconate SXS); drycleaning compositions (Richonate SXS); germicides (Naxonate 4L, 5L, G); metal processing surfactants (Lakeway SXS; Witconate SXS)

Sodium xylenesulfonate (cont'd.)

PROPERTIES:

Form:

Liquid (Carsosulf SXS Liquid; Conco SXS; Eltesol SX30; Lakeway SXS; Lankrosol SXS-30; Naxonate 4L, 5L; Manro SXS30, SXS40; Pilot SXS-40, SXS-96; Reworyl NXS40; Sterling 2XS; Sulfotex SXS; Surco SXS; Ultra SXS Liquid; Ultrawet 40SX; Witconate SXS Liquid)
Clear liquid (Alkatrope SX-40; Hartotrope SXS40; Polystep A-2; Stepanate X; SXS)
Pellets (Eltesol SX93, SX Pellets)
Powder (Eltesol SX93; Hartotrope SXS Powder; Naxonate G; Pilot SXS-96; Ultra SXS Powder; Witconate SXS Powder)

Color:

Water-white (Conco SXS)
White (Eltesol SX Pellets)
Off-white (Eltesol SX93)
Cream (Naxonate G)
Pale straw (Lankrosol SXS-30)
Pale yellow (Eltesol SX30; Lakeway SXS; Manro SXS30, SXS40)
Gardner 1 (Richonate SXS)
Klett 40 (Witconate SXS Liquid)
Klett 50 max. (Naxonate 4L)
Klett 60 max. (Naxonate 5L)

Composition:

30% active (Eltesol SX30; Lankrosol SXS-30; Manro SXS30)
39% active (Richonate SXS)
39–42% active (Surco SXS)
40% active (Carsosulf SXS Liquid; Conco SXS; Hartotrope SXS40; Manro SXS40; Naxonate 4L; Pilot SXS-40; Polystep A-2; Stepanate X; Sterling 2XS; Witconate SXS Liquid)
40% conc. (ESI-Terge SXS; Ninex 303; Norfox SXS-40; Reworyl NXS40; Siponate SXS; Sulfotex SXS; Ultra SXS Liquid)
41% active (Alkatrope SX-40; SXS)
41% conc. (Lakeway SXS)
41.5% conc. (Ultrawet 40SX)
50% active (Naxonate 5L)
88% active (Eltesol SX Pellets)
90% active (Hartotrope SXS Powder)
90% conc. (Ultra SXS Powder)
91% active (Witconate SXS Powder)
93% active (Eltesol SX93)
93% conc. (Manro SXS93)
95% active (Naxonate G)
96% active (Pilot SXS-96)
96% conc. (Norfox SXS-96)

Solubility:
Sol. in water (Lankrosol SXS-30; Stepanate X)
Ionic Nature:
Anionic
Sp.gr.:
1.128 (20 C) (Lankrosol SXS-30)
1.15 (20 C) (Manro SXS30)
1.17 (SXS); (25/20 C) (Surco SXS); (30 C) (Manro SXS40)
1.18–1.22 (Carsosulf SXS-Liquid)
Density:
0.32 g/cm³ (Witconate SXS Powder)
0.5 g/cm³ (20 C) (Eltesol SX Pellets)
1.2 g/cm³ (Witconate SXS Liquid)
9.80 lb/gal (Richonate SXS; SXS)
9.83 lb/gal (Stepanate X)
10.0 lb/gal (Lakeway SXS)
Visc.:
10 cps (Brookfield) (Lakeway SXS)
2.7 cs (20 C) (Lankrosol SXS-30)
6.0 (SXS)
B.P.:
212 F (SXS)
M.P.:
31 F (SXS)
Pour Pt.:
< 0 C (Lankrosol SXS-30)
Flash Pt.:
> 200 F (COC) (Lankrosol SXS-30; Surco SXS)
Cloud Pt.:
< –5 C (Lakeway SXS)
pH:
7.0 (1% aq.) (Lankrosol SXS-30)
7.0–9.0 (Naxonate 4L, 5L; Stepanate X)
7.0–10.0 (Eltesol SX30; Manro SXS40; Surco SXS)
7.5–10.5 (Carsosulf SXS Liquid)
8.0–9.5 (Richonate SXS)
8.0–10.0 (Manro SXS30)
8.5 (Lakeway SXS)
8.6 (Witconate SXS Liquid)
9.0–10.5 (Eltesol SXS93, SX Pellets)
9.6 (Witconate SXS Powder)
TOXICITY/HANDLING:
Skin irritant on prolonged contact with conc. form (Lankrosol SXS-30; Witconate SXS

Sodium xylenesulfonate (cont'd.)

Liquid, SXS Powder)
Protective gloves and goggles should be worn (Lankrosol SXS-30)
Use normal safety precautions (Witconate SXS Liquid, SXS Powder)
Avoid contact with skin and eyes (Hartotrope SXS40)

STORAGE/HANDLING:

Storage and transfer in stainless steel is recommended; store above 25 C to avoid crystallization; if crystallization occurs, heating will reliquefy (Witconate SXS Liquid, SXS Powder)
Stainless steel, glass, resin, or polyethylene-lined equipment (Lankrosol SXS-30)

STD. PKGS.:

Drums, T/T (Hartotrope SXS40)
Tankwagons (Naxonate 5L)
200 Leverpak (Naxonate G)
45-gal drums or road tankers (Manro SXS30, SXS40)
55-gal lined drums or tank wagons (Naxonate 4L)
550-lb bung-type drums (Alkatrope SX-40)

Sorbitan monoisostearate

SYNONYMS:

1,4-Anhydro-D-glucitol, 6-isooctadecanoate
Anhydrosorbitol monoisostearate
D-Glucitol, 1,4-anhydro-, 6-isooctadecanoate
Sorbitan isostearate (CTFA)
Sorbitan, monoisooctadecanoate

EMPIRICAL FORMULA:

$C_{24}H_{46}O_6$

TRADENAME EQUIVALENTS:

Arlacel 987 [ICI United States]
Crill 6 [Croda]
Montane 70 [Seppic]
Nikkol SI-10R, SI-10T, SI-15R, SI-15T [Nikko]
S-Maz 67 [Mazer]

CATEGORY:

Dispersant, emulsifier, wetting agent, surfactant, antistat, softener, lubricant, defoamer, opacifier, coemulsifier

APPLICATIONS:

Cosmetic industry preparations: (Nikkol SI-10R, SI-10T, SI-15R, SI-15T; S-Maz 67); creams and lotions (Arlacel 987)
Food applications: (S-Maz 67)
Household products: (S-Maz 67)

Industrial applications: dyes and pigments (Crill 6); industrial oils (S-Maz 67); textile/leather processing (S-Maz 67)
Pharmaceutical applications: (Nikkol SI-10R, SI-10T, SI-15R, SI-15T)

PROPERTIES:

Form:
Liquid (Arlacel 987; Montane 70; Nikkol SI-10R, SI-10T, SI-15R, SI-15T; S-Maz 67)
Viscous liquid (Crill 6)

Color:
Pale yellow (Crill 6; S-Maz 67)

Composition:
100% conc. (Arlacel 987; Montane 70; Nikkol SI-10R, SI-10T, SI-15R, SI-15T; S-Maz 67)

Solubility:
Miscible in certain proportions with ethanol (S-Maz 67)
Partly sol. in isopropyl myristate (Crill 6)
Sol. in min. oil (Crill 6; S-Maz 67)
Miscible in certain proportions with naphtha (S-Maz 67)
Partly sol. in oleyl alcohol (Crill 6)
Sol. in olive oil (Crill 6)
Miscible in certain proportions with toluol (S-Maz 67)
Sol. in veg. oil (S-Maz 67)
Disp. in water (S-Maz 67)

Ionic Nature:
Nonionic (Montane 70; Nikkol SI-10R, SI-10T, SI-15R, SI-15T; S-Maz 67)

Sp.gr.:
1.0 (S-Maz 67)

Visc.:
1000 cps (S-Maz 67)

HLB:
4.5 (Nikkol SI-15R, SI-15T)
4.7 (Crill 6; S-Maz 67)
5.0 (Nikkol SI-10R, SI-10T)

Acid No.:
10 max. (S-Maz 67)

Saponification No.:
143–153 (Crill 6)
146–157 (S-Maz 67)

Hydroxyl No.:
230–255 (S-Maz 67)

Sorbitan monopalmitate

SYNONYMS:

1,4-Anhydro-D-glucitol, 6-hexadecanoate
D-Glucitol, 1,4-anhydro-, 6-hexadecanoate
Sorbitan palmitate (CTFA)

EMPIRICAL FORMULA:

$C_{22}H_{42}O_6$

STRUCTURE:

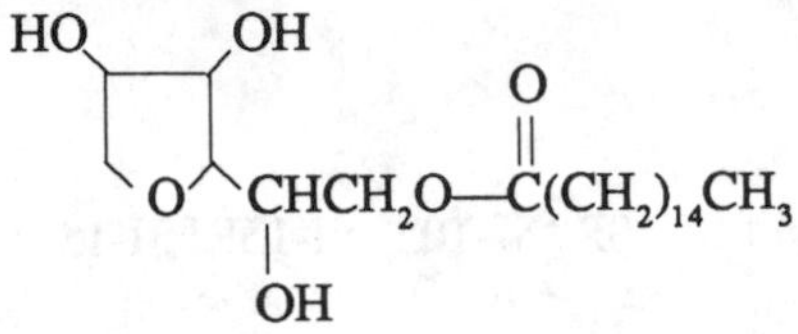

HO OH O
O CHCH$_2$O—C(CH$_2$)$_{14}$CH$_3$
OH

CAS No.:

26266-57-9

TRADENAME EQUIVALENTS:

Ahco FP-67 [ICI United States]
Arlacel 40 [ICI United States]
Armotan MP [Akzo]
Crill 2 [Croda]
Emasol P-10 [Kao]
Emsorb 2510 [Emery]
Glycomul P [Glyco]
Hodag SMP [Hodag]
Kuplur SMP [BASF Wyandotte]
Liposorb P [Lipo]
Lonzest SMP [Lonza]
Montane 40 [Seppic]
Newcol 40 [Nippon Nyukazai]
Nikkol SP-10 [Nikko]
Nissan Nonionic PP-40R [Nippon Oil & Fats]
Radiamuls 135, SORB 2135 [Oleofina SA]
Radiasurf 7135 [Oleofina SA]
S-Maz 40 [Mazer]
Sorbax SMP [Chemax]
Sorbon S-40 [Toho]
Span 40 [ICI United States]

CATEGORY:

Solubilizer, emulsifier, coemulsifier, dispersant, lubricant, thickener, antistat, corrosion inhibitor, softener, defoamer, opacifier, stabilizer, surfactant, cosolvent, wetting agent, viscosity reducer, mold release agent, slip agent, antiblock agent, superfatting aid, antifog aid, detergent

APPLICATIONS:

Cosmetic industry preparations: (Crill 2; Emsorb 2510; Glycomul P; Nissan Nonionic PP-40R; Radiasurf 7135; S-Maz 40; Sorbax SMP; Span 40)

Farm products: herbicides (Crill 2); insecticides/pesticides (Crill 2; Radiasurf 7135)

Food applications: (Crill 2; Lonzest SMP; Nissan Nonionic PP-40R; S-Maz 40; Span 40); food emulsifying (Glycomul P; Nikkol SP-10; Radiamuls 135, SORB 2135); food packaging (Crill 2)

Household detergents: (S-Maz 40; Radiasurf 7135; Span 40)

Industrial applications: (Glycomul P; S-Maz 40); construction (Radiasurf 7135); dyes and pigments (Emasol P-10; Radiasurf 7135); industrial processing (Hodag SMP; Newcol 40; Radiasurf 7135; S-Maz 40); lubricating/cutting oils (Emsorb 2510; Radiasurf 7135); metalworking (Crill 2); paint mfg. (Crill 2); paper mfg. (Radiasurf 7135); petroleum industry (Crill 2); plastics (Crill 2; Radiasurf 7135); polishes and waxes (Crill 2); printing inks (Crill 2); textile/leather processing (Ahco FP-67; Crill 2; Lonzest SMP; Nissan Nonionic PP-40R; Radiasurf 7135; S-Maz 40; Sorbax SMP; Span 40)

Industrial cleaners: drycleaning compositions (Radiasurf 7135); metal processing surfactants (Crill 2; Radiasurf 7135)

Pharmaceutical applications: (Crill 2; Glycomul P; Nissan Nonionic PP-40R; Radiasurf 7135; Span 40)

PROPERTIES:

Form:

Solid (Ahco FP-67; Emasol P-10; Emsorb 2510; Hodag SMP; Lonzest SMP; Montane 40; Newcol 40; Nikkol SP-10; Radiasurf 7135; Sorbax SMP; Sorbon S-40; Span 40)

Beads (Arlacel 40; Glycomul P; Liposorb P)

Flake (Liposorb P; Radiamuls 135, SORB 2135; S-Maz 40)

Powder (Radiamuls 135, SORB 2135)

Wax (Kuplur SMP)

Waxy solid (Armotan MP; Nissan Nonionic PP-40R)

Hard waxy solid (Crill 2)

Color:

White (Radiasurf 7135)

Off-white (Kuplur SMP)

Cream (Arlacel 40; Glycomul P)

Pale tan (Crill 2)

Tan (Liposorb P; S-Maz 40; Span 40)

Gardner 7 max. (Nissan Nonionic PP-40R)

Gardner 8 (Emsorb 2510)

Composition:

98.5% active (Kuplur SMP)

100% active (Arlacel 40; Liposorb P; Span 40)

100% conc. (Ahco FP-67; Armotan MP; Emasol P-10; Emsorb 2510; Glycomul P; Hodag SMP; Lonzest SMP; Montane 40; Newcol 40; Nikkol SP-10; Nissan

Sorbitan monopalmitate (*cont'd.*)

Nonionic PP-40R; S-Maz 40; Sorbon S-40)

Solubility:

Partly sol. hot in acetone (Glycomul P; S-Maz 40)
Disp. in butyl stearate (Emsorb 2510)
Partly sol. in ethanol (Crill 2); partly sol. hot (Glycomul P; S-Maz 40)
Sol. hot in ethyl acetate (Glycomul P)
Sol. in isopropanol (Arlacel 40); (@ 1%) (Span 40); cloudy (@ 10%) (Radiasurf 7135)
Sol. cloudy in min. oil (@ 10%) (Radiasurf 7135); partly sol. hot (Glycomul P; S-Maz 40)
Miscible hot with naphtha in certain proportions (S-Maz 40)
Partly sol. in oleic acid (Crill 2)
Partly sol. in oleyl alcohol (Crill 2)
Partly sol. in olive oil (Crill 2)
Sol. hazy in perchloroethylene (@ 1%) (Span 40)
Partly sol. in propylene glycol (Crill 2)
Sol. in Stoddard solvent (Emsorb 2510)
Partly sol. hot in toluol (Glycomul P; S-Maz 40)
Sol. cloudy in trichlorethylene (@ 10%) (Radiasurf 7135)
Sol. in veg. oil (Glycomul P); partly sol. hot (S-Maz 40)
Disp. in water (Kuplur SMP); (@ 10%) (Radiasurf 7135)
Sol. in xylene (@ 1%) (Span 40)

Ionic Nature:

Nonionic (Ahco FP-67; Arlacel 40; Armotan MP; Crill 2; Emasol P-10; Emsorb 2510; Glycomul P; Kuplur SMP; Liposorb P; Lonzest SMP; Montane 40; Nikkol SP-10; Nissan Nonionic PP-40R; S-Maz 40; Sorbon S-40)

Sp.gr.:

0.943 (98.9 C) (Radiasurf 7135)
≈ 1.0 (Arlacel 40)
1.01 (Kuplur SMP)

Visc.:

51.70 cps (98.9 C) (Radiasurf 7135)

M.P.:

45–46 C (S-Maz 40)
46 C (Crill 2)
48 C (Emsorb 2510)
≈ 50 C (Radiamuls 135, SORB 2135)
54.5 C (Radiasurf 7135)

Pour Pt.:

48 C (Span 40)

Solidification Pt.:

45–51 C (Nissan Nonionic PP-40R)

Flash Pt.:

214 C (COC) (Radiasurf 7135)

> 300 F (Arlacel 40)

Fire Pt.:

> 300 F (Arlacel 40)

HLB:

≈ 5.3 (Radiamuls 135)
5.3 avg. (Radiasurf 7135)
6.3 (Radiamuls SORB 2135)
6.6 (Emsorb 2510)
6.7 (Ahco FP-67; Arlacel 40; Crill 2; Emasol P-10; Glycomul P; Hodag SMP; Kuplur SMP; Montane 40; S-Maz 40; Nikkol SP-10; Nissan Nonionic PP-40R; Sorbax SMP; Span 40)
6.7 ± 1 (Liposorb P)

Acid No.:

7.0 max. (Nissan Nonionic PP-40R; Radiamuls 135; Radiasurf 7135)
7.5 max. (Glycomul P; Kuplur SMP; S-Maz 40)

Iodine No.:

1.0 max. (Radiamuls 135; Radiasurf 7135)

Saponification No.:

139–150 (Glycomul P)
139–151 (Liposorb P)
140–150 (Crill 2; Kuplur SMP; S-Maz 40)
140–160 (Radiasurf 7135)

Hydroxyl No.:

272–306 (Glycomul P; Liposorb P)
275–305 (Kuplur SMP; S-Maz 40)

STD. PKGS.:

25-kg net paper bags (Radiamuls SORB 2135)
25-kg net multiply paper bags or bulk (Radiasurf 7135)
190-kg net bung drums (Radiamuls 135)
55-gal (450 lb net) steel drums (Kuplur SMP)

Sorbitan monostearate

SYNONYMS:

1,4-Anhydro-D-glucitol, 6-octadecanoate
Anhydrosorbitol monostearate
D-Glucitol, 1,4-anhydro-, 6-octadecanoate
Sorbitan, monooctadecanoate
Sorbitan stearate (CTFA)

EMPIRICAL FORMULA:

$C_{24}H_{46}O_6$

Sorbitan monostearate (cont'd.)

STRUCTURE:

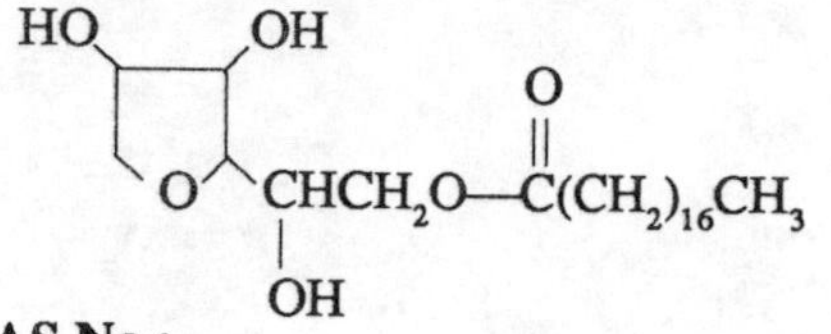

HO OH

O

O $CHCH_2O—C(CH_2)_{16}CH_3$

OH

CAS No.:

1338-41-6

TRADENAME EQUIVALENTS:

Ahco 909 [ICI United States]
Alkamuls SMS [Alkaril]
Arlacel 60 [ICI United States]
Armotan MS [Akzo Chemie Italia]
Crill 3 [Croda]
Drewmulse SMS [PVO Int'l.]
Drewsorb 60 [PVO Int'l.]
Durtan 60, 60K [Durkee Ind. Foods]
Emasol S-10 [Kao]
Emasol Super S-10F [Kao]
Emsorb 2505 [Emery]
Emultex SMS [La Tessilchimica]
Glycomul S [Glyco]
Grindtek SMS [Grinsted]
Hodag SMS [Hodag]
Ionet S-60C [Sanyo]
Kuplur SMS [BASF Wyandotte]
Liposorb S [Lipo]
Lonzest SMS [Lonza]
Montane 70 [Seppic]
Nikkol SS-10 [Nikko]
Nissan Nonion SP-60•R [Nippon Oil & Fats]
Radiamuls 145, SORB 2145 [Oleofina S.A.]
Radiasurf 7145 [Oleofina S.A.]
S-Maz 60, 60K [Mazer]
Soprofor S/60 [Geronazzo SpA]
Sorbax SMS [Chemax]
Sorbon S-60 [Toho]
Sorgen 50 [Dai-ichi Kogyo Seiyaku Co.]
Span 60 [ICI United States]

CATEGORY:

Emulsifier, solubilizer, dispersant, wetting agent, detergent, viscosity control agent, germicide, lubricant, surfactant, coupling agent, antistat, emulsion stabilizer, thickener, plasticizer, antifoaming agent, softener, anticorrosive agent, descouring aid, superfatting and bodying aid, antifog aid, opacifier

APPLICATIONS:

Cosmetic industry preparations: (Armotan MS; Crill 3; Drewmulse SMS; Emsorb 2505; Emultex SMS; Glycomul S; Ionet S-60C; Nissan Nonion SP-60•R; Radiasurf 7145; S-Maz 60; Sorbax SMS; Sorgen 50; Span 60); colorants (Emasol S-10); creams and lotions (Crill 3; Drewmulse SMS; Durtan 60; Ionet S-60C); ointments (Ionet S-60C); perfumery (Drewmulse SMS); shampoos (Drewmulse SMS); shaving preparations (Drewmulse SMS)

Farm products: herbicides (Crill 3); insecticides/pesticides (Crill 3; Ionet S-60C; Radiasurf 7145)

Food applications: (Crill 3; Lonzest SMS; Nissan Nonion SP-60•R; S-Maz 60; Sorgen 50; Span 60); dairy products (Radiamuls 145, SORB 2145); flavors (Drewmulse SMS; Radiamuls SORB 2145); food emulsifying (Drewsorb 60; Durtan 60, 60K; Emasol Super S-10F; Glycomul S; Radiamuls 145, SORB 2145)

Household products: (Emsorb 2505; S-Maz 60); cleaners (Span 60); detergent additive (Radiasurf 7145)

Industrial applications: (Alkamuls SMS; Glycomul S); concrete (Radiasurf 7145); corrosion inhibitors (Hodag SMS; Ionet S-60C; Lonzest SMS; Radiasurf 7145); dyes and pigments (Emasol S-10; Ionet S-60C; Radiasurf 7145; Span 60); industrial processing (Hodag SMS; S-Maz 60); lubricating/cutting oils (Radiasurf 7145; S-Maz 60); metalworking (Ionet S-60C); paper mfg. (Alkamuls SMS; Emsorb 2505; Radiasurf 7145); petroleum industry (Radiasurf 7145); plastics (Alkamuls SMS; Durtan 60; Radiasurf 7145; Span 60); polishes, waxes, oils (Crill 3; Emsorb 2505); silicone products (Alkamuls SMS; Crill 3); textile/leather processing (Ahco 909; Alkamuls SMS; Crill 3; Emsorb 2505; Emultex SMS; Ionet S-60C; Lonzest SMS; Nissan Nonion SP-60•R; Radiasurf 7145; S-Maz 60; Soprofor S/60; Sorbax SMS; Span 60)

Industrial cleaners: drycleaning compositions (Radiasurf 7145); metal processing surfactants (Crill 3; Radiasurf 7145)

Pharmaceutical applications: (Armotan MS; Crill 3; Drewmulse SMS; Glycomul S; Nissan Nonion SP-60•R; Radiasurf 7145; Sorgen 50); topical ointments (Drewmulse SMS); vitamins (Drewmulse SMS; Radiamuls 145, SORB 2145)

PROPERTIES:

Form:

Liquid (Montane 70)

Solid (Ahco 909; Arlacel 60; Drewmulse SMS; Emasol S-10; Emsorb 2505; Hodag SMS; Ionet S-60C; Radiasurf 7145; S-Maz 60K; Soprofor S/60; Sorbax SMS; Sorbon S-60; Span 60)

Beads (Drewsorb 60; Durtan 60, 60K; Glycomul S; Liposorb S)

Flake (Alkamuls SMS; Emultex SMS; Liposorb S; Lonzest SMS; Nikkol SS-10; Radiamuls 145, SORB 2145; S-Maz 60; Sorgen 50)

Needle-like (Armotan MS)

Powder (Emasol Super S-10F; Grindtek SMS; Radiamuls 145, SORB 2145)

Sorbitan monostearate *(cont'd.)*

Waxy solid (Kuplur SMS; Nissan Nonion SP-60•R)
Hard waxy solid (Crill 3)

Color:

White (Radiasurf 7145)
Cream (Alkamuls SMS; Armotan MS; Drewsorb 60; Durtan 60K; Glycomul S; Kuplur SMS; Liposorb S; S-Maz 60)
Pale tan (Crill 3)
Tan (Grindtek SMS; Span 60)
Gardner 4 (Emsorb 2505)
Gardner 5 max. (Nissan Nonion SP-60•R)

Odor:

Low (Crill 3)

Composition:

98.5% conc. min. (Kuplur SMS)
> 99% active (Armotan MS)
100% active (Ahco 909; Arlacel 60; Durtan 60K; Emasol S-10, Super S-10F; Emultex SMS; Glycomul S; Hodag SMS; Ionet S-60C; Liposorb S; Lonzest SMS; Montane 70; Nikkol SS-10; Nissan Nonion SP-60•R; S-Maz 60; Soprofor S/60; Sorbon S-60; Sorgen 50; Span 60)

Solubility:

Sol. in aromatic solvent (@ 10%) (Alkamuls SMS)
Disp. in butyl stearate (@ 5%) (Emsorb 2505)
Sol. warm in ethanol (Nissan Nonion SP-60•R); partly sol. warm (Grindtek SMS); partly sol. hot (S-Maz 60); poorly sol. hot (Glycomul S)
Poorly sol. hot in ethyl acetate (Glycomul S)
Slightly sol. in ethyl ether (Nissan Nonion SP-60•R)
Disp. in glycerol trioleate (@ 5%) (Emsorb 2505)
Sol. in isopropanol (@ 1%) (Span 60)
Sol. warm in kerosene (Nissan Nonion SP-60•R)
Sol. warm in methanol (Nissan Nonion SP-60•R)
Partly sol. hot in min. oil (S-Maz 60); poorly sol. hot (Glycomul S); insol. (@ 5%) (Emsorb 2505)
Sol. in oils (S-Maz 60, 60K)
Partly sol. in oleic acid (Crill 3)
Partly sol. in oleyl alcohol (Crill 3)
Partly sol. in olive oil (Crill 3)
Partly sol. warm in paraffin oil (Grindtek SMS)
Sol. in perchloroethylene (@ 10%) (Alkamuls SMS); sol. hazy (@ 1%) (Span 60)
Sol. in solvents (S-Maz 60, 60K)
Insol. in Stoddard solvent (@ 5%) (Emsorb 2505)
Sol. warm in tetrachloromethan (Nissan Nonion SP-60•R)
Sol. warm in toluene (Grindtek SMS)
Partly sol. hot in toluol (S-Maz 60); poorly sol. hot Glycomul S)

Sol. cloudy in trichlorethylene (@ 10%) (Radiasurf 7145)
Sol. in veg. oil (Glycomul S); partly sol. hot (S-Maz 60)
Disp. in water (Drewsorb 60; S-Maz 60, 60K); disp. (@ 10%) (Radiasurf 7145); disp. warm (Nissan Nonion SP-60•R); disp. @ 50 C (Kuplur SMS); insol. (@ 5%) (Emsorb 2505)
Partly sol. warm in white spirit (Grindtek SMS)
Sol. warm in xylene (Nissan Nonion SP-60•R); sol. hazy (@ 1%) (Span 60); insol. (@ 5%) (Emsorb 2505)

Ionic Nature:
Nonionic (Ahco 909; Arlacel 60; Drewmulse SMS; Durtan 60, 60K; Emasol S-10, Super S-10F; Emultex SMS; Glycomul S; Hodag SMS; Ionet S-60C; Kuplur SMS; Liposorb S; Lonzest SMS; Montane 70; Nikkol SS-10; Nissan Nonion SP-60•R; Radiamuls 145; Radiasurf 7145; S-Maz 60; Soprofor S/60; Sorbon S-60; Sorgen 50; Span 60)

M.W.:
510 avg. (Radiasurf 7145)

Sp.gr.:
0.943 (98.9 C) (Radiasurf 7145)
1.01 (Kuplur SMS)

Density:
1.0 g/ml (Alkamuls SMS)

Visc.:
48.20 cps (98.9 C) (Radiasurf 7145)

M.P.:
50 C (Emsorb 2505)
50–53 C (S-Maz 60)
50–60 C (Radiamuls 145, SORB 2145)
51–54 C (Crill 3)
51–58 C (Armotan MS)
58 C (Radiasurf 7145)
121–127 F (Durtan 60K)

Pour Pt.:
53 C (Span 60)

Solidification Pt.:
49–55 C (Nissan Nonion SP-60•R)

Flash Pt.:
232 C (COC) (Radiasurf 7145)
480 F (Emsorb 2505)

HLB:
4.5 (Emultex SMS)
4.7 (Ahco 909; Alkamuls SMS; Arlacel 60; Crill 3; Drewmulse SMS; Drewsorb 60; Durtan 60, 60K; Emasol S-10, Super S-10F; Glycomul S; Hodag SMS; Ionet S-60C; Kuplur SMS; Nikkol SS-10; Nissan Nonion SP-60•R; S-Maz 60; Sorbax SMS;

Sorbitan monostearate (cont'd.)

Sorgen 50)
4.7 ± 1 (Liposorb S)
5.0 (Radiamuls 145, SORB 2145; Radiasurf 7145)
5.2 (Emsorb 2505; Grindtek SMS)

Acid No.:
7 max. (Nissan Nonion SP-60•R; Radiamuls 145; Radiasurf 7145)
10 max. (Glycomul S; Kuplur SMS; S-Maz 60)

Iodine No.:
1 max. (Radiasurf 7145)
60–75 (Radiamuls 145)

Saponification No.:
146–158 (Glycomul S; Radiasurf 7145)
147–157 (Alkamuls SMS; Crill 3; Drewmulse SMS; Drewsorb 60; Durtan 60; Kuplur SMS; Liposorb S; S-Maz 60)

Hydroxyl No.:
235–260 (Alkamuls SMS; Durtan 60; Kuplur SMS; Liposorb S; S-Maz 60)
236–260 (Glycomul S)

STD. PKGS.:
25-kg net multiply paper bags (Radiamuls 145, SORB 2145)
25-kg net multiply paper bags or bulk (Radiasurf 7145)
55-gal (200 lb net) fiber drums (Kuplur SMS)

Soya amine

SYNONYMS:
Amines, soya alkyl
Soyamine (CTFA)

STRUCTURE:
RNH_2
where R represents the soya radical

CAS No.:
61790-18-9

TRADENAME EQUIVALENTS:
Armeen SD [Armak]
Kemamine P-997 [Humko Sheffield] (tech.)
Kemamine P-997D [Humko Sheffield] (distilled)
Lilamin 115 [Lilachim S.A.]
Lilamin 115D [Lilachim S.A.] (distilled)

CATEGORY:
Dispersant, emulsifier, flotation agent, intermediate, corrosion inhibitor, acid scavenger, release agent, lubricant, bactericidal

APPLICATIONS:

Industrial applications: (Armeen SD; Kemamine P-997, P-997D); dyes and pigments (Lilamin 115, 115D); fuel oil additives (Kemamine P-997, P-997D; Lilamin 115, 115D); petroleum industry (Lilamin 115, 115D); plastics (Kemamine P-997, P-997D); printing inks (Lilamin 115, 115D); road building (Lilamin 115, 115D); rubber (Kemamine P-997, P-997D; Lilamin 115, 115D)

PROPERTIES:

Form:

Liquid (Lilamin 115)
Paste (Armeen SD; Kemamine P-997, P-997D)

Color:

Gardner 2 max. (Kemamine P-997D)
Gardner 3 (Armeen SD)
Gardner 3 max. (Kemamine P-997)
Gardner 5 max. (Lilamin 115)

Composition:

93% conc. (Kemamine P-997)
95% active (Lilamin 115)
97% conc. (Kemamine P-997D)

Solubility:

Sol. in acetone (Armeen SD)
Sol. in carbon tetrachloride (Armeen SD)
Sol. in chloroform (Armeen SD)
Sol. in ethanol (Armeen SD)
Sol. in isopropanol (Armeen SD)
Sol. in methanol (Armeen SD)
Sol. in common organic solvents (Kemamine P-997, P-997D)
Sol. in toluene (Armeen SD)

Ionic Nature:

Cationic (Kemamine P-997, P-997D)

M.W.:

270 (Lilamin 115)

Sp.gr.:

0.81 (38/4 C) (Armeen SD)

Visc.:

46.2 SSU (35 C) (Armeen SD)

M.P.:

81–86 F (Armeen SD)

Pour Pt.:

70 F (Armeen SD)

Flash Pt.:

305 F (Armeen SD)

Soya amine (cont'd.)

Fire Pt.:
345 F (Armeen SD)
Cloud Pt.:
85 F (Armeen SD)
Iodine No.:
90 min. (Kemamine P-997, P-997D; Lilamin 115)

Stearyl amine

SYNONYMS:
1-Octadecanamine
Octadecylamine
n-Octadecylamine
Octadecylamine, primary
Stearamine (CTFA)
Stearyl primary amine
EMPIRICAL FORMULA:
$C_{18}H_{39}N$
STRUCTURE:
$CH_3(CH_2)_{16}CH_2NH_2$
CAS No.:
124-30-1
TRADENAME EQUIVALENTS:
Adogen 142 (D) [Sherex]
Amine 18D [Kenobel]
Armeen 18, 18D [Armak]
Crodamine 1.18D [Croda Universal]
Kemamine P-990 [Humko Sheffield] (tech.)
Kemamine P-990D [Humko Sheffield] (distilled)
Lilamin 142 [Lilachim S.A.]
Lilamin 142S [Lilachim S.A.] (distilled)
Nissan Amine AB [Nippon Oil & Fats]
CATEGORY:
Corrosion inhibitor, anticaking agent, dispersant, emulsifier, flotation agent, intermediate, acid scavenger, mold release agent, lubricant, bactericidal, germicide, wetting agent, softener, flushing aid
APPLICATIONS:
Farm products: (Nissan Amine AB)
Industrial applications: (Kemamine P-990, P-990D); asphalt (Nissan Amine AB); ceramics (Nissan Amine AB); concrete (Nissan Amine AB); dyes and pigments (Lilamin 142, 142D; Nissan Amine AB); grease additive (Nissan Amine AB); ore

flotation (Nissan Amine AB); organic synthesis (Nissan Amine AB); petroleum industry (Adogen 142 (D); Lilamin 142, 142D); Kemamine P-990, P-990D); plastics (Kemamine P-990, P-990D); printing inks (Lilamin 142, 142D); road building (Lilamin 142, 142D); rubber (Kemamine P-990, P-990D; Lilamin 142, 142D; Nissan Amine AB); textile/leather processing (Nissan Amine AB); water treatment (Nissan Amine AB)

PROPERTIES:

Form:

Solid (Armeen 18, 18D; Crodamine 1.18D; Kemamine P-990, P-990D; Lilamin 142)
Flake (Nissan Amine AB)
Waxy solid (Nissan Amine AB)

Color:

White (Nissan Amine AB)
Gardner 1 max. (Kemamine P-990D)
Gardner 3 max. (Kemamine P-990; Lilamin 142)

Composition:

93% conc. (Kemamine P-990)
95% active (Lilamin 142)
97% conc. (Kemamine P-990D)
98% conc. (Amine 18D)
98% primary amine min. (Nissan Amine AB)
100% conc. (Adogen 142 (D); Crodamine 1.18D)

Solubility:

Slightly sol. in acetone (Armeen 18)
Sol. in carbon tetrachloride (Armeen 18, 18D)
Sol. in chloroform (Armeen 18, 18D)
Sol. in ethanol (Armeen 18, 18D)
Sol. in isopropanol (Armeen 18, 18D)
Slightly sol. in kerosene (Armeen 18, 18D)
Slightly sol. in methanol (Armeen 18D)
Sol. in common organic solvents (Kemamine P-990, P-990D)
Sol. in toluene (Armeen 18, 18D)

Ionic Nature:

Cationic (Amine 18D; Crodamine 1.18D; Kemamine P-990, P-990D)

M.W.:

273 (Lilamin 142)

Sp.gr.:

0.791 (60/4 C) (Armeen 18D)
0.792 (60/4 C) (Armeen 18)

Visc.:

43.7 SSU (45 C) (Armeen 18D)
45.6 SSU (55 C) (Armeen 18)

Stearyl amine (cont'd.)

M.P.:
122–133 F (Armeen 18, 18D)
Pour Pt.:
110 F (Armeen 18D)
115 F (Armeen 18)
Solidification Pt.:
47–53 C (Nissan Amine AB)
Flash Pt.:
300 F (Armeen 18D)
320 F (Armeen 18)
Fire Pt.:
350 F (Armeen 18D)
365 F (Armeen 18)
Iodine No.:
2.0 max. (Kemamine P-990, P-990D; Nissan Amine AB)
3.0 max. (Lilamin 142)
STD. PKGS.:
15-kg paper bag, 13-kg can, 160-kg drum (Nissan Amine AB)

Sucrose distearate (CTFA)

SYNONYMS:
α-D-Glucopyranoside, β-D-fructofuranosyl, dioctadecanoate
EMPIRICAL FORMULA:
$C_{48}H_{90}O_{13}$
CAS No.:
27195-16-0
TRADENAME EQUIVALENTS:
Crodesta F-10, F-50 [Croda]
Ryoto Sugar Ester S-570, S-770 [Ryoto Co.]
CATEGORY:
Dispersant, wetting agent, emulsifier
APPLICATIONS:
Cosmetic industry preparations: (Crodesta F-10, F-50); toiletries (Crodesta F-10, F-50)
Food applications: food emulsifying (Ryoto Sugar Ester S-570, S-770)
Pharmaceutical applications: (Crodesta F-10, F-50)
PROPERTIES:
Form:
Creamy powder (Crodesta F-10)
Powder (Crodesta F-50; Ryoto Sugar Ester S-570, S-770)

Color:
White (Crodesta F-10, F-50)
Composition:
100% active, 3% monoester (Crodesta F-10)
100% active, 29% monoester (Crodesta F-50)
100% conc. (Ryoto Sugar Ester S-570, S-770)
Ionic Nature:
Nonionic (Crodesta F-10, F-50; Ryoto Sugar Ester S-570, S-770)
M.P.:
60–68 C (Crodesta F-10)
74–78 C (Crodesta F-50)
HLB:
> 3.0 (Crodesta F-10)
5.0 (Ryoto Sugar Ester S-570)
6.5 (Crodesta F-50)
7.0 (Ryoto Sugar Ester S-770)
Acid No.:
5.0 max. (Crodesta F-10, F-50)
Iodine No.:
1.0 max. (Crodesta F-10, F-50)
Saponification No.:
93–153 (Crodesta F-50)
140–200 (Crodesta F-10)
Hydroxyl No.:
80–130 (Crodesta F-10)
419–469 (Crodesta F-50)

Sucrose monococoate

SYNONYMS:
Sucrose cocoate (CTFA)
TRADENAME EQUIVALENTS:
Crodesta SL-40 [Croda]
CATEGORY:
Dispreant, wetting agent, emulsifier
APPLICATIONS:
Cleansers: skin cleansers (Crodesta SL-40)
Cosmetic industry preparations: (Crodesta SL-40); shampoos (Crodesta SL-40); toiletries (Crodesta SL-40)
Pharmaceutical applications: (Crodesta SL-40)

Sucrose monococoate (cont'd.)

PROPERTIES:
Form:
Liquid (Crodesta SL-40)
Color:
Amber (Crodesta SL-40)
Composition:
40% conc., 55% monoester (Crodesta SL-40)
Solubility:
Sol. in glycols (Crodesta SL-40)
Sol. in water (Crodesta SL-40)
Ionic Nature:
Nonionic (Crodesta SL-40)
HLB:
15.0 (Crodesta SL-40)
Acid No.:
5.0 max. (Crodesta SL-40)
Iodine No.:
1.0 max. (Crodesta SL-40)

Sucrose monostearate

SYNONYMS:
β-D-Fructofuranosyl-α-D-glucopyranoside, monooctadecanoate
α-D-Glucopyranoside, β-D-fructofuranosyl, monooctadecanoate
Sucrose stearate (CTFA)
EMPIRICAL FORMULA:
$C_{30}H_{56}O_{12}$
CAS No.:
25168-73-4
TRADENAME EQUIVALENTS:
Crodesta F110, F160 [Croda]
Ryoto Sugar Ester S-1570, S-1670 [Ryoto Co. Ltd.]
CATEGORY:
Dispersant, wetting agent, emulsifier, thickening agent, suspending agent
APPLICATIONS:
Cosmetic industry preparations: (Crodesta F110, F160); toiletries (Crodesta F110, F160)
Food applications: food emulsifying (Ryoto Sugar Ester S-1570, S-1670)
Pharmaceutical applications: (Crodesta F110, F160)

PROPERTIES:

Form:

Powder (Crodesta F110, F160; Ryoto Sugar Ester S-1570, S-1670)

Color:

White (Crodesta F110, F160)

Composition:

100% active, 52% monoester (Crodesta F110)
100% active, 75% monoester (Crodesta F160)
100% conc. (Ryoto Sugar Ester S-1570, S-1670)

Solubility:

Sol. in glycols (Crodesta F160)
Sol. in water (Crodesta F160)

Ionic Nature:

Nonionic (Crodesta F110, F160; Ryoto Sugar Ester S-1570, S-1670)

HLB:

12.0 (Crodesta F110)
14.5 (Crodesta F160)
15.0 (Ryoto Sugar Ester S-1570)
16.0 (Ryoto Sugar Ester S-1670)

Acid No.:

5.0 max. (Crodesta F110, F160)

Iodine No.:

1.0 max. (Crodesta F110, F160)

Saponification No.:

75–153 (Crodesta F160)
85–145 (Crodesta F110)

Hydroxyl No.:

475–525 (Crodesta F110)
545–595 (Crodesta F160)

Tallowamidopropyl hydroxysultaine (CTFA)

SYNONYMS:

Quaternary ammonium compounds, (3-tallowamidopropyl) (2-hydroxy-3-sulfopropyl) dimethyl, hydroxide, inner salt

(3-Tallowamidopropyl) (2-hydroxy-3-sulfopropyl) dimethyl quaternary ammonium compounds, hydroxide, inner salt

STRUCTURE:

$$RC(=O)—NH—(CH_2)_3—N^+(CH_3)_2—CH_2CH(OH)CH_2SO_3^-$$

where RCO^- represents the tallow acid radical

TRADENAME EQUIVALENTS:

Mirataine TABS [Miranol]

CATEGORY:

Dispersant

APPLICATIONS:

Cleansers: liquid hand soap (Mirataine TABS)

Household detergents: soap and detergent bars (Mirataine TABS)

Industrial cleaners: (Mirataine TABS); lime soap dispersant (Mirataine TABS)

PROPERTIES:

Form:

Clear to slightly hazy gel; thins to a pumpable liquid at 75 C (Mirataine TABS)

Composition:

41% solids (Mirataine TABS)

Ionic Nature:

Amphoteric (Mirataine TABS)

pH:

8.3 (Mirataine TABS)

Biodegradable: (Mirataine TABS)

TOXICITY/HANDLING:

Severe primary skin irritant; may cause mild conjunctival irritation (Mirataine TABS)

Tallow amine (CTFA)

SYNONYMS:

Amines, tallow alkyl
Tallow primary amine

STRUCTURE:

RNH_2
where R represents the tallow fatty radicals

CAS No.:

61790-33-8

TRADENAME EQUIVALENTS:

Adogen 170 (D) [Sherex]
Amine BG [Kenobel]
Amine BGD [Kenobel] (distilled)
Arosurf MG-170 [Sherex]
Crodamine 1.T [Croda]
Kemamine P-974 [Humko Sheffield] (tech.)
Kemamine P-974D [Humko Sheffield] (distilled)
Lilamin 170 [Lilachim S.A.]
Lilamin 170D [Lilachim S.A.] (distilled)
Radiamine 6170, 6171 [Oleofina]
Tomah Tallow Amine [Tomah] (tech.)

CATEGORY:

Dispersant, emulsifier, corrosion inhibitor, flotation agent, intermediate, acid scavenger, mold release agent, lubricant, bactericidal, antistripping aid, flushing agent

APPLICATIONS:

Cosmetic industry preparations: (Radiamine 6170, 6171); beauty products (Lilamin 170, 170D); creams (Lilamin 170, 170 D)

Farm applications: herbicides (Crodamine 1.T)

Industrial applications: (Kemamine P-974, P-974D; Tomah Tallow Amine); chemical synthesis (Radiamine 6170, 6171); clays (Lilamin 170, 170D); dyes and pigments (Crodamine 1.T; Lilamin 170, 170D; Radiamine 6170, 6171); metalworking (Crodamine 1.T); ore flotation (Arosurf MG-170; Crodamine 1.T; Kemamine P-974, P-974D; Radiamine 6170, 6171; Tomah Tallow Amine); petroleum industry (Adogen 170 (D); Lilamin 170, 170D); Kemamine P-974, P-974D); plastics (Crodamine 1.T; Kemamine P-974, P-974D); printing inks (Lilamin 170, 170D); road building (Lilamin 170, 170D); rubber (Crodamine 1.T; Kemamine P-974, P-974D; Lilamin 170, 170D; Radiamine 6170, 6171); textile/leather processing (Crodamine 1.T; Radiamine 6170, 6171)

PROPERTIES:

Form:

Liquid (Crodamine 1.T)
Paste (Amine BG, BGD; Arosurf MG-170; Kemamine P-974, P-974D; Lilamin 170; Radiamine 6170, 6171)
Solid (Tomah Tallow Amine)

Tallow amine (cont'd.)

Color:
Gardner 1 max. (Kemamine P-974D)
Gardner 3 max. (Kemamine P-974; Lilamin 170)

Composition:
93% conc. (Kemamine P-974)
95% active (Lilamin 170)
97% conc. (Amine BGD; Kemamine P-974D)
98% conc. (Amine BG)
100% conc. (Adogen 170 (D); Crodamine 1.T; Radiamine 6170, 6171; Tomah Tallow Amine)

Solubility:
Sol. in common organic solvents (Kemamine P-974, P-974D)

Ionic Nature:
Cationic (Amine BG, BGD; Crodamine 1.T; Kemamine P-974, P-974D; Lilamin 170; Radiamine 6170, 6171; Tomah Tallow Amine)

M.W.:
266 (Lilamin 170)

Sp. Gr.:
0.799 (60 C) (Lilamin 170)

Visc.:
3.35 cps (60 C) (Lilamin 170)

M.P.:
32 C (Lilamin 170)

Flash Pt.:
155 C (OC) (Lilamin 170)

Iodine No.:
38 min. (Kemamine P-974, P-974D; Lilamin 170

Tallow amine, hydrogenated

SYNONYMS:
Amines, hydrogenated tallow alkyl
Hydrogenated tallow amine (CTFA)
Primary hydrogenated tallow amine

STRUCTURE:
$R{-}NH_2$
where R represents the hydrogenated tallow radical

CAS No.:
61788-45-2

TRADENAME EQUIVALENTS:

Adogen 140 (D) [Sherex]
Amine HBG [Kenobel]
Amine HBGD [Kenobel] (distilled)
Armeen HT, HTD [Armak]
Arosurf MG-140 [Sherex]
Crodamine 1.HT [Croda]
Kemamine P-970 [Humko Sheffield] (tech.)
Kemamine P-970D [Humko Sheffield] (distilled)
Lilamin 140 [Lilachim S.A.]
Lilamin 140D [Lilachim S.A.] (distilled)
Nissan Amine ABT [Nippon Oil & Fats]
Radiamine 6140 [Oleofina S.A.]
Radiamine 6141 [Oleofina S.A.] (distilled)

CATEGORY:

Corrosion inhibitor, dispersant, emulsifier, flotation reagent, intermediate, anticaking agent, acid scavenger, mold release agent, lubricant, flushing agent, germicide, bactericidal, wetting agent, softener

APPLICATIONS:

Cosmetic industry preparations: (Armeen HT, HTD; Radiamine 6140, 6141)
Farm products: (Nissan Amine ABT); herbicides (Crodamine 1.HT)
Industrial applications: asphalt (Nissan Amine ABT); ceramics (Nissan Amine ABT); chemical synthesis (Kemamine P-970, P-970D; Nissan Amine ABT; Radiamine 6140, 6141); clays (Lilamin 140, 140D); concrete (Nissan Amine ABT); dyes and pigments (Crodamine 1.HT; Lilamin 140, 140D; Nissan Amine ABT); grease/ lubricant additive (Lilamin 140, 140D; Nissan Amine ABT); industrial processing (Armeen HT, HTD); metalworking (Crodamine 1.HT; Nissan Amine ABT); ore flotation (Arosurf MG-140; Crodamine 1.HT; Kemamine P-970, P-970D; Nissan Amine ABT; Radiamine 6140, 6141); petroleum industry (Adogen 140 (D); Kemamine P-970, P-970D; Lilamin 140, 140D); plastics (Crodamine 1.HT; Kemamine P-970, P-970D); printing inks (Lilamin 140, 140D); road building (Lilamin 140, 140D); rubber (Crodamine 1.HT; Kemamine P-970, P-970D; Lilamin 140, 140D; Nissan Amine ABT; Radiamine 6140, 6141); textile/leather processing (Crodamine 1.HT; Nissan Amine ABT; Radiamine 6140, 6141); water treatment (Nissan Amine ABT)

PROPERTIES:

Form:

Liquid (Crodamine 1.HT)
Flake (Nissan Amine ABT)
Solid (Amine HBG, HBGD; Armeen HT, HTD; Arosurf MG-140; Kemamine P-970, P-970D; Lilamin 140; Radiamine 6140, 6141)
Waxy solid (Nissan Amine ABT)

Tallow amine, hydrogenated (*cont'd.*)

Color:
White (Arosurf MG-140; Nissan Amine ABT)
Gardner 1 max. (Kemamine P-970D)
Gardner 2 (Armeen HTD)
Gardner 3 max. (Kemamine P-970; Lilamin 140)
Gardner 9 (Armeen HT)

Composition:
93% conc. (Kemamine P-970)
95% active (Lilamin 140)
97% conc. (Amine HBGD; Kemamine P-970D)
98% conc. (Amine HBG; Armeen HT)
98% primary amine min. (Nissan Amine ABT)
100% conc. (Adogen 140 (D); Crodamine 1.HT; Radiamine 6140, 6141)

Solubility:
Sol. in carbon tetrachloride (Armeen HT, HTD)
Sol. in chloroform (Armeen HT, HTD)
Sol. in ethanol (Armeen HT, HTD)
Sol. in isopropanol (Armeen HT, HTD)
Slightly sol. in kerosene (Armeen HT, HTD)
Sol. in methanol (Armeen HT, HTD)
Sol. in common organic solvents (Kemamine P-970, P-970D)
Sol. in toluene (Armeen HT, HTD)

Ionic Nature:
Cationic (Amine HBG, HBGD; Armeen HT, HTD; Crodamine 1.HT; Kemamine P-970, P-970D; Lilamin 140, 140D; Radiamine 6140, 6141)

M.W.:
266 (Lilamin 140)

Sp.gr.:
0.974 (60/4 C) (Armeen HTD)
0.795 (60 C) (Lilamin 140); (60/4 C) (Armeen HT)

Visc.:
44.1 SSU (55 C) (Armeen HTD)
47.5 SSU (55 C) (Armeen HT)
4.40 cps (60 C) (Lilamin 140)

M.P.:
46 C (Lilamin 140)
70–120 F (Armeen HTD)
79–136 F (Armeen HT)

Pour Pt.:
100 F (Armeen HTD)
110 F (Armeen HT)

Solidification Pt.:
40–46 C (Nissan Amine ABT)

Flash Pt.:
158 C (OC) (Lilamin 140)
315 F (Armeen HTD)
320 F (Armeen HT)
Fire Pt.:
360 F (Armeen HTD)
365 F (Armeen HT)
Cloud Pt.:
110 F (Armeen HTD)
115 F (Armeen HT)
Iodine No.:
3.0 max. (Kemamine P-970, P-970D; Lilamin 140; Nissan Amine ABT)
STD. PKGS.:
15-kg paper bag, 14-kg can, 160-kg drum (Nissan Amine ABT)

Tallow trimethyl ammonium chloride

SYNONYMS:
Quaternary ammonium compounds, tallow alkyl trimethyl, chlorides
Tallowtrimonium chloride (CTFA)
Trimethyl tallow ammonium chloride
STRUCTURE:

$$\left[R{-}N(CH_3){-}CH_3 \atop \quad | \; CH_3 \right]^+ Cl^-$$

$$\left[\begin{array}{c} CH_3 \\ | \\ R{-}N{-}CH_3 \\ | \\ CH_3 \end{array}\right]^+ Cl^-$$

CAS No.:
8030-78-2; 74791-05-2
TRADENAME EQUIVALENTS:
Adogen 471 [Sherex]
Arquad T-27W, T-50 [Armak]
Jet Quat T-50 [Jetco]
Radiaquat 6471 [Oleofina S.A.]
CATEGORY:
Dispersant, antistat, emulsifier, foaming agent, wetting agent, corrosion inhibitor, softening agent, dyeing aid, bactericide
APPLICATIONS:
Cosmetic industry preparations: hair conditioner (Arquad T-50)
Industrial applications: construction (Jet Quat T-50); latex applications (Arquad T-50); mineral flotation (Arquad T-50); paper mfg. (Arquad T-27W); petroleum industry (Arquad T-27W; Jet Quat T-50); plastics (Arquad T-27W); textile/leather

Tallow trimethyl ammonium chloride (*cont'd.*)

processing (Arquad T-27W, T-50; Jet Quat T-50); water and sewage treatment (Arquad T-27W, T-50)

Industrial cleaners: metal processing surfactants (Arquad T-27W, T-50)

PROPERTIES:

Form:

Liquid (Adogen 471; Arquad T-27W, T-50; Radiaquat 6471)

Color:

Gardner 3 max. (Arquad T-27W)
Gardner 6 max. (Adogen 471)
Gardner 8 max. (Arquad T-50)

Composition:

26–29% quaternary in 49% aq. isopropanol (Arquad T-27W)
49–52% quaternary (Adogen 471)
50% active in aq. isopropanol (Arquad T-50)
50% conc. (Radiaquat 6471)

Solubility:

Sol. in water (@ 20 C) (Arquad T-27W)

Ionic Nature:

Cationic (Adogen 471; Arquad T-27W, T-50; Jet Quat T-50; Radiaquat 6471)

M.W.:

339 (Adogen 471)
340 (Arquad T-50)
343 (of active) (Arquad T-27W)

Sp.gr.:

0.88 (Arquad T-50)

Flash Pt.:

58 F (PM) (Adogen 471)
< 80 F (Arquad T-50)

HLB:

14.2 (Arquad T-27W, T-50)

Stability:

Good (Arquad T-50)

pH:

5.0–8.0 (10% aq.) (Arquad T-27W)

Biodegradable: (Arquad T-50)

TOXICITY/HANDLING:

Skin irritant, severe eye irritant; protective clothing, goggles, gloves should be worn (Arquad T-50)

STORAGE/HANDLING:

Avoid contact with strong oxidizing agents, anionics; avoid temps. > 70 C; will kill bacteria in biological disposal systems; flammable (Arquad T-50)

STD. PKGS.:

180-kg net bung-type steel drums (Arquad T-50)

Tetrapotassium pyrophosphate (CTFA)

SYNONYMS:
Diphosphoric acid, tetrapotassium salt
Potassium pyrophosphate
TKPP
EMPIRICAL FORMULA:
$H_4O_7P_2 \cdot 4K$
STRUCTURE:
$K_4P_2O_7$
CAS No.:
7320-34-5
TRADENAME EQUIVALENTS:
Empiphos 4KP [Albright & Wilson/Marchon]
CATEGORY:
Dispersant, stabilizer, clarifying agent, sequestering agent
APPLICATIONS:
Household detergents: built detergents (Empiphos 4KP); heavy-duty liquid cleaner (Empiphos 4KP); liquid soaps (Empiphos 4KP)
Industrial applications: emulsion paints (Empiphos 4KP); rubber (Empiphos 4KP); water treatment (Empiphos 4KP)
PROPERTIES:
Solubility:
Sol. 191 g/100 g water @ 25 C (Empiphos 4KP)
Bulk Density:
75 lb/ft^3 (Empiphos 4KP)
pH:
10.3 (Empiphos 4KP)

Tetrasodium dicarboxyethyl stearyl sulfosuccinamate(CTFA)

SYNONYMS:
L-Aspartic acid, N-(3-carboxy-1-oxosulfopropyl)-N-octadecyl, tetrasodium salt
N-(3-Carboxy-1-oxosulfopropyl)-N-octadecyl-L-aspartic acid, tetrasodium salt
N-(1,2-Dicarboxyethyl) N-alkyl (C_{18}) sulfosuccinamate
Tetrasodium N-(1,2-dicarboxyethyl)-N-octadecyl sulfosuccinate
EMPIRICAL FORMULA:
$C_{26}H_{47}NO_{10}S \cdot 4Na$
STRUCTURE:

```
                          O
                          ||
NaOOCCH2CH—N—CCH2CHCOONa
          |     |         |
      NaOOC   C18H37    SO3Na
```

Tetrasodium dicarboxyethyl stearyl sulfosuccinamate (cont'd.)

CAS No.:

3401-73-8; 37767-39-8

TRADENAME EQUIVALENTS:

Aerosol 22 [American Cyanamid]
Alconate 2CSA [Alcolac]
Monawet SNO-35 [Mona]
Rewopol B2003 [Rewo]

CATEGORY:

Emulsifier, dispersant, surfactant, wetting agent, solubilizer, viscosity depressant, antigelling agent, stabilizer, mild detergent, foaming agent

APPLICATIONS:

Cleansers: (Aerosol 22)
Cosmetic industry preparations: (Aerosol 22; Monawet SNO-35)
Farm products: agricultural oils/sprays (Aerosol 22; Alconate 2CSA; Monawet SNO-35); insecticides/pesticides (Alconate 2CSA)
Household detergents: (Aerosol 22; Alconate 2CSA; Rewopol B2003)
Industrial applications: carpet backing (Rewopol B2003); latex applications (Aerosol 22; Rewopol B2003); paper mfg. (Rewopol B2003); polymers/polymerization (Aerosol 22; Alconate 2CSA; Monawet SNO-35; Rewopol B2003); textile/leather processing (Aerosol 22; Monawet SNO-35)
Industrial cleaners: (Aerosol 22; Alconate 2CSA; Monawet SNO-35); metal processing surfactants (Aerosol 22)

PROPERTIES:

Form:

Liquid (Alconate 2CSA)
Clear liquid (Monawet SNO-35)
Clear to cloudy liquid (Aerosol 22)

Color:

Light amber (Monawet SNO-35)
Gardner 8 (Aerosol 22)

Composition:

35% active (Aerosol 22; Alconate 2CSA)
35% solids (Monawet SNO-35)

Solubility:

Excellent sol. in calcium, alkaline, and other electrolyte sol'ns. (Alconate 2CSA)
Good sol. in high electrolyte salt sol'ns. (Monawet SNO-35)
Excellent sol. in water (Aerosol 22); sol. in water (Monawet SNO-35)

Ionic Nature:

Anionic (Aerosol 22; Alconate 2CSA; Monawet SNO-35; Rewopol B2003)

M.W.:

653 avg. (Monawet SNO-35)

Tetrasodium dicarboxyethyl stearyl sulfosuccinamate (cont'd.)

Sp.gr.:
1.12 (Aerosol 22)
1.12–1.16 (Monawet SNO-35)
Density:
9.4 lb/gal (Aerosol 22)
9.5 lb/gal (Monawet SNO-35)
Visc.:
53 cps (Aerosol 22)
16–18 s (#2 Zahn cup) (Monawet SNO-35)
F.P.:
Separates below 10 C (Aerosol 22)
45 ± 5 F (Monawet SNO-35)
M.P.:
> 200 C (Aerosol 22)
Acid No.:
2.0 max. (Monawet SNO-35)
Iodine No.:
0.5 max. (Monawet SNO-35)
Stability:
High alkaline stability; high calcium tolerance (Monawet SNO-35)
pH:
7.0–8.0 (Monawet SNO-35)
Surface Tension:
41 dynes/cm (Aerosol 22)
43 dynes/cm (0.1%) (Monawet SNO-35)
Biodegradable: (Aerosol 22; Monawet SNO-35)
STD. PKGS.:
55-gal Liquipaks or bulk tankwagons (Monawet SNO-35)

Tetrasodium pyrophosphate

SYNONYMS:
Diphosphoric acid, tetrasodium salt
Sodium pyrophosphate
TSPP
EMPIRICAL FORMULA:
$H_4O_7P_2 \cdot 4Na$
STRUCTURE:
$Na_4P_2O_7$

Tetrasodium pyrophosphate (*cont'd.*)

CAS No.:
7722-88-5

TRADENAME EQUIVALENTS:
Generically sold by:
Monsanto, Olin

CATEGORY:
Dispersant, deflocculant, emulsifier, sequestrant

APPLICATIONS:
Food applications: food additives (generic)
Household detergents: detergent builder (generic)
Industrial applications: cement slurries (generic); clays (generic); dyes and pigments (generic); paint mfg. (generic); paper coating (generic); petroleum industry (generic); printing inks deinking (generic); rubber (generic); textile/leather processing (generic); water treatment (generic; Olin)
Industrial cleaners: metal processing surfactants (generic); textile scouring (generic)

PROPERTIES:

Form:
Granular (generic; Olin)
Powder (generic; Olin)
Regular (Olin)

Color:
White (generic)

Composition:
95% conc. (Olin)

Solubility:
Decomposed by alcohols (generic)
Sol. in water (generic); sol. 6 g/100 g water @ 24 C (Olin)

Sp.gr.:
2.45 (generic)

Density:
56 lb/ft³ (regular, Olin)
58 lb/ft³ (granular, Olin)
59 lb/ft³ (powdered, Olin)

M.P.:
880 C (generic)

pH:
10.3 (1% sol'n.) (Olin)

TOXICITY/HANDLING:
Toxic by inhalation (generic)
Toxic by oral ingestion; may produce eye, skin, and mucous membrane irritation; wear protective goggles, clothing; use adequate ventilation to control airborne dust (Olin)

STD. PKGS.:
100-lb net moistureproof paper bags, 125- and 350-lb net Leverpak drums (Olin)

SYNONYMS:
Benzene, methyl-
Methylbenzene
Phenylmethane
Toluol (obsolete)
EMPIRICAL FORMULA:
C_7H_8
STRUCTURE:
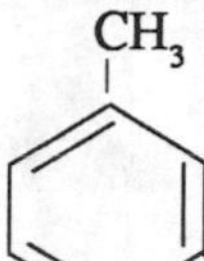

CAS No.:
108-88-3
TRADENAME EQUIVALENTS:
Amsco Toluene [Union Chem.] (nitration grade)
Shell Toluene [Shell]
Shell Toluene (nitration grade) [Shell]
Toluol [Ashland, Shell, Union Oil]
CATEGORY:
Solvent, diluent, thinner, adhesive solvent
APPLICATIONS:
Automotive industry: (Shell Toluene)
Household detergents: detergent mfg. (generic)
Industrial applications: adhesives/cements (Shell Toluene (nitration grade); Toluol); aviation gasoline (generic); chemical mfg. (generic; Shell Toluene); explosives (generic); paint mfg. (generic; Shell Toluene); plastics (generic); resins and gums (generic); rubber (generic; Shell Toluene (nitration grade); Toluol)
Industrial cleaners: drycleaning compositions (Shell Toluene)
PROPERTIES:
Form:
Liquid (generic; Shell Toluene (nitration grade); Toluol)
Clear liquid (Amsco Toluene)
Color:
Colorless (generic; Amsco Toluene; Shell Toluene (nitration grade); Toluol)
Odor:
Benzene-like (generic)
Characteristic (Shell Toluene (nitration grade))
Typical (Toluol)
Light aromatic (Amsco Toluene)
Composition:
99.9+% aromatics (Shell Toluene)

99.9+ vol.% toluene (Amsco Toluene)

Solubility:

Sol. in alcohols (generic)
Sol. in benzene (generic)
Sol. in ether (generic)
Misc. with most common organic solvents (Amsco Toluene)
Insol. in water (generic)

Sp.gr.:

0.866 (20/4 C) (generic)
0.869–0.8973 (Shell Toluene (nitration grade))
0.870–0.875 (Toluol)
0.871 (Amsco Toluene); (60/60 F) (Shell Toluene)

Density:

7.25 lb/gal (Amsco Toluene); (60 F) (Shell Toluene)

Visc.:

0.69 cSt (20 C) (Amsco Toluene)

F.P.:

–94.5 C (generic)

B.P.:

109–111 C (Toluol)
110.3 C (initial) (Amsco Toluene)
110.6 ± 0.1 (Shell Toluene (nitration grade))
110.7 C (generic)

Flash Pt.:

4.4 C (CC) (generic)
41 F (CC) (Shell Toluene)
45 F (Toluol); (TCC) (Amsco Toluene)

Ref. Index:

1.4933 (20 C) (Amsco Toluene)
1.497 (20 C) (generic)

TOXICITY/HANDLING:

Toxic by ingestion, inhalation and skin absorption; TLV: 100 ppm in air (generic)
100 ppm TLV; 150 ppm STEL (Amsco Toluene)

STORAGE/HANDLING:

Flammable, dangerous fire risk; explosive limits in air 1.27–5% (generic)
DOT flammable liquid (Amsco Toluene)

STD. PKGS.:

55-gal drums, 5-gal cans, tank trucks, tank cars, barges (Amsco Toluene)
Drum, tank car, tank truck, barge (Toluol)

Trichloroethane (CTFA)

SYNONYMS:
Ethane, 1,1,1-trichloro-
Methylchloroform
1,1,1-Trichloroethane

EMPIRICAL FORMULA:
$C_2H_3Cl_3$

STRUCTURE:
CH_3CCl_3

CAS No.:
71-55-6

TRADENAME EQUIVALENTS:
Aerothene TT [Dow] (inhibited)
Chlorothene NU [C.P. Hall]
Chlorothene SM Solvent [Dow] (inhibited)
Dowclene LS [Dow] (inhibited)
Prelete Defluxer [Dow] (inhibited with alcohols)
Tri-Ethane [PPG Industries] (inhibited)
Sold generically by:
Ethyl Corp.

CATEGORY:
Solvent, vapor depressant, fumigant

APPLICATIONS:
Degreasers: (generic; Aerothene TT; Chlorothene SM Solvent; Tri-Ethane)
Farm products: insecticides/pesticides (generic)
Industrial applications: adhesives (Chlorothene NU; Tri-Ethane); aerosols (Aerothene TT; Tri-Ethane); dyes and pigments (Tri-Ethane); electronics (Aerothene TT); industrial formulation (generic; Chlorothene SM Solvent; Tri-Ethane); lubricating/cutting oils (Tri-Ethane); paints and coatings (Aerothene TT; Chlorothene NU); printing inks (Tri-Ethane); textile/leather processing (generic; Tri-Ethane)
Industrial cleaners: cold cleaning (generic; Tri-Ethane); leather/suede cleaning (Dowclene LS); post-solder cleaning (Prelete Defluxer); maintenance cleaning (Tri-Ethane); solvent cleaners (generic)

PROPERTIES:

Form:
Clear liquid (Chlorothene NU; Tri-Ethane)
Liquid (generic)

Color:
Colorless (generic; Chlorothene NU)
Water-white (Tri-Ethane)

Solubility:
Sol. in alcohols (generic)
Sol. in ether (generic)

Trichloroethane (cont'd.)

Sol. in most organic solvents (Aerothene TT; Chlorothene SM Solvent; Dowclene LS; Prelete Defluxer)

Very low sol. in water (Aerothene TT; Chlorothene SM Solvent; Dowclene LS; Prelete Defluxer); 0.07 g/100 g (Tri-Ethane); insol. (generic)

M.W.:

133.42 (Tri-Ethane)

Sp.gr.:

1.304 (Tri-Ethane)
1.32 (Chlorothene NU)
1.325 (generic)

Density:

10.84 lb/gal (Tri-Ethane)

Visc.:

0.735 cps (Tri-Ethane)

F.P.:

–45 C (Tri-Ethane)
–38 C (generic)

B.P.:

74.1 C (Tri-Ethane)
75 C (generic)

Flash Pt.:

None (generic); (TOC) (Tri-Ethane)

Fire Pt.:

None (TOC) (Tri-Ethane)

Ref. Index:

1.4379 (Tri-Ethane)

pH:

7.0 (Tri-Ethane)

Specific Heat:

0.258 cal/g °C (20 C) (Tri-Ethane)

Dielectric Constant:

8.3 (1000 cps) (Tri-Ethane)

TOXICITY/HANDLING:

Low toxicity (Chlorothene NU)

Fatal vapor concs. can occur in confined or poorly ventilated areas; ingestion of substantial quantities could be fatal; eye contact may result in pain and irritation; prolonged/repeated skin contact can cause irritation and dermatitis; wear protective goggles and clothing (Tri-Ethane)

Irritant to eyes and tissue (generic)

STORAGE/HANDLING:

Nonflammable (generic; Chlorothene NU; Tri-Ethane)

Has no flash pt. but high vapor concs. in air can be ignited by a high-energy source; at temps. above 700 C the liquid or vapor will form toxic and corrosive gases; do not permit contact with strong alkalis; do not store in aluminum, magnesium, or their alloys (Tri-Ethane)

STD. PKGS.:

Drums (Chlorothene NU)

55-gal drums, tank trucks, tank cars (Tri-Ethane)

Tripropylene glycol monomethyl ether

SYNONYMS:

POP (3) methyl ether

PPG (3) methyl ether

PPG-3 methyl ether (CTFA)

STRUCTURE:

$CH_3(OCHCH_2)_nOH$ (with CH_3 on the CH)

where avg. $n = 3$

CAS No.:

37286-64-9 (generic)

RD No.: 977067-03-0

TRADENAME EQUIVALENTS:

Arcosolv TPM [Arco]

Poly-Solv TPM [Olin]

CATEGORY:

Solvent, chemical intermediate, solubilizer, plasticizer, extraction solvent, coupling agent, dispersant

APPLICATIONS:

Cosmetic industry preparations: (Arcosolv TPM; Poly-Solv TPM); nail polish (Poly-Solv TPM)

Farm products: (Arcosolv TPM); agricultural oils/sprays (Poly-Solv TPM); insect repellents (Poly-Solv TPM)

Household detergents: hard surface cleaner (Poly-Solv TPM)

Industrial applications: adhesives (Poly-Solv TPM); brake fluids (Poly-Solv TPM); dyes and pigments (Poly-Solv TPM); paints and coatings (Arcosolv TPM; Poly-Solv TPM); plastics (Poly-Solv TPM); polishes and waxes (Poly-Solv TPM); printing inks (Arcosolv TPM; Poly-Solv TPM); textile/leather processing (Poly-Solv TPM)

Industrial cleaners: solvent cleaners (Arcosolv TPM)

PROPERTIES:

Form:

Liquid (Arcosolv TPM; Poly-Solv TPM)

Tripropylene glycol monomethyl ether *(cont'd.)*

Color:
APHA 15 max. (Arcosolv TPM; Poly-Solv TPM)
Odor:
Mild, characteristic (Poly-Solv TPM)
Mild, pleasant, characteristic (Arcosolv TPM)
Composition:
99.85% conc. min. (Arcosolv TPM)
Solubility:
Misc. with many organic solvents (Arcosolv TPM; Poly-Solv TPM)
Completely sol. in water (Arcosolv TPM); misc. (Poly-Solv TPM)
M.W.:
206.28 (Poly-Solv TPM)
206.3 (Arcosolv TPM)
Sp.gr.:
0.962–0.965 (Arcosolv TPM)
0.969 (20/20 C) (Poly-Solv TPM)
Density:
8.03 lb/gal (Arcosolv TPM)
8.06 lb/gal (Poly-Solv TPM)
Visc.:
5.8 cstk (Arcosolv TPM)
6.1 cP (20 C) (Poly-Solv TPM)
F.P.:
–79 C (Arcosolv TPM)
–78 C (Poly-Solv TPM)
B.P.:
242 C (760 mm) (Poly-Solv TPM)
242.4 C (760 mm Hg) (Arcosolv TPM)
Flash Pt.:
114 C (PMCC) (Arcosolv TPM)
121 C (COC) (Poly-Solv TPM)
Ref. Index:
1.428 (Poly-Solv TPM)
1.430 (20 C) (Arcosolv TPM)
Surface Tension:
29.0 dynes/cm (Arcosolv TPM)
Conductivity:
3.54 K $\times 10^4$ cal/cm^{2}s°C/cm (60 C) (Arcosolv TPM)
Specific Heat:
0.51 cal/g/°C (Arcosolv TPM)
TOXICITY/HANDLING:
Relatively low acute toxicity; may cause mild adverse effects on skin and eye contact; avoid ingestion, eye contact, and prolonged skin contact and inhalation of vapors

(Arcosolv TPM)
Mild primary skin and eye irritant; avoid contact with skin and inhalation of vapors (Poly-Solv TPM)

STORAGE/HANDLING:
Store in carbon steel vessels; avoid contact with air when storing for long periods of time (Arcosolv TPM)

STD. PKGS.:
Drums, tank car, tank truck (Arcosolv TPM)
55-gal (440 lb net) drums, tank cars, tank trucks (Poly-Solv TPM)

Tris (hydroxymethyl) amino methane

SYNONYMS:
2-Amino-2-(hydroxymethyl)-1,3-propanediol
1,3-Propanediol, 2-amino-2-(hydroxymethyl)-
THAM
Tri (hydroxymethyl) amino methane
Tris amine buffer
Tromethamine (CTFA)

EMPIRICAL FORMULA:
$C_4H_{11}NO_3$

STRUCTURE:

$$\begin{array}{c} CH_2OH \\ | \\ H_2N\text{—}C\text{—}CH_2OH \\ | \\ CH_2OH \end{array}$$

CAS No.:
77-86-1

TRADENAME EQUIVALENTS:
Tris Amino (40% Concentrate) [Angus]
Tris Amino (Crystals) [Angus]

CATEGORY:
Dispersant, solubilizer, neutralizer, emulsifier (in soap form), absorbent

APPLICATIONS:
Industrial applications: dyes and pigments (Tris Amino (Crystals), (40% Conc.)); paints and coatings (Tris Amino (Crystals), (40% Conc.)); polymers (Tris Amino (Crystals), (40% Conc.))

PROPERTIES:

Form:
Crystalline solid (generic)

Tris (hydroxymethyl) amino methane (*cont'd.*)

Color:
White (generic)
Composition:
40% conc. (Tris Amino (40% Conc.))
Solubility:
Sol. in water 80 g/100 cc (20 C) (generic)
M.W.:
121.14 (generic)
B.P.:
219–220 C (10 mm) (generic)
M.P.:
171–172 C (generic)
TOXICITY/HANDLING:
Irritant to eyes and skin (generic)
STORAGE/HANDLING:
Combustible (generic)

Xylene (CTFA)

SYNONYMS:

Benzene, dimethyl-

Commercial mixture of 3 isomers, *o*-, *m*-, and *p*-xylene, predominantly *m*- and *p*-Dimethylbenzene

EMPIRICAL FORMULA:

C_8H_{10}

STRUCTURE:

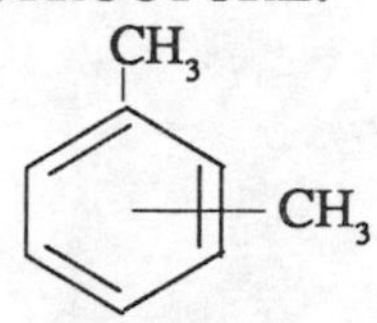

CAS No.:

1330-20-7

TRADENAME EQUIVALENTS:

Amsco Xylene [Union Chem.] (5 degree quality)

Shell Xylene [Shell]

Xylol [Ashland, Shell, Union Oil]

CATEGORY:

Solvent

APPLICATIONS:

Automotive industry: (Shell Xylene)

Industrial applications: adhesives/cements (generic; Xylol); aviation gasoline (generic); chemical mfg. (generic; Shell Xylene); paints/lacquers/enamels (generic; Shell Xylene); resins (generic); rubber (generic; Xylol)

Industrial cleaners: drycleaning compositions (Shell Xylene)

PROPERTIES:

Form:

Clear liquid (generic; Amsco Xylene; Xylol)

Color:

Colorless (Amsco Xylene; Xylol)

Odor:

Light aromatic (Amsco Xylene)

Typical (Xylol)

Composition:

97.89 vol.% total C_8 aromatics (Amsco Xylene)

99.5 vol.% aromatics (Shell Xylene)

Xylene (cont'd.)

Solubility:
Sol. in alcohols (generic)
Sol. in ether (generic)
Misc. with most common organic solvents (Amsco Xylene)
Insol. in water (generic)

Sp.gr.:
0.86 (generic)
0.86–0.88 (Xylol)
0.871 (Amsco Xylene); (60/60 F) (Shell Xylene)

Density:
7.25 lb/gal (Amsco Xylene); (60 F) (Shell Xylene)

Visc.:
0.77 cSt (20 C) (Amsco Xylene)

B.P.:
137–144 C (Xylol)
137.7 C (Amsco Xylene)
81 F (Shell Xylene)

Flash Pt.:
27.2–46.1 C (TOC) (generic)
81 F (Amsco Xylene)

Ref. Index:
1.4949 (20 C) (Amsco Xylene)

TOXICITY/HANDLING:
Toxic (Xylol)
Toxic by ingestion and inhalation; TLV: 100 ppm in air (generic)
100 ppm TLV; 150 ppm STEL (Amsco Xylene)

STORAGE/HANDLING:
Flammable (Xylol)
Flammable, moderate fire risk (generic)
DOT flammable liquid (Amsco Xylene)

STD. PKGS.:
5-gal cans, 55-gal drums, tank trucks, tank cars, barges (Amsco Xylene)
Drums, tank car, tank truck barge (Xylol)

TRADENAME PRODUCTS AND GENERIC EQUIVALENTS

Ablunol NP6 [Taiwan Surfactant]—POE (6) nonyl phenyl ether
Ablunol NP8 [Taiwan Surfactant]—POE (8) nonyl phenyl ether
Ablunol NP9 [Taiwan Surfactant]—POE (9) nonyl phenyl ether
Ablunol NP15 [Taiwan Surfactant]—POE (15) nonyl phenyl ether
Ablunol NP20 [Taiwan Surfactant]—POE (20) nonyl phenyl ether
Actiflo 68, 70 [Central Soya]—Lecithin
Adinol CT95 [Croda]—Sodium methyl cocoyl taurate
Adinol T, T35 [Croda]—Sodium methyl oleoyl taurate
Adogen 140 (D) [Sherex]—Tallow amine, hydrogenated
Adogen 142 (D) [Sherex]—Stearyl amine
Adogen 160 (D) [Sherex]—Coconut amine
Adogen 170 (D) [Sherex]—Tallow amine
Adogen 471 [Sherex]—Tallow trimethyl ammonium chloride
AE-1214/3 [Procter & Gamble]—POE (3) lauryl ether
Aerosol 18, 19 [Amer. Cyanamid]—Disodium stearyl sulfosuccinamate
Aerosol 22 [Amer. Cyanamid]—Tetrasodium dicarboxyethyl stearyl sulfosuccinamate
Aerosol A-102 [Amer. Cyanamid]—Disodium deceth-6 sulfosuccinate
Aerosol A-103 [Amer. Cyanamid, Cyanamid B.V.]—Disodium nonoxynol-10 sulfosuccinate
Aerosol A-196 [Amer. Cyanamid]—Sodium dicyclohexyl sulfosuccinate
Aerosol A-268 [Amer. Cyanamid]—Disodium isodecyl sulfosuccinate
Aerosol AY, AY-65, AY-100 [Amer. Cyanamid]—Sodium diamyl sulfosuccinate
Aerosol GPG [Amer. Cyanamid]—Sodium dioctyl sulfosuccinate
Aerosol IB-45 [Amer. Cyanamid]—Sodium diisobutyl sulfosuccinate
Aerosol MA-80 [Amer. Cyanamid]—Sodium dihexyl sulfosuccinate
Aerosol OT-70PG, OT-75, OT-100, OT-B, OTS [Amer. Cyanamid]—Sodium dioctyl sulfosuccinate
Aerosol TR-70 [Amer. Cyanamid]—Sodium ditridecyl sulfosuccinate
Aerothene MM [Dow]—Methylene chloride (inhibited)
Aerothene TT [Dow]—Trichloroethane (inhibited)
Ahco FP-67 [ICI United States]—Sorbitan monopalmitate
Alcolec 439-C [American Lecithin]—Lecithin
Alcolec 495, 619-B, 621, 628G, 634, 638, 650, 658, 662 [American Lecithin]—Lecithin
Alcolec 4135 [American Lecithin]—Lecithin
Alcolec Extra A [American Lecithin]—Lecithin
Alcolec Granules [American Lecithin]—Lecithin
Alcolec HS-3 [American Lecithin]—Lecithin
Alcolec Powder [American Lecithin]—Lecithin
Alcolec RCX-1 [American Lecithin]—Lecithin

Alcolec S [American Lecithin]—Lecithin
Alcolec Z-3 [American Lecithin]—Lecithin (hydroxylated)
Alconate 2CH [Alcolac]—Sodium dicyclohexyl sulfosuccinate
Alconate 2CSA [Alcolac]—Tetrasodium dicarboxyethyl stearyl sulfosuccinamate
Alconate D-6 [Alcolac]—Disodium deceth-6 sulfosuccinate
Alcopol O-60%, O-70PG, O-100% [Allied Colloids Ltd.]—Sodium dioctyl sulfosuccinate
Alcopol OB [Allied Colloids Ltd.]—Sodium diisobutyl sulfosuccinate
Alcopol OD [Allied Colloids]—Sodium ditridecyl sulfosuccinate
Alcopol OS [Allied Colloids Ltd.]—Sodium dihexyl sulfosuccinate
Aldo DGDO [Glyco]—Decaglycerol decaoleate
Aldo HGDS [Glyco]—Hexaglyceryl distearate
Alfol 4 [Continental Oil]—*n*-Butyl alcohol
Alipal CO-433 [GAF]—Sodium nonoxynol-4 sulfate
Alipal CO-436 [GAF]—Ammonium nonoxynol-4 sulfate
Alkamuls 200-DL [Alkaril]—POE (4) dilaurate
Alkamuls 400-DL [Alkaril]—POE (8) dilaurate
Alkamuls 600-DL [Alkaril]—POE (12) dilaurate
Alkasurf CO-5 [Alkaril]—POE (5) castor oil
Alkasurf CO-10 [Alkaril]—POE (10) castor oil
Alkasurf CO-25 [Alkaril]—POE (25) castor oil
Alkasurf CO-30 [Alkaril]—POE (30) castor oil
Alkasurf CO-40M [Alkaril]—POE (40) castor oil
Alkasurf NP-1 [Alkaril]—POE (1) nonyl phenyl ether
Alkasurf NP-4 [Alkaril]—POE (4) nonyl phenyl ether
Alkasurf NP-5 [Alkaril]—POE (5) nonyl phenyl ether
Alkasurf NP-6 [Alkaril]—POE (6) nonyl phenyl ether
Alkasurf NP-8 [Alkaril]—POE (8) nonyl phenyl ether
Alkasurf NP-9 [Alkaril]—POE (9) nonyl phenyl ether
Alkasurf NP-10 [Alkaril]—POE (10) nonyl phenyl ether
Alkasurf NP-11 [Alkaril]—POE (11) nonyl phenyl ether
Alkasurf NP-12 [Alkaril]—POE (12) nonyl phenyl ether
Alkasurf NP-15, NP-15 80% [Alkaril]—POE (15) nonyl phenyl ether
Alkasurf NP-20, NP-20 70% [Alkaril]—POE (20) nonyl phenyl ether
Alkasurf OP-1 [Alkaril]—POE (1) octyl phenyl ether
Alkasurf OP-5 [Alkaril]—POE (5) octyl phenyl ether
Alkasurf OP-6 [Alkaril]—POE (6) octyl phenyl ether
Alkasurf OP-8 [Alkaril]—POE (8) octyl phenyl ether (7–8 moles EO)
Alkasurf OP-10 [Alkaril]—POE (10) octyl phenyl ether
Alkasurf OP-12 [Alkaril]—POE (12) octyl phenyl ether (12–13 moles EO)
Alkasurf OP-30, OP-30 70% [Alkaril]—POE (30) octyl phenyl ether
Alkasurf OP-40 [Alkaril]—POE (40) octyl phenyl ether

Alkasurf OP-70-50% [Alkaril]—POE (70) octyl phenyl ether
Alkasurf SA-2 [Alkaril]—POE (2) stearyl ether
Alkasurf SA-20 [Alkaril]—POE (20) stearyl ether
Alkasurf SS-O-40, SS-O-60, SS-O-75 [Alkaril]—Sodium dioctyl sulfosuccinate
Alkasurf SS-TA [Alkaril]—Disodium stearyl sulfosuccinamate
Alkasurf TDA-15 [Alkaril]—POE (15) tridecyl ether
Alkaterge E [IMC]—Ethyl hydroxymethyl oleyl oxazoline
Alkatrope SX-40 [Alkaril]—Sodium xylenesulfonate
Alrowet D-65 [Ciba-Geigy]—Sodium dioctyl sulfosuccinate
Amasperse N [Amer. Color & Chem.]—Sodium lignosulfonate
Amberlig [Reed Lignin]—Calcium lignosulfonate
Amine 18D [Kenobel]—Stearyl amine
Amine BG [Kenobel]—Tallow amine
Amine BGD [Kenobel]—Tallow amine (distilled)
Amine HBG [Kenobel]—Tallow amine, hydrogenated
Amine HBGD [Kenobel]—Tallow amine, hydrogenated (distilled)
Amine KK [Kenobel]—Coconut amine
AMP, AMP-95 [IMC]—Aminomethyl propanol
Amsco Heptane [Union Oil]—Heptane
Arcosolv DPM [Arco]—Dipropylene glycol monomethyl ether
Arcosolv PM [Arco]—Propylene glycol monomethyl ether
Arcosolv TPM [Arco]—Tripropylene glycol monomethyl ether
Arkopal N-040 [Amer. Hoechst]—POE (4) nonyl phenyl ether
Arkopal N-060 [Amer. Hoechst]—POE (6) nonyl phenyl ether
Arkopal N-080 [Amer. Hoechst]—POE (8) nonyl phenyl ether
Arkopal N-090 [Amer. Hoechst]—POE (9) nonyl phenyl ether
Arkopal N-100 [Amer. Hoechst]—POE (10) nonyl phenyl ether
Arkopal N-110 [Amer. Hoechst]—POE (11) nonyl phenyl ether
Arkopal N-130 [Amer. Hoechst]—POE (13) nonyl phenyl ether
Arkopal N-150 [Amer. Hoechst]—POE (15) nonyl phenyl ether
Arlacel 40 [ICI United States]—Sorbitan monopalmitate
Arlacel 987 [ICI United States]—Sorbitan monoisostearate
Arlasolve 200 [ICI United States]—POE (20) isocetyl ether
Arlatone T [ICI United States]—POE (40) sorbitan peroleate
Armeen 18, 18D [Armak]—Stearyl amine
Armeen C, CD [Armak]—Coconut amine
Armeen HT, HTD [Armak]—Tallow amine, hydrogenated
Armeen SD [Armak]—Soya amine
Armeen Z, Z-9 [Armak]—Cocaminobutyric acid
Armotan MP [Akzo]—Sorbitan monopalmitate
Arosurf MG-140 [Sherex]—Tallow amine, hydrogenated
Arosurf MG-160 [Sherex]—Coconut amine

Arosurf MG-170 [Sherex]—Tallow amine
Arquad T-27W, T-50 [Armak]—Tallow trimethyl ammonium chloride
Arylene M40, M60 [Hart]—Sodium dioctyl sulfosuccinate
Astrowet B-45 [Alco]—Sodium diisobutyl sulfosuccinate
Astrowet H-80 [Alco]—Sodium dihexyl sulfosuccinate
Astrowet O-70-PG, O-75 [Alco]—Sodium dioctyl sulfosuccinate
Atcowet W [Bostik South]—Sodium dioctyl sulfosuccinate
Atlas G-100 [ICI Specialty Chem.]—Dimethyl isosorbide
Avirol SO-70P [Henkel]—Sodium dioctyl sulfosuccinate
Borax [U.S. Borax]—Sodium borate (decahydrate)
Brij 72 [ICI United States]—POE (2) stearyl ether
Brij 76 [ICI United States]—POE (10) stearyl ether (with BHA and citric acid)
Brij 78 [ICI United States]—POE (20) stearyl ether (with BHA and citric acid)
Brij 721 [ICI United States]—POE (21) stearyl ether (with BHA and citric acid)
Butyl Carbitol [Union Carbide]—Diethylene glycol monobutyl ether
Butyl Cellosolve [Union Carbide]—Ethylene glycol monobutyl ether
Butyl Dioxitol [Shell]—Diethylene glycol monobutyl ether
Butyl Oxitol [Shell]—Ethylene glycol monobutyl ether
Calgolac [Merck]—Sodium hexametaphosphate
Calgon [Merck]—Sodium hexametaphosphate
Caprol 6G2S [Capital City]—Hexaglyceryl distearate
Caprol 10G100 [Stokely-Van Camp]—Decaglycerol decaoleate
Captex 200 [Capital City]—Propylene glycol dicaprylate/dicaprate
Captex 300 [Capital City]—Caprylic/capric triglyceride
Carbitol Solvent [Union Carbide]—Diethylene glycol monoethyl ether
Carsonon N-4 [Carson]—POE (4) nonyl phenyl ether
Carsonon N-6 [Carson]—POE (6) nonyl phenyl ether
Carsonon N-8 [Carson]—POE (8) nonyl phenyl ether
Carsonon N-9 [Carson]—POE (9) nonyl phenyl ether (9–10 EO)
Carsonon N-10 [Carson]—POE (10) nonyl phenyl ether (10–11 EO)
Carsonon N-11 [Lonza]—POE (11) nonyl phenyl ether
Carsonon N-12 [Lonza]—POE (12) nonyl phenyl ether
Carsonon TD-10 [Carson]—POE (10) tridecyl ether
Carsonon TD-11 [Carson]— POE (11) tridecyl ether
Carsosulf SXS Liquid [Carson]—Sodium xylenesulfonate
Castorwax [CasChem]—Hydrogenated castor oil
Castorwax MP-70, MP-80 [CasChem]—Hydrogenated castor oil
Cedepal CA-210 [Domtar]—POE (1) octyl phenyl ether (1.5 EO)
Cedepal CA-520 [Domtar]—POE (5) octyl phenyl ether
Cedepal CA-630 [Domtar]—POE (9) octyl phenyl ether
Cedepal CA-720 [Domtar]—POE (12) octyl phenyl ether (12.5 EO)
Cedepal CA-890, CA-897 [Domtar]—POE (40) octyl phenyl ether

Cedepal CO-210 [Domtar]—POE (1) nonyl phenyl ether (1.5 EO)
Cedepal CO-430 [Domtar]—POE (4) nonyl phenyl ether
Cedepal CO-530 [Domtar]—POE (6) nonyl phenyl ether
Cedepal CO-610 [Domtar]—POE (8) nonyl phenyl ether
Cedepal CO-630 [Domtar]—POE (9) nonyl phenyl ether
Cedepal CO-710 [Domtar]—POE (10) nonyl phenyl ether (10–11 EO)
Cedepal CO-730 [Domtar]—POE (15) nonyl phenyl ether
Cellosolve [Union Carbide]—Ethylene glycol monoethyl ether
Cellosolve Acetate [Union Carbide]—Ethylene glycol monoethyl ether acetate
Centrocap Series [Central Soya]—Lecithin
Centrol Series [Central Soya]—Lecithin
Centrolene Series [Central Soya]—Lecithin (hydroxylated)
Centrolex Series [Central Soya]—Lecithin
Centromix Series [Central Soya]—Lecithin
Centrophase 31 [Central Soya/Chemurgy]—Lecithin
Centrophase HR [Central Soya]—Lecithin (heat-resistant)
Centrophil Series [Central Soya]—Lecithin
Cenwax G [Union Camp]—Hydrogenated castor oil
Cenwax ME [Union Camp]—Methyl hydroxystearate
Chemal TDA-9 [Chemax]—POE (9) tridecyl ether
Chemax CO-5 [Chemax]—POE (5) castor oil
Chemax CO-25 [Chemax]—POE (25) castor oil
Chemax CO-30 [Chemax]—POE (30) castor oil
Chemax CO-40 [Chemax]—POE (40) castor oil
Chemax DNP-150 [Chemax]—POE (150) dinonyl phenyl ether
Chemax DOSS-70, DOSS-75E [Chemax]—Sodium dioctyl sulfosuccinate
Chemax HCO-200/50 [Chemax]—POE (200) hydrogenated castor oil
Chemax NP-4 [Chemax]—POE (4) nonyl phenyl ether
Chemax NP-6 [Chemax]—POE (6) nonyl phenyl ether
Chemax NP-9 [Chemax]—POE (9) nonyl phenyl ether
Chemax NP-10 [Chemax]—POE (10) nonyl phenyl ether
Chemax NP-15 [Chemax]—POE (15) nonyl phenyl ether
Chemax NP-20 [Chemax]—POE (20) nonyl phenyl ether
Chemax OP-3 [Chemax]—POE (3) octyl phenyl ether
Chemax OP-5 [Chemax]—POE (5) octyl phenyl ether
Chemax OP-7 [Chemax]—POE (7) octyl phenyl ether
Chemax OP-9 [Chemax]—POE (9) octyl phenyl ether
Chemax OP-40/70 [Chemax]—POE (40) octyl phenyl ether
Chemcol NPE-40 [Chemform]—POE (4) nonyl phenyl ether
Chemcol NPE-60 [Chemform]—POE (6) nonyl phenyl ether
Chemcol NPE-100 [Chemform]—POE (10) nonyl phenyl ether
Chemcol NPE-200 [Chemform]—POE (20) nonyl phenyl ether

Chemeen HT-50 [Chemax]—POE (50) tallow amine
Chlorothene NU [C.P. Hall]—Trichloroethane
Chlorothene SM Solvent [Dow]—Trichloroethane (inhibited)
Cithrol 2DL [Croda Chem. Ltd.]—POE (4) dilaurate
Cithrol 4DL [Croda]—POE (8) dilaurate
Cithrol 6DL [Croda Chem. Ltd.]—POE (12) dilaurate
Clearate Special Extra, WDF [W.A. Cleary]—Lecithin
Commercial Heptane [Phillips]—Heptane
Complemix 100 [Amer. Cyanamid]—Sodium dioctyl sulfosuccinate
Conco AAS-50M [Continental]—Ammonium dodecylbenzenesulfonate
Conco NI-43 [Continental Chem.]—POE (4) nonyl phenyl ether
Conco NI-60 [Continental Chem.]—POE (6) nonyl phenyl ether
Conco NI-90 [Continental Chem.]—POE (9) nonyl phenyl ether
Conco NI-100 [Continental Chem.]—POE (9) nonyl phenyl ether (9–10 EO)
Conco NI-110 [Continental Chem.]—POE (10) nonyl phenyl ether (10–11 EO)
Conco NI-150 [Continental Chem.]—POE (15) nonyl phenyl ether
Conco NI-185 [Continental Chem.]—POE (20) nonyl phenyl ether
Conco SXS [Continental]—Sodium xylenesulfonate
Condanol SBDO60 [Dutton & Reinisch]—Sodium dioctyl sulfosuccinate
CPH-79-N [C.P. Hall]—POE (8) dilaurate
Cremophor NP14 [BASF AG]—POE (14) nonyl phenyl ether
Cremophor RH40 [BASF AG]—POE (40) hydrogenated castor oil
Cremophor RH60 [BASF]—POE (60) hydrogenated castor oil
Cremophor RH410 [BASF-Wyandotte]—POE (40) hydrogenated castor oil
Crill 2 [Croda]—Sorbitan monopalmitate
Crill 6 [Croda]—Sorbitan monoisostearate
Crillet 6 [Croda]—POE (20) sorbitan monoisostearate
Crodamet 1.015 [Croda]—POE (15) oleyl amine (primary amine)
Crodamet 1.02 [Croda]—POE (2) oleyl amine (primary amine)
Crodamet 1.05 [Croda]—POE (5) oleyl amine (primary amine)
Crodamine 1.HT [Croda]—Tallow amine, hydrogenated
Crodamine 1.T [Croda]—Tallow amine
Crodamine 1.18D [Croda Universal]—Stearyl amine
Crodesta A-10, A-20 [Croda]—Acetylated sucrose distearate
Crodesta F-10, F-50 [Croda]—Sucrose distearate
Crodesta F110, F160 [Croda]—Sucrose monostearate
Crodesta SL-40 [Croda]—Sucrose monococoate
Croduret 40 [Croda Chem. Ltd.]—POE (40) hydrogenated castor oil
Croduret 60 [Croda Chem. Ltd.]—POE (60) hydrogenated castor oil
Croduret 100 [Croda Chem. Ltd.]—POE (100) hydrogenated castor oil
Croduret 200 [Croda Chem. Ltd.]—POE (200) hydrogenated castor oil
Crystamet [Stauffer]—Sodium metasilicate (pentahydrate)

Cyclopol SBDO [Cyclo]—Sodium dioctyl sulfosuccinate
Cyclopol SBG-280 [Cyclo]—Disodium oleamido PEG-2 sulfosuccinate
Darvan #2 [Vanderbilt]—Sodium lignosulfonate
Dehybor [U.S. Borax]—Sodium borate (anhyd.)
Dehydol LS2 [Henkel KGaA]—POE (2) lauryl ether
Dehydol LS3 [Henkel KGaA]—POE (3) lauryl ether
Denwet CM [Graden]—Sodium dioctyl sulfosuccinate
Dequest 2000 [Monsanto]—Aminotrimethylene phosphonic acid
Dequest 2006 [Monsanto]—Pentasodium aminotrimethylene phosphonate
Deriphat 154 [Henkel]—Disodium tallowiminodipropionate
Dioxitol-High Gravity, -Low Gravity [Shell]—Diethylene glycol monoethyl ether
Dowanol DB [Dow]—Diethylene glycol monobutyl ether
Dowanol DE [Dow]—Diethylene glycol monoethyl ether
Dowanol DPM [Dow, Dow Chem. Europe SA]—Dipropylene glycol monomethyl ether
Dowanol EB [Dow]—Ethylene glycol monobutyl ether
Dowanol EPH [Dow]—Ethylene glycol monophenyl ether
Dowanol PM [Dow]—Propylene glycol monomethyl ether
Dowclene LS [Dow]—Trichloroethane (inhibited)
Drewfax 528 [Drew Produtos]—Sodium dioctyl sulfosuccinate
Drewmulse 6-2-S [PVO Int'l.]—Hexaglyceryl distearate
Drewmulse 10-8-O [PVO Int'l.]—Decaglycerol octaoleate
Drewmulse 10-10-O [PVO Int'l.—Decaglycerol decaoleate
Drewmulse 10-10-S [PVO Int'l.]—Decaglycerol decastearate
Drewpol 6-2-S [PVO Int'l.]—Hexaglyceryl distearate
Drewpol 10-8-O [PVO Int'l.]—Decaglycerol octaoleate
Drewpol 10-10-O [PVO Int'l.]—Decaglycerol decaoleate
Drewpol 10-10-S [PVO Int'l.]—Decaglycerol decastearate
Drymet [Stauffer]—Sodium metasilicate (anhydrous)
Dynasperse A, B [Reed Lignin]—Sodium lignosulfonate (modified)
Dyqex [Georgia-Pacific]—Sodium lignosulfonate
Edenol 302 [Henkel/Canada]—Propylene glycol dicaprylate/dicaprate
Ektasolve DB [Eastman]—Diethylene glycol monobutyl ether
Ektasolve DE [Eastman]—Diethylene glycol monoethyl ether
Ektasolve DM [Eastman]—Diethylene glycol monomethyl ether
Ektasolve EB [Eastman]—Ethylene glycol monobutyl ether
Ektasolve EE [Eastman]—Ethylene glycol monoethyl ether
Ektasolve EE Acetate [Eastman]—Ethylene glycol monoethyl ether acetate
Eltesol ACS60 [Albright & Wilson/Detergents]—Ammonium cumenesulfonate
Eltesol AX40 [Albright & Wilson/Marchon]—Ammonium xylenesulfonate
Eltesol ST34, ST40, ST90 [Albright & Wilson/Marchon]—Sodium toluenesulfonate
Eltesol ST Pellets [Albright & Wilson/Detergents]—Sodium toluenesulfonate
Eltesol SX30, SX93 [Albright & Wilson/Marchon]—Sodium xylenesulfonate

Eltesol SX Pellets [Albright & Wilson/Marchon]—Sodium xylenesulfonate
Emasol P-10 [Kao]—Sorbitan monopalmitate
Emcol 4100M [Witco/Organics]—Disodium myristamido MEA-sulfosuccinate
Emcol 4161L [Witco/Organics]—Disodium oleamido PEG-2 sulfosuccinate
Emcol 4500, 4560 [Witco/Organics]—Sodium dioctyl sulfosuccinate
Emcol 4600 [Witco/Organics]—Sodium ditridecyl sulfosuccinate
Emcol CC-9 [Witco/Organics]—POP (9) methyl diethyl ammonium chloride
Emcol CC-36 [Witco/Organics]—POP (25) methyl diethyl ammonium chloride
Emcol CC-42 [Witco/Organics]—POP (40) methyl diethyl ammonium chloride
Emerest 2622 [Emery]—POE (4) dilaurate
Emerest 2652 [Emery]—POE (8) dilaurate
Emerest 2704 [Emery]—POE (4) dilaurate
Emerest 2706 [Emery]—POE (8) dilaurate
Emerest 2717 [Emery]—POE (100) monostearate
Emery 6705 [Emery]—Ethylene glycol monophenyl ether
Emphos CS-1361 [Witco/Organics]—Sodium nonoxynol-9 phosphate
Empilan KB2 [Albright & Wilson/Marchon]—POE (2) lauryl ether
Empilan KB3, KC3 [Albright & Wilson/Marchon]—POE (3) lauryl ether
Empilan KM20 [Albright & Wilson/Marchon]—POE (20) cetyl/stearyl ether
Empilan NP9 [Albright & Wilson/Marchon]—POE (9) nonyl phenyl ether
Empimin MA [Albright & Wilson/Detergents]—Sodium dihexyl sulfosuccinate
Empimin OP-45, O-70, OT [Albright & Wilson]—Sodium dioctyl sulfosuccinate
Empiphos 4KP [Albright & Wilson/Marchon]—Tetrapotassium pyrophosphate
Empiphos STP, STP/D, STP Gran M, STP/L16 [Albright & Wilson/Phosphates Div.]—
Pentasodium triphosphate
Emsorb 2510 [Emery]—Sorbitan monopalmitate
Emulphogene BC-720 [GAF]—POE (10) tridecyl ether
Emulphogene BC-840 [GAF]—POE (15) tridecyl ether
Emulphor EL-620 [GAF]—POE (30) castor oil
Emulphor EL-719 [GAF]—POE (40) castor oil
Emulphor VN-430 [GAF]—POE (5) monooleate
ESI-Terge SXS [Emulsion Systems]—Sodium xylenesulfonate
Ethoduomeen T/13 [Akzo]—POE (3) tallow aminopropylamine
Ethofat 60/15 [Armak]—POE (5) monostearate
Ethofat 60/20 [Armak]—POE (10) monostearate
Ethofat C/15 [Armak]—POE (5) monococoate
Ethofat C/25 [Armak]—POE (15) monococoate
Ethofat O/15 [Armak]—POE (5) monooleate
Ethofat O/20 [Armak]—POE (10) monooleate
Ethomeen 18/20 [Armak]—POE (10) stearyl amine
Ethomeen 18/25 [Armak]—POE (15) stearyl amine (tertiary amine)
Ethomeen O/12 [Armak]—POE (2) oleyl amine (tertiary amine)

Ethomeen O/15 [Armak]—POE (5) oleyl amine (tertiary amine)
Ethomeen O/25 [Armak]—POE (15) oleyl amine (tertiary amine)
Ethomeen T/60 [Armak]—POE (50) tallow amine
Ethomid HT/60 [Armak]—POE (50) hydrogenated tallow amide
Ethomid O/15 [Armak]—POE (5) oleyl amide
Ethylan 20 [Lankro Chem. Ltd.]—POE (20) nonyl phenyl ether
Ethylan NP1 [Lankro Chem. Ltd.]—POE (1) nonyl phenyl ether
Etocas 10 [Croda]—POE (10) castor oil
Etocas 30 [Croda]—POE (30) castor oil
Etocas 40 [Croda]—POE (40) castor oil
Etocas 100 [Croda Surfactants]—POE (100) castor oil
Eumulgin B-1 [Henkel]—POE (12) cetyl/stearyl ether
Eumulgin B-2 [Henkel]—POE (20) cetyl/stearyl ether
Eumulgin HRE40 [Henkel KGaA]—POE (40) hydrogenated castor oil
Eumulgin HRE60 [Henkel KGaA]—POE (60) hydrogenated castor oil
Eumulgin O5 [Henkel]—POE (5) oleyl ether
Eumulgin RO-40 [Henkel/Canada]—POE (40) hydrogenated castor oil
Finsolv SB [Finetex]—Isostearyl benzoate
Finsolv TN [Finetex]—C12–15 alcohols benzoate
Fosfodril [FMC]—Sodium hexametaphosphate
Fostex AMP [Henkel]—Aminotrimethylene phosphonic acid
Fostex P [Henkel]—Etidronic acid
Gafac RA-600 [GAF]—POE (4) decyl ether phosphate
Gafac RS-610 [GAF]—POE (6) tridecyl ether phosphate
Gemtex PA-70, PA-75, PA-85P, PAX-60, SC-40, SC-70, SC-75 [Finetex]—Sodium dioctyl sulfosuccinate
Geropon AY [Geronazzo SpA]—Sodium diamyl sulfosuccinate
Geropon CYA/45 [Geronazzo SpA]—Sodium diisobutyl sulfosuccinate
Glutrin [Reed Lignin]—Calcium lignosulfonate
Glycol Ether EB [Ashland, Shell]—Ethylene glycol monobutyl ether
Glycomul P [Glyco]—Sorbitan monopalmitate
Goulac [Reed Lignin]—Calcium lignosulfonate
Gradonic N-95 [Graden]—POE (9) nonyl phenyl ether (9.5 EO)
Hartotrope AXS40 [Hart Chem. Ltd.]—Ammonium xylenesulfonate
Hartotrope STS40, STS Powder [Hart Chem.]—Sodium toluenesulfonate
Hartotrope SXS40, SXS Powder [Hart]—Sodium xylenesulfonate
Hetoxamate MO-5 [Heterene]—POE (5) monooleate
Hetoxamate SA-5 [Heterene]—POE (5) monostearate
Hetoxamine O2 [Heterene]—POE (2) oleyl amine
Hetoxamine O5 [Heterene]—POE (5) oleyl amine
Hetoxamine O15 [Heterene]—POE (15) oleyl amine
Hetoxamine ST-15 [Heterene]—POE (15) stearyl amine (tertiary amine)

Hetoxide C-25 [Heterene]—POE (25) castor oil
Hetoxide C-30 [Heterene]—POE (30) castor oil
Hetoxide C-40 [Heterene]—POE (40) castor oil
Hetoxide HC-40 [Heterene]—POE (40) hydrogenated castor oil
Hetoxide HC-60 [Heterene]—POE (60) hydrogenated castor oil
Hetoxide NP-4 [Heterene]—POE (4) nonyl phenyl ether
Hetoxide NP-9 [Heterene]—POE (9) nonyl phenyl ether
Hetoxol CS-15, CSA-15 [Heterene]—POE (15) cetyl/stearyl ether
Hetoxol CS-20 [Heterene]—POE (20) cetyl/stearyl ether
Hetoxol L3N [Heterene]—POE (3) lauryl ether
Hetoxol OL5 [Heterene]—POE (5) oleyl ether
Hetoxol STA-2 [Heterene]—POE (2) stearyl ether
Hetoxol STA-10 [Heterene]—POE (10) stearyl ether
Hetoxol STA-20 [Heterene]—POE (20) stearyl ether
Hetoxol TD-9 [Heterene]—POE (9) tridecyl ether
Hetsulf 50A [Heterene]—Ammonium dodecylbenzenesulfonate
Hetsulf IPA [Heterene]—Monoisopropanolamine dodecylbenzenesulfonate
Hexaphos [FMC]—Sodium hexametaphosphate
Hipochem EK-18 [High Point]—Sodium dioctyl sulfosuccinate
Hodag 22L [Hodag]—POE (4) dilaurate
Hodag CC-22, CC-22-S [Hodag]—Propylene glycol dicaprylate/dicaprate
Hodag CC-33, CC-33-F [Hodag]—Caprylic/capric triglyceride
Hodag SMP [Hodag]—Sorbitan monopalmitate
Hodag SVO-629 [Hodag]—Hexaglyceryl distearate (veg. grade)
Hostacerin O-5 [Amer. Hoechst]—POE (5) oleyl ether
Hostacerin T-3 [Amer. Hoechst]—POE (3) cetyl/stearyl ether
Hostapal N-040 [Amer. Hoechst]—POE (4) nonyl phenyl ether
Hostapal N-060 [Amer. Hoechst]—POE (6) nonyl phenyl ether
Hostapal N-090 [Amer. Hoechst]—POE (9) nonyl phenyl ether
Hostapal N-100 [Amer. Hoechst]—POE (10) nonyl phenyl ether
Hostapon CT Paste [Hoechst AG]—Sodium methyl cocoyl taurate
Hostapon KA Pdr. Hi Conc. [Amer. Hoechst]—Sodium cocoyl isethionate
Hostapon KA Special [Hoechst AG]—Sodium cocoyl isethionate
Hostapon T [Hoechst AG]—Sodium methyl oleoyl taurate
Hostapon T Paste 33, T Powder H.C. [Amer. Hoechst]—Sodium methyl oleoyl taurate
Hyonic NP-40 [Diamond Shamrock]—POE (4) nonyl phenyl ether
Hyonic NP-60 [Diamond Shamrock]—POE (6) nonyl phenyl ether
Hyonic NP-90 [Diamond Shamrock]—POE (9) nonyl phenyl ether
Hyonic NP-100 [Diamond Shamrock]—POE (10) nonyl phenyl ether
Hyonic NP-110 [Diamond Shamrock]—POE (11) nonyl phenyl ether
Hyonic NP-120 [Diamond Shamrock]—POE (12) nonyl phenyl ether
Hyonic OP-70 [Diamond Shamrock]—POE (7) octyl phenyl ether

Hyonic OP-80 [Diamond Shamrock]—POE (8) octyl phenyl ether
Hyonic OP-100 [Diamond Shamrock]—POE (9) octyl phenyl ether
Hyonic OP-407 [Diamond Shamrock]—POE (40) octyl phenyl ether
Hyonic OP-705 [Diamond Shamrock]—POE (70) octyl phenyl ether
Hyonic PE-100 [Diamond Shamrock]—POE (10) nonyl phenyl ether
Iconol DNP-150 [BASF Wyandotte]—POE (150) dinonyl phenyl ether
Iconol NP-1.5 [BASF Wyandotte]—POE (1) nonyl phenyl ether (1.5 EO)
Iconol NP-4 [BASF Wyandotte]—POE (4) nonyl phenyl ether
Iconol NP-5 [BASF Wyandotte]—POE (5) nonyl phenyl ether
Iconol NP-6 [BASF Wyandotte]—POE (6) nonyl phenyl ether
Iconol NP-9 [BASF Wyandotte]—POE (9) nonyl phenyl ether
Iconol NP-10 [BASF Wyandotte]—POE (10) nonyl phenyl ether
Iconol NP-15 [BASF Wyandotte]—POE (15) nonyl phenyl ether
Iconol NP-20 [BASF Wyandotte]—POE (20) nonyl phenyl ether
Iconol OP-5 [BASF Wyandotte]—POE (5) octyl phenyl ether
Iconol OP-7 [BASF Wyandotte]—POE (7) octyl phenyl ether
Iconol OP-10 [BASF Wyandotte]—POE (10) octyl phenyl ether
Iconol OP-30 [BASF Wyandotte]—POE (30) octyl phenyl ether
Iconol OP-40, O-40-70% [BASF Wyandotte]—POE (40) octyl phenyl ether
Iconol TDA-9 [BASF Wyandotte]—POE (9) tridecyl ether
Iconol TDA-10 [BASF Wyandotte]—POE (10) tridecyl ether
Igepal CA-210 [GAF]—POE (1) octyl phenyl ether (1.5 EO)
Igepal CA-420 [GAF]—POE (3) octyl phenyl ether
Igepal CA-520 [GAF]—POE (5) octyl phenyl ether
Igepal CA-620 [GAF]—POE (7) octyl phenyl ether
Igepal CA-630 [GAF]—POE (9) octyl phenyl ether
Igepal CA-720 [GAF]—POE (12) octyl phenyl ether (12.5 EO)
Igepal CA-887 [GAF]—POE (30) octyl phenyl ether
Igepal CA-890, CA-897 [GAF]—POE (40) octyl phenyl ether
Igepal CO-210 [GAF]—POE (1) nonyl phenyl ether (1.5 EO)
Igepal CO-430 [GAF]—POE (4) nonyl phenyl ether
Igepal CO-520 [GAF]—POE (5) nonyl phenyl ether
Igepal CO-530 [GAF]—POE (6) nonyl phenyl ether
Igepal CO-610 [GAF]—POE (8) nonyl phenyl ether
Igepal CO-620 [GAF]—POE (8) nonyl phenyl ether (8.5 EO)
Igepal CO-630 [GAF]—POE (9) nonyl phenyl ether
Igepal CO-660 [GAF]—POE (10) nonyl phenyl ether
Igepal CO-710 [GAF]—POE (10) nonyl phenyl ether (10–11 EO)
Igepal CO-720 [GAF]—POE (12) nonyl phenyl ether
Igepal CO-730 [GAF]—POE (15) nonyl phenyl ether
Igepal CO-850 [GAF]—POE (20) nonyl phenyl ether
Igepal DM-970 [GAF]—POE (150) dinonyl phenyl ether

Igepon AC-78 [GAF]—Sodium cocoyl isethionate
Igepon T-33, T-43, T-51, T-77 [GAF]—Sodium methyl oleoyl taurate
Igepon TC-42 [GAF]—Sodium methyl cocoyl taurate
Igepon TM-43 [GAF]—Sodium methyl oleoyl taurate
Imwitor 308 [Dynamit-Nobel]—Glyceryl monocaprylate
Imwitor 310 [Dynamit-Nobel]—Glyceryl monocaprate
Imwitor 742 [Dynamit-Nobel]—Caprylic/capric glycerides
Incrocas 10 [Croda Surfactants]—POE (10) castor oil
Incrocas 30 [Croda Surfactants]—POE (30) castor oil
Incrocas 100 [Croda Surfactants]—POE (100) castor oil
Incropol CS-12 [Croda Surfactants]—POE (12) cetyl/stearyl ether
Incropol CS-20 [Croda Surfactants]—POE (20) cetyl/stearyl ether
Incropol L-2 [Croda]—POE (2) lauryl ether
Incropol L-7, L-7-90 [Croda Surfactants Inc.]—POE (7) lauryl ether
Industrol COH-200 [BASF Wyandotte]—POE (200) hydrogenated castor oil
Industrol DL-9 [BASF Wyandotte]—POE (8) dilaurate
Industrol DT-13 [BASF Wyandotte]—POE (12) ditallate
Industrol MS-5 [BASF Wyandotte]—POE (5) monostearate
Jet Quat T-50 [Jetco]—Tallow trimethyl ammonium chloride
Kemamine P-650 [Humko Sheffield]—Coconut amine (tech.)
Kemamine P-650D [Humko Sheffield]—Coconut amine (distilled)
Kemamine P-970 [Humko Sheffield]—Tallow amine, hydrogenated (tech.)
Kemamine P-970D [Humko Sheffield]—Tallow amine, hydrogenated (distilled)
Kemamine P-974 [Humko Sheffield]—Tallow amine (tech.)
Kemamine P-974D [Humko Sheffield]—Tallow amine (distilled)
Kemamine P-990 [Humko Sheffield]—Stearyl amine (tech.)
Kemamine P-990D [Humko Sheffield]—Stearyl amine (distilled)
Kemamine P-997 [Humko Sheffield]—Soya amine (tech.)
Kemamine P-997D [Humko Sheffield]—Soya amine (distilled)
Kessco PEG 200 Dilaurate [Armak]—POE (4) dilaurate
Kessco PEG 300 Dilaurate [Armak]—POE (6) dilaurate
Kessco PEG 400 Dilaurate [Armak]—POE (8) dilaurate
Kessco PEG 600 Dilaurate [Armak]—POE (12) dilaurate
Kessco PEG 1540 Distearate [Armak]—POE (32) distearate
Kessco PEG 6000 Monostearate [Armak]—POE (150) monostearate
Kuplur SMP [BASF Wyandotte]—Sorbitan monopalmitate
LA-55-3 [Hefti Ltd.]—POE (3) lauryl ether
Lakeway SXS [Bofors Lakeway]—Sodium xylenesulfonate
Lamacit 877 [Grunau]—POE (14) nonyl phenyl ether
Lankropol KMA [Lankro]—Sodium dihexyl sulfosuccinate
Lankropol KO2, KO Special [Lankro]—Sodium dioctyl sulfosuccinate
Lankropol ODS/LS, ODS/PT [Lankro]—Disodium stearyl sulfosuccinamate

Lankrosol SXS-30 [Lankro]—Sodium xylenesulfonate
Lexin K [American Lecithin]—Lecithin
Lexol GT855, GT865 [Inolex]—Caprylic/capric triglyceride
Lexol PG855, PG865 [Inolex]—Propylene glycol dicaprylate/dicaprate
Lignosite 401 [Georgia-Pacific]—Calcium lignosulfonate
Lignosite 431, 458, 823, 854 [Georgia-Pacific/Bellingham Div.]—Sodium lignosulfonate
Lignosite 1840, L [Georgia-Pacific]—Calcium lignosulfonate
Lignosol AXD [Reed]—Sodium lignosulfonate (acidic desugarized, partially desulfonated)
Lignosol B, BD, LC [Reed Ltd.]—Calcium lignosulfonate
Lignosol D-10, D-30 [Reed]—Sodium lignosulfonate
Lignosol DXD [Reed]—Sodium lignosulfonate (alkaline desugarized, partially desulfonated)
Lignosol FTA [Reed]—Sodium lignosulfonate (desugarized, desulfonated)
Lignosol HCX [Reed]—Sodium lignosulfonate (alkaline sugar-free)
Lignosol NSX-110 [Reed Ltd.]—Sodium lignosulfonate
Lignosol NSX-120 [Reed]—Sodium lignosulfonate (modified)
Lignosol SF, SFL [Reed Ltd.] (desugarized)—Calcium lignosulfonate
Lignosol SFS [Reed]—Sodium lignosulfonate (sugar-free)
Lignosol SFX [Reed]—Sodium lignosulfonate (desugarized)
Lignosol SFX-65, XD, X [Reed Ltd.]—Sodium lignosulfonate
Lignosol WT [Reed]—Sodium lignosulfonate
Lilamin 115 [Lilachim S.A.]—Soya amine
Lilamin 115D [Lilachim S.A.]—Soya amine (distilled)
Lilamin 140 [Lilachim S.A.]—Tallow amine, hydrogenated
Lilamin 140D [Lilachim S.A.]—Tallow amine, hydrogenated (distilled)
Lilamin 142 [Lilachim S.A.]—Stearyl amine
Lilamin 142S [Lilachim S.A.]—Stearyl amine (distilled)
Lilamin 160 [Lilachim S.A.]—Coconut amine
Lilamin 160D [Lilachim S.A.]—Coconut amine (distilled)
Lilamin 170 [Lilachim S.A.]—Tallow amine
Lilamin 170D [Lilachim S.A.]—Tallow amine (distilled)
Lipal 10TD [PVO]—POE (10) tridecyl ether
Lipal 15CSA [PVO]—POE (15) cetyl/stearyl ether
Lipal 20SA [PVO]—POE (20) stearyl ether
Lipal 400DL [PVO]—POE (8) dilaurate
Lipocol O-2 [Lipo]—POE (5) oleyl ether
Lipocol S-2 [Lipo]—POE (2) stearyl ether
Lipocol S-10 [Lipo]—POE (10) stearyl ether
Lipocol S-20 [Lipo]—POE (20) stearyl ether
Lipocol SC-4 [Lipo]—POE (4) cetyl/stearyl ether

Lipocol SC-15 [Lipo]—POE (15) cetyl/stearyl ether
Lipocol SC-20 [Lipo]—POE (20) cetyl/stearyl ether
Liponate GC [Lipo]—Caprylic/capric triglyceride
Liponate PC [Lipo]—Propylene glycol dicaprylate/dicaprate
Lipopeg 2DL [Lipo]—POE (4) dilaurate
Lipopeg 4DL [Lipo]—POE (8) dilaurate
Lipopeg 100-S [Lipo]—POE (100) monostearate
Liposorb P [Lipo]—Sorbitan monopalmitate
Lonzest SMP [Lonza]—Sorbitan monopalmitate
Lorapal HV5 [Dutton & Reinisch]—POE (5) nonyl phenyl ether
Lorapal HV8 [Dutton & Reinisch]—POE (8) nonyl phenyl ether
Lorapal HV9 [Dutton & Reinisch]—POE (9) nonyl phenyl ether
Lorapal HV10 [Dutton & Reinisch]—POE (10) nonyl phenyl ether
Lubrol N13 [ICI Ltd. Organics Div.]—POE (13) nonyl phenyl ether
Mackanate CM [McIntyre]—Disodium cocoyl monoethanolamide sulfosuccinate
Mackanate DOS-70M5, DOS-75 [McIntyre]—Sodium dioctyl sulfosuccinate
Mackanate OD [McIntyre]—Disodium oleamido PEG-2 sulfosuccinate
Mackazoline C [McIntyre]—Cocoyl imidazoline
Macol CSA-2 [Mazer]—POE (2) cetyl/stearyl ether
Macol CSA-4 [Mazer]—POE (4) cetyl/stearyl ether
Macol CSA-10 [Mazer]—POE (10) cetyl/stearyl ether
Macol CSA-15 [Mazer]—POE (15) cetyl/stearyl ether
Macol CSA-20 [Mazer]—POE (20) cetyl/stearyl ether
Macol LA-790 [Mazer]—POE (7) lauryl ether
Macol NP-4 [Mazer]—POE (4) nonyl phenyl ether
Macol NP-6 [Mazer]—POE (6) nonyl phenyl ether
Macol NP-9.5 [Mazer]—POE (9) nonyl phenyl ether (9.5 EO)
Macol NP-11 [Mazer]—POE (11) nonyl phenyl ether
Macol NP-20 [Mazer]—POE (20) nonyl phenyl ether
Macol OA-5 [Mazer]—POE (5) oleyl ether
Macol SA-2 [Mazer]—POE (2) stearyl ether
Macol SA-10 [Mazer]—POE (10) stearyl ether
Macol SA-15 [Mazer]—POE (15) stearyl ether
Macol SA-20 [Mazer]—POE (20) stearyl ether
Macol SA-40 [Mazer]—POE (40) stearyl ether
Macol TD-10 [Mazer]—POE (10) tridecyl ether
Makon 4 [Stepan]—POE (4) nonyl phenyl ether
Makon 6 [Stepan]—POE (6) nonyl phenyl ether
Makon 8 [Stepan]—POE (8) nonyl phenyl ether
Makon 10 [Stepan]—POE (10) nonyl phenyl ether
Makon 12 [Stepan]—POE (12) nonyl phenyl ether
Makon 14 [Stepan]—POE (14) nonyl phenyl ether

Makon 20 [Stepan]—POE (20) nonyl phenyl ether
Makon OP-9 [Stepan]—POE (9) octyl phenyl ether (9–10 moles EO)
Manoxol OT, OT60%, OT/B [Manchem Ltd.]—Sodium dioctyl sulfosuccinate
Manoxol OT/P [Manchem Ltd.]—Sodium dioctyl sulfosuccinate (pharmaceutical grade)
Manro STS40 [Manro]—Sodium toluenesulfonate
Manro SXS30, SXS40, SXS93 [Manro]—Sodium xylenesulfonate
Mapeg 200DL [Mazer]—POE (4) dilaurate
Mapeg 400DL [Mazer]—POE (8) dilaurate
Mapeg 600 DOT [Mazer]—POE (12) ditallate
Mapeg 600DL [Mazer]—POE (12) dilaurate
Mapeg 1540 DS [Mazer]—POE (32) distearate
Mapeg 6000 MS [Mazer]—POE (150) monostearate
Mapeg S-100 [Mazer]—POE (100) monostearate
Maracell E [Amer. Can]—Sodium lignosulfonate (partially desulfonated)
Marasperse C-21 [Amer. Can]—Calcium lignosulfonate
Marasperse CB [Amer. Can]—Sodium lignosulfonate
Marasperse CBOS-3 [Reed]—Sodium lignosulfonate (partially desulfonated)
Marasperse CBX-2 [Reed]—Sodium lignosulfonate
Marasperse N-22 [Amer. Can]—Sodium lignosulfonate
Marlinat DF8 [Chemische Werke Huls AG]—Sodium dioctyl sulfosuccinate
Marlipal MG [Chemische Werke Huls AG]—POE (7) lauryl ether
Mazol PGO-108 [Mazer]—Decaglycerol octaoleate
Mazol PGO-1010 [Mazer]—Decaglycerol decaoleate
Merpasol DIO60, DIO75 [Kempen]—Sodium dioctyl sulfosuccinate
Merpoxen NO15 [Elektrochemische Fabrik Kempen]—POE (1) nonyl phenyl ether (1.5 EO)
Merpoxen NO40 [Elektrochemische Fabrik Kempen]—POE (4) nonyl phenyl ether
Merpoxen NO60 [Elektrochemische Fabrik Kempen]—POE (6) nonyl phenyl ether
Merpoxen NO80 [Elektrochemische Fabrik Kempen]—POE (8) nonyl phenyl ether
Merpoxen NO90 [Elektrochemische Fabrik Kempen]—POE (9) nonyl phenyl ether
Merpoxen NO95 [Elektrochemische Fabrik Kempen]—POE (9) nonyl phenyl ether (9.5 EO)
Merpoxen NO100 [Elektrochemische Fabrik Kempen]—POE (10) nonyl phenyl ether
Merpoxen NO110 [Elektrochemische Fabrik Kempen]—POE (11) nonyl phenyl ether
Merpoxen NO120 [Elektrochemische Fabrik Kempen]—POE (12) nonyl phenyl ether
Merpoxen NO150 [Elektrochemische Fabrik Kempen]—POE (15) nonyl phenyl ether
Merpoxen NO200 [Elektrochemische Fabrik Kempen]—POE (20) nonyl phenyl ether
Merpoxen RO300 [Elektrochemische Fabrik Kempen]—POE (30) castor oil
Merpoxen RO400 [Elektrochemische Fabrik Kempen]—POE (40) castor oil
Methyl Carbitol [Union Carbide]—Diethylene glycol monomethyl ether
Metso 20 [PQ Corp.]—Sodium metasilicate
Metso Beads 2048 [PQ Corp.]—Sodium metasilicate (anhydrous)

Metso Pentabead 20 [PQ Corp.]—Sodium metasilicate (pentahydrate)
Miglyol 810, 812 [Dynamit-Nobel]—Caprylic/capric triglyceride
Miglyol 840 [Dynamit-Nobel]—Propylene glycol dicaprylate/dicaprate
Miramine CC [Miranol]—Cocoyl imidazoline
Mirataine COB [Miranol]—Coco/oleamidopropyl betaine
Mirataine T2C [Miranol]—Disodium tallowiminodipropionate
Mirataine TABS [Miranol]—Tallowamidopropyl hydroxysultaine
Monamate OPA-30, OPA-100 [Mona]—Disodium oleamido PEG-2 sulfosuccinate
Monateric TDB-35 [Mona]—Disodium tallowiminodipropionate
Monawet MB-45 [Mona]—Sodium diisobutyl sulfosuccinate
Monawet MM-80 [Mona]—Sodium dihexyl sulfosuccinate
Monawet MO-65-150, MO-65PEG, MO-70, MO-70-150, MO-70E, MO-70R, MO-70RP, MO-70S, MO-75E, MO-84R2W, MO85P [Mona]—Sodium dioctyl sulfosuccinate
Monawet MT-70, MT-70E, MT-80H2W [Mona]—Sodium ditridecyl sulfosuccinate
Monawet SNO-35 [Mona]—Tetrasodium dicarboxyethyl stearyl sulfosuccinamate
Monawet TD-30 [Mona]—Disodium deceth-6 sulfosuccinate
Monazoline C [Mona]—Cocoyl imidazoline
Montane 40 [Seppic]—Sorbitan monopalmitate
Montane 70 [Seppic]—Sorbitan monoisostearate
Montanox 70 [Seppic]—POE (20) sorbitan monoisostearate
Myritol 318 [Henkel, Henkel KGaA]—Caprylic/capric triglyceride
Myrj 59 [ICI]—POE (100) monostearate
Na-Cumene Sulfonate 40, Powder [Chemische Werke Huls AG]—Sodium cumenesulfonate
Na-Toluene Sulfonate 30, 40 [Chemische Werke Huls AG]—Sodium toluenesulfonate
Nansa AS40 [Albright & Wilson/Marchon]—Ammonium dodecylbenzenesulfonate
Naxonate 4AX [Nease]—Ammonium xylenesulfonate
Naxonate 4L, 5L, G [Nease]—Sodium xylenesulfonate
Naxonate 4ST, ST [Ruetgers Nease]—Sodium toluenesulfonate
Naxonate 6AC [Nease]—Ammonium cumenesulfonate
Naxonate 45SC, SC [Nease]—Sodium cumenesulfonate
Nekal WT-27 [GAF]—Sodium dioctyl sulfosuccinate
Neobee M-5, O [PVO]—Caprylic/capric triglyceride
Neobee M-20 [PVO]—Propylene glycol dicaprylate/dicaprate
Neodol 91-6 [Shell]—Pareth-91-6
Neutronyx 600 [Onyx]—POE (9) nonyl phenyl ether (9.5 EO)
Neutronyx 656 [Onyx]—POE (11) nonyl phenyl ether
Neutronyx S-60 [Onyx]—Ammonium nonoxynol-4 sulfate
Newcol 40 [Nippon Nyukazai]—Sorbitan monopalmitate
Newcol 210 [Nippon Nyukazai]—Ammonium dodecylbenzenesulfonate
Nikkol BL-2 [Nikko]—POE (2) lauryl ether

Nikkol CMT-30 [Nikko]—Sodium methyl cocoyl taurate
Nikkol CO-10 [Nikko]—POE (10) castor oil
Nikkol CO-40TX [Nikko]—POE (40) castor oil
Nikkol HCO-40 [Nikko]—POE (40) hydrogenated castor oil
Nikkol HCO-50 [Nikko]—POE (50) hydrogenated castor oil
Nikkol HCO-60 [Nikko]—POE (60) hydrogenated castor oil
Nikkol HCO-100 [Nikko]—POE (100) hydrogenated castor oil
Nikkol MYS-10 [Nikko]—POE (10) monostearate
Nikkol NP-7.5 [Nikko]—POE (8) nonyl phenyl ether (7.5 EO)
Nikkol NP-10 [Nikko]—POE (10) nonyl phenyl ether
Nikkol NP-15 [Nikko]—POE (15) nonyl phenyl ether
Nikkol OP-3 [Nikko]—POE (3) octyl phenyl ether
Nikkol OP-10 [Nikko]—POE (10) octyl phenyl ether
Nikkol OP-30 [Nikko]—POE (30) octyl phenyl ether
Nikkol PEN-4612 [Nikko]—POP (6) POE (12) tetradecyl ether
Nikkol PEN-4620 [Nikko]—POP (6) POE (20) tetradecyl ether
Nikkol PEN-4630 [Nikko]—POP (6) POE (30) tetradecyl ether
Nikkol SI-10R, SI-10T, SI-15R, SI-15T [Nikko]—Sorbitan monoisostearate
Nikkol SP-10 [Nikko]—Sorbitan monopalmitate
Nikkol TI-10 [Nikko]—POE (20) sorbitan monoisostearate
Ninex 303 [Stepan/Europe]—Sodium xylenesulfonate
Nissan Amine AB [Nippon Oil & Fats]—Stearyl amine
Nissan Amine ABT [Nippon Oil & Fats]—Tallow amine, hydrogenated
Nissan Nonion HS-240 [Nippon Oil & Fats]—POE (40) octyl phenyl ether
Nissan Nonion HS-270 [Nippon Oil & Fats]—POE (70) octyl phenyl ether
Nissan Nonion NS-206 [Nippon Oil & Fats]—POE (6) nonyl phenyl ether
Nissan Nonion NS-212 [Nippon Oil & Fats]—POE (12) nonyl phenyl ether
Nissan Nonionic PP-40R [Nippon Oil & Fats]—Sorbitan monopalmitate
Nissan Rapisol B-30, B-80 [Nippon Oil & Fats]—Sodium dioctyl sulfosuccinate
Nopalcol 6-DTW [Diamond Shamrock]—POE (12) ditallate
Norfox NP-1 [Norman, Fox & Co.]—POE (1) nonyl phenyl ether (1.5 EO)
Norfox NP-4 [Norman, Fox & Co.]—POE (4) nonyl phenyl ether
Norfox NP-6 [Norman, Fox & Co.]—POE (6) nonyl phenyl ether
Norfox NP-9 [Norman, Fox & Co.]—POE (9) nonyl phenyl ether
Norfox NP-11 [Norman, Fox & Co.]—POE (11) nonyl phenyl ether
Norfox SXS-40, SXS-96 [Norman, Fox & Co.]—Sodium xylenesulfonate
Norlig 12 [Amer. Can]—Sodium lignosulfonate
Norlig A [Amer. Can]—Calcium lignosulfonate
NP-55-40 [Hefti Ltd.]—POE (4) nonyl phenyl ether
NP-55-50 [Hefti Ltd.]—POE (5) nonyl phenyl ether
NP-55-60 [Hefti Ltd.]—POE (6) nonyl phenyl ether
NP-55-80 [Hefti Ltd.]—POE (8) nonyl phenyl ether

NP-55-90 [Hefti Ltd.]—POE (9) nonyl phenyl ether
NP-55-95 [Hefti Ltd.]—POE (9) nonyl phenyl ether (9.5 EO)
NP-55-120 [Hefti Ltd.]—POE (12) nonyl phenyl ether
NP-55-150 [Hefti Ltd.]—POE (15) nonyl phenyl ether
Nutrol 100 [Clough Chem. Co. Ltd.]—POE (9) octyl phenyl ether
Nutrol 600 [Clough Chem. Ltd.]—POE (9) nonyl phenyl ether
Nutrol 611 [Clough Chem. Ltd.]—POE (8) nonyl phenyl ether
Nutrol 622 [Clough Chem. Ltd.]—POE (4) nonyl phenyl ether
Nutrol 640 [Clough Chem. Ltd.]—POE (15) nonyl phenyl ether
Nutrol 656 [Clough Chem. Ltd.]—POE (11) nonyl phenyl ether
Octapol 100 [Sanyo]—POE (10) octyl phenyl ether
Octapol 300 [Sanyo]—POE (30) octyl phenyl ether
Octapol 400 [Sanyo]—POE (40) octyl phenyl ether
Orzan LS, LS-50, SL-50 [ITT Rayonier]—Sodium lignosulfonate
Orzan S [Crown Zellerbach]—Sodium lignosulfonate
Oxitol [Shell]—Ethylene glycol monoethyl ether
PB-92 [Reed Lignin]—Calcium lignosulfonate (modified)
Peganol NP1.5 [Borg-Warner]—POE (1) nonyl phenyl ether (1.5 EO)
Peganol NP4 [Borg-Warner]—POE (4) nonyl phenyl ether
Peganol NP5 [Borg-Warner]—POE (5) nonyl phenyl ether
Peganol NP6 [Borg-Warner]—POE (6) nonyl phenyl ether
Peganol NP9 [Borg-Warner]—POE (9) nonyl phenyl ether
Peganol NP10 [Borg-Warner]—POE (10) nonyl phenyl ether
Peganol NP12 [Borg-Warner]—POE (12) nonyl phenyl ether
Peganol NP15 [Borg-Warner]—POE (15) nonyl phenyl ether
Peganol NP20 [Borg-Warner]—POE (20) nonyl phenyl ether
Peganol OP3 [GAF]—POE (3) octyl phenyl ether
Peganol OP6 [GAF]—POE (6) octyl phenyl ether
Peganol OP8 [GAF]—POE (8) octyl phenyl ether
Peganol OP10 [GAF]—POE (10) octyl phenyl ether
Peganol OP30, OP 30 70% [GAF]—POE (30) octyl phenyl ether
Peganol OP40, OP40 70% [GAF]—POE (40) octyl phenyl ether
Peganol OP70, OP70 70% [GAF]—POE (70) octyl phenyl ether
Pegosperse 100MR [Glyco]—Diethylene glycol monoricinoleate
Pegosperse 200DL [Glyco]—POE (4) dilaurate
Pegosperse 400DL [Glyco]—POE (8) dilaurate
Pegosperse 400DTR [Glyco]—POE (8) ditriricinoleate
Pegosperse 1500 DO [Glyco]—POE 1500 dioleate
Penetron Conc. [Hart Prod. Corp.]—Sodium dioctyl sulfosuccinate
Pentex 99 [Colloids Inc.]—Sodium dioctyl sulfosuccinate
Perk [Stauffer Chem.]—Perchloroethylene (dry cleaning grade, industrial grade)
Perkare [PPG]—Perchloroethylene (activated; with detergent and antistat)

Phenyl Cellosolve [Union Carbide]—Ethylene glycol monophenyl ether
Pilot SXS-40, SXS-96 [Pilot]—Sodium xylenesulfonate
Plurafac D-25 [BASF Wyandotte]—PPG-6 Pareth-28-11
Polyaldo DGDO [Glyco]—Decaglycerol decaoleate
Polyaldo HGDS [Lonza]—Hexaglyceryl distearate
Polybor 3 [U.S. Borax]—Sodium borate
Polyfon F, H, O, T [Westvaco]—Sodium lignosulfonate
Polyphos [Olin]—Sodium hexametaphosphate
Poly-Solv DB [Olin]—Diethylene glycol monobutyl ether
Poly-Solv DE (High Gravity), DE (Low Gravity) [Olin]—Diethylene glycol monoethyl ether
Poly-Solv DM [Olin]—Diethylene glycol monomethyl ether
Poly-Solv DPM [Olin]—Dipropylene glycol monomethyl ether
Poly-Solv EB [Olin]—Ethylene glycol monobutyl ether
Poly-Solv EE [Olin]—Ethylene glycol monoethyl ether
Poly-Solv MPM [Olin]—Propylene glycol monomethyl ether
Poly-Solv TPM [Olin]—Tripropylene glycol monomethyl ether
Polystep A-2 [Stepan]—Sodium xylenesulfonate
Polystep F-1 [Stepan]—POE (4) nonyl phenyl ether
Polystep F-2 [Stepan]—POE (6) nonyl phenyl ether
Polystep F-3 [Stepan]—POE (8) nonyl phenyl ether
Polystep F-4 [Stepan]—POE (10) nonyl phenyl ether
Polystep F-5 [Stepan]—POE (12) nonyl phenyl ether
Polystep F-7 [Stepan]—POE (15) nonyl phenyl ether
Polystep F-8 [Stepan]—POE (20) nonyl phenyl ether
PPG Perchlor [PPG]—Perchloroethylene (dry cleaning grade, degreasing/general solvent grade)
PPG Perchlor HD [PPG]—Perchloroethylene (heavy-duty grade)
Prelete Defluxer [Dow]—Trichloroethane (inhibited with alcohols)
Protowet 4337 [Procter Chem. Co.]—Sodium dicyclohexyl sulfosuccinate
Punctilious Pure Ethyl Alcohol [U.S. Industrial Chem.]—Ethyl alcohol
Pyrobor [Kerr-McGee]—Sodium borate (anhyd.)
Quimipol ENF80 [Quimigal-Quimica de Portugal]—POE (8) nonyl phenyl ether
Radiamine 6140 [Oleofina S.A.]—Tallow amine, hydrogenated
Radiamine 6141 [Oleofina S.A.]—Tallow amine, hydrogenated (distilled)
Radiamine 6160 [Oleofina S.A.]—Coconut amine
Radiamine 6161 [Oleofina S.A.]—Coconut amine (distilled)
Radiamine 6170, 6171 [Oleofina]—Tallow amine
Radiamuls 135, SORB 2135 [Oleofina SA]—Sorbitan monopalmitate
Radiaquat 6471 [Oleofina S.A.]—Tallow trimethyl ammonium chloride
Radiasurf 7135 [Oleofina SA]—Sorbitan monopalmitate
Raylig [ITT Rayonier]—Sodium lignosulfonate

Raymix [ITT Rayonier]—Sodium lignosulfonate (desugared)
Reax 15B [Westvaco]—Sodium lignosulfonate (modified, Kraft)
Reax 45A, 45L, 80C, 81A, 82, 83A, 85A [Westvaco]—Sodium lignosulfonate (modified)
Reax 88B [Westvaco]—Sodium lignosulfonate (sugar-free, modified Kraft)
Reax SR-1, SR-7 [Westvaco]—Sodium lignosulfonate (modified Kraft)
Renex 31 [ICI United States]—POE (15) tridecyl ether
Renex 647 [Atlas Chem. Ind. NV]—POE (4) nonyl phenyl ether
Renex 648 [ICI United States]—POE (5) nonyl phenyl ether
Renex 649 [Atlas Chem. Industries NV]—POE (20) nonyl phenyl ether
Renex 678 [ICI United States]—POE (15) nonyl phenyl ether
Renex 679 [Atlas Chem. Industries NV]—POE (13) nonyl phenyl ether
Renex 682 [ICI United States]—POE (12) nonyl phenyl ether
Renex 688 [ICI United States]—POE (8) nonyl phenyl ether
Renex 690 [ICI United States]—POE (10) nonyl phenyl ether
Renex 697 [ICI United States]—POE (6) nonyl phenyl ether
Renex 698 [ICI United States]—POE (9) nonyl phenyl ether (9–9.5 EO)
Renex 750 [ICI Specialty Chem.]—POE (10) octyl phenyl ether
Rewopal HV4 [Rewo Chem. Werke GmbH]—POE (4) nonyl phenyl ether
Rewopal HV5 [Dutton & Reinisch]—POE (5) nonyl phenyl ether
Rewopal HV6 [Rewo Chem. Werke GmbH]—POE (6) nonyl phenyl ether
Rewopal HV8 [Dutton & Reinisch]—POE (8) nonyl phenyl ether
Rewopal HV9 [Dutton & Reinisch]—POE (9) nonyl phenyl ether
Rewopal HV10 [Dutton & Reinisch]—POE (10) nonyl phenyl ether
Rewopol B2003 [Rewo]—Tetrasodium dicarboxyethyl stearyl sulfosuccinamate
Rewopol NEHS40, SBDO70 [Rewo]—Sodium dioctyl sulfosuccinate
Rewopol SBC 212, SBC 212/G [Rewo Chemische Werke GmbH]—Disodium cocoyl monoethanolamide sulfosuccinate
Rewopol SBF18 [Rewo]—Disodium stearyl sulfosuccinamate
Reworyl ACS60 [Rewo Chem. Werke GmbH]—Ammonium cumenesulfonate
Reworyl NCS40 [Rewo Chem. Werke GmbH]—Sodium cumenesulfonate
Reworyl NTS40 [Rewo Chem. Werke GmbH]—Sodium toluenesulfonate
Reworyl NXS40 [Rewo]—Sodium xylenesulfonate
Rexol 25/1 [Hart Chem. Ltd.]—POE (1) nonyl phenyl ether
Rexol 25/20 [Hart Chem. Ltd.]—POE (20) nonyl phenyl ether
Rexol 45/1 [Hart Chem. Ltd.]—POE (1) octyl phenyl ether
Rexol 45/3 [Hart Chem. Ltd.]—POE (3) octyl phenyl ether
Rexol 45/5 [Hart Chem. Ltd.]—POE (5) octyl phenyl ether
Rexol 45/7 [Hart Chem. Ltd.]—POE (7) octyl phenyl ether
Rexol 45/10 [Hart Chem. Ltd.]—POE (10) octyl phenyl ether
Rexol 45/12 [Hart Chem. Ltd.]—POE (12) octyl phenyl ether
Rexol 45/307 [Hart Chem. Ltd.]—POE (30) octyl phenyl ether
Rexol 45/407 [Hart Chem. Ltd.]—POE (40) octyl phenyl ether

Richonate SXS [Richardson]—Sodium xylenesulfonate
RS-55-100 [Hefti Ltd.]—POE (100) monostearate
Ryoto Sugar Ester S-570, S-770 [Ryoto Co.]—Sucrose distearate
Ryoto Sugar Ester S-1570, S-1670 [Ryoto Co. Ltd.]—Sucrose monostearate
S-Maz 40 [Mazer]—Sorbitan monopalmitate
S-Maz 67 [Mazer]—Sorbitan monoisostearate
Schercopol CMS-Na [Scher]—Disodium cocoyl monoethanolamide sulfosuccinate
Schercozoline C [Scher]—Cocoyl imidazoline
Simulsol 72 [Seppic]—POE (2) stearyl ether
Siponic L-7-90 [Alcolac]—POE (7) lauryl ether
ST-55-2 [Hefti Ltd.]—POE (2) stearyl ether
Sanmorin OT70 [Sayno Chem. Ind. Ltd.]—Sodium dioctyl sulfosuccinate
Santone 10-10-O [Durkee]—Decaglycerol decaoleate
Scher PEG 200 Dilaurate [Scher]—POE (4) dilaurate
Scher PEG 400 Dilaurate [Scher]—POE (8) dilaurate
Schercomid HT-60 [Scher]—POE (50) hydrogenated tallow amide
Schercopol DOS-70, DOS-PG-85 [Scher]—Sodium dioctyl sulfosuccinate
Secosol DOS/70 [Stepan Europe]—Sodium dioctyl sulfosuccinate
Serdet DNK30 [Servo B.V.]—Sodium nonoxynol-4 sulfate
Serdox NNP4 [Servo B.V.]—POE (4) nonyl phenyl ether
Serdox NNP5 [Servo B.V.]—POE (5) nonyl phenyl ether
Serdox NNP6 [Servo B.V.]—POE (6) nonyl phenyl ether
Serdox NNP8.5 [Servo B.V.]—POE (8) nonyl phenyl ether (8.5 EO)
Serdox NNP9 [Servo B.V.]—POE (9) nonyl phenyl ether
Serdox NNP10 [Servo B.V.]—POE (10) nonyl phenyl ether
Serdox NNP11 [Servo B.V.]—POE (11) nonyl phenyl ether
Serdox NNP12 [Servo B.V.]—POE (12) nonyl phenyl ether
Serdox NNP13 [Servo B.V.]—POE (13) nonyl phenyl ether
Serdox NNP15 [Servo B.V.]—POE (15) nonyl phenyl ether
Serdox NNP20/70 [Servo B.V.]—POE (20) nonyl phenyl ether
Serdox NOP9 [Servo B.V.]—POE (9) octyl phenyl ether
Serdox NOP30/70 [Servo B.V.]—POE (30) octyl phenyl ether
Serdox NOP40/70 [Servo B.V.]—POE (40) octyl phenyl ether
Sermul EA54 [Servo B.V.]—Sodium nonoxynol-4 sulfate
Sermul EN20/70 [Servo B.V.]—POE (20) nonyl phenyl ether
Simulsol 76 [Seppic]—POE (10) stearyl ether
Simulsol 78 [Seppic]—POE (20) stearyl ether
Simulsol M59 [Seppic]—POE (100) monostearate
Siponate SXS [Alcolac]—Sodium xylenesulfonate
Siponic E-2 [Alcolac]—POE (4) cetyl/stearyl ether
Siponic E-3 [Alcolac]—POE (6) cetyl/stearyl ether
Siponic E-5 [Alcolac]—POE (10) cetyl/stearyl ether

Siponic E-7 [Alcolac]—POE (15) cetyl/stearyl ether
Siponic E-10 [Alcolac]—POE (20) cetyl/stearyl ether
Siponic F-90 [Alcolac]—POE (9) octyl phenyl ether
Siponic F-300 [Alcolac]—POE (30) octyl phenyl ether
Siponic F-400 [Alcolac]—POE (40) octyl phenyl ether
Siponic F-707 [Alcolac]—POE (70) octyl phenyl ether
Siponic NP4 [Alcolac]—POE (4) nonyl phenyl ether
Siponic NP6 [Alcolac]—POE (6) nonyl phenyl ether
Siponic NP9 [Alcolac]—POE (9) nonyl phenyl ether
Siponic NP10.5 [Alcolac]—POE (10) nonyl phenyl ether
Siponic NP15 [Alcolac]—POE (15) nonyl phenyl ether
Siponic OP1.5 [Alcolac]—POE (1) octyl phenyl ether (1.5 EO)
Siponic OP7 [Alcolac]—POE (7) octyl phenyl ether
Siponic OP10 [Alcolac]—POE (10) octyl phenyl ether
Siponic OP30 [Alcolac]—POE (30) octyl phenyl ether
Siponic OP40 [Alcolac]—POE (40) octyl phenyl ether
Siponic TD-9-90 [Alcolac]—POE (9) tridecyl ether
Softisan 378 [Dynamit-Nobel]—Caprylic/capric triglyceride
Solusol 75%, 84%, 85%, 100% [Amer. Cyanamid]—Sodium dioctyl sulfosuccinate
Sorbax SMP [Chemax]—Sorbitan monopalmitate
Sorbon S-40 [Toho]—Sorbitan monopalmitate
Soyalec DBF, DBP, SBF, SBP, UBF, WDF-FG [Canada Packers]—Lecithin
Span 40 [ICI United States]—Sorbitan monopalmitate
ST-55-20 [Hefti Ltd.]—POE (20) stearyl ether
Standamul 302 [Henkel]—Propylene glycol dicaprylate/dicaprate
Standamul 318 [Henkel]—Caprylic/capric triglyceride
Standamul B-1 [Henkel]—POE (12) cetyl/stearyl ether
Standamul B-2 [Henkel]—POE (20) cetyl/stearyl ether
Standamul O-5 [Henkel]—POE (5) oleyl ether
Standapol SH-100, SH-135 Special [Henkel]—Disodium oleamido PEG-2 sulfosuccinate
Stantex T-14 [Henkel]—Sodium dioctyl sulfosuccinate
Stauffer MC+ [Stauffer]—Methylene chloride
Steinapal HV5 [Dutton & Reinisch]—POE (5) nonyl phenyl ether
Steinapal HV8 [Dutton & Reinisch]—POE (8) nonyl phenyl ether
Steinapal HV9 [Dutton & Reinisch]—POE (9) nonyl phenyl ether
Steinapal HV10 [Dutton & Reinisch]—POE (10) nonyl phenyl ether
Stepanate AM [Stepan]—Ammonium xylenesulfonate
Stepanate C-S [Stepan]—Sodium cumenesulfonate
Stepanate T [Stepan]—Sodium toluenesulfonate
Stepanate X [Stepan]—Sodium xylenesulfonate
Sterling 2XS [Canada Packers]—Sodium xylenesulfonate

Sterox ND [Monsanto]—POE (4) nonyl phenyl ether
Sterox NE [Monsanto]—POE (5) nonyl phenyl ether
Sterox NF [Monsanto]—POE (6) nonyl phenyl ether
Sterox NG [Monsanto]—POE (6) nonyl phenyl ether (6.5 EO)
Sterox NJ [Monsanto]—POE (9) nonyl phenyl ether (9.5 EO)
Sterox NK [Monsanto]—POE (10) nonyl phenyl ether (10.5 EO)
Sterox NL [Monsanto]—POE (11) nonyl phenyl ether
Sterox NM [Monsanto]—POE (12) nonyl phenyl ether (12.5 EO)
Sulfotex SXS [Henkel]—Sodium xylenesulfonate
Surco SXS [Onyx/Millmaster]—Sodium xylenesulfonate
Surfactol 318 [NL Ind.]—POE (5) castor oil
Surfactol 365 [NL Ind.]—POE (40) castor oil
Surfonic N-10 [Jefferson]—POE (1) nonyl phenyl ether
Surfonic N-40 [Jefferson]—POE (4) nonyl phenyl ether
Surfonic N-60 [Jefferson]—POE (6) nonyl phenyl ether
Surfonic N-60 [Jefferson]—POE (6) nonyl phenyl ether
Surfonic N-85 [Texaco]—POE (8) nonyl phenyl ether
Surfonic N-95 [Jefferson]—POE (9) nonyl phenyl ether (9.5 EO)
Surfonic N-95 [Jefferson]—POE (9) nonyl phenyl ether (9.5 EO)
Surfonic N-100, N-102 [Jefferson]—POE (10) nonyl phenyl ether
Surfonic N-100, N-102 [Jefferson]—POE (10) nonyl phenyl ether
Surfonic N-120 [Jefferson]—POE (12) nonyl phenyl ether
Surfonic N-150 [Jefferson]—POE (15) nonyl phenyl ether
Surfonic N-200 [Jefferson]—POE (20) nonyl phenyl ether
SXS [Continental]—Sodium xylenesulfonate
Synperonic NP4 [ICI Petrochemicals Div.]—POE (4) nonyl phenyl ether
Synperonic NP5 [ICI Petrochem. Div.]—POE (5) nonyl phenyl ether
Synperonic NP6 [ICI Petrochem. Div.]—POE (6) nonyl phenyl ether
Synperonic NP8 [ICI Petrochem. Div.]—POE (8) nonyl phenyl ether
Synperonic NP9 [ICI Petrochem. Div.]—POE (9) nonyl phenyl ether
Synperonic NP10 [ICI Petrochem. Div.]—POE (10) nonyl phenyl ether
Synperonic NP12 [ICI Petrochem. Div.]—POE (12) nonyl phenyl ether
Synperonic NP13 [ICI Petrochem. Div.]—POE (13) nonyl phenyl ether
Synperonic NP15 [ICI Petrochem. Div.]—POE (15) nonyl phenyl ether
Synperonic NP20 [ICI Petrochem. Div.]—POE (20) nonyl phenyl ether
Synperonic OP10 [ICI Petrochem. Div.]—POE (10) octyl phenyl ether
Synperonic OP11 [ICI Petrochem. Div.]—POE (11) octyl phenyl ether
Tagat R40 [Th. Goldschmidt AG]—POE (40) hydrogenated castor oil
Tagat R60 [Th. Goldschmidt AG]—POE (60) hydrogenated castor oil
Tauranol M, ML, MS [Finetex]—Sodium methyl oleoyl taurate
Tauranol WS, WS Conc., WS H.P., WSP [Finetex]—Sodium methyl cocoyl taurate
T-Det C-40 [Thompson-Hayward]—POE (40) castor oil

T-Det D-150 [Thompson-Hayward]—POE (150) dinonyl phenyl ether
T-Det N-1.5 [Thompson-Hayward]—POE (1) nonyl phenyl ether (1.5 EO)
T-Det N-4 [Thompson-Hayward]—POE (4) nonyl phenyl ether
T-Det N-6 [Thompson-Hayward]—POE (6) nonyl phenyl ether
T-Det N-8 [Thompson-Hayward]—POE (8) nonyl phenyl ether
T-Det N-9.5 [Thompson-Hayward]—POE (9) nonyl phenyl ether (9.5 EO)
T-Det N-10.5 [Thompson-Hayward]—POE (10) nonyl phenyl ether (10.5 EO)
T-Det N-12 [Thompson-Hayward]—POE (12) nonyl phenyl ether
T-Det N-20 [Thompson-Hayward]—POE (20) nonyl phenyl ether
T-Det O-4 [Thompson-Hayward]—POE (3) octyl phenyl ether (3–4 moles EO)
T-Det O-6 [Thompson-Hayward]—POE (6) octyl phenyl ether (6–7 moles EO)
T-Det O-8 [Thompson-Hayward]—POE (8) octyl phenyl ether (7–8 moles EO)
T-Det O-9 [Thompson-Hayward]—POE (9) octyl phenyl ether
T-Det O-40, O-407 [Thompson-Hayward]—POE (40) octyl phenyl ether
Temsperse S002, S003 [Temfibre]—Sodium lignosulfonate (modified)
Tergenol G, S Liquid, Slurry [Hart Prod.]—Sodium methyl oleoyl taurate
Tergitol NP-4 [Union Carbide]—POE (4) nonyl phenyl ether
Tergitol NP-5 [Union Carbide]—POE (5) nonyl phenyl ether
Tergitol NP-6 [Union Carbide]—POE (6) nonyl phenyl ether
Tergitol NP-8 [Union Carbide]—POE (8) nonyl phenyl ether
Tergitol NP-9 [Union Carbide]—POE (9) nonyl phenyl ether
Tergitol NP-10 [Union Carbide]—POE (10) nonyl phenyl ether (10.5 EO)
Tergitol NP-13 [Union Carbide]—POE (13) nonyl phenyl ether
Tergitol NP-14 [Union Carbide]—POE (4) nonyl phenyl ether
Tergitol NP-15 [Union Carbide]—POE (15) nonyl phenyl ether
Tergitol NPX [Union Carbide]—POE (10) nonyl phenyl ether (10.5 EO)
Tergitol TP-9 [Union Carbide]—POE (9) nonyl phenyl ether
Teric 13A9 [ICI Australia Ltd.]—POE (9) tridecyl ether
Teric GN5 [ICI Australia Ltd.]—POE (5) nonyl phenyl ether (tech.)
Teric GN8 [ICI Australia Ltd.]—POE (8) nonyl phenyl ether (tech.)
Teric GN9 [ICI Australia Ltd.]—POE (9) nonyl phenyl ether (tech.)
Teric GN10 [ICI Australia Ltd.]—POE (10) nonyl phenyl ether (tech.)
Teric GN12 [ICI Australia Ltd.]—POE (12) nonyl phenyl ether (tech.)
Teric GN13 [ICI Australia Ltd.]—POE (13) nonyl phenyl ether (tech.)
Teric GN15 [ICI Australia Ltd.]—POE (15) nonyl phenyl ether (tech.)
Teric N2 [ICI Australia Ltd.]—POE (2) nonyl phenyl ether
Teric N4 [ICI Australia Ltd.]—POE (4) nonyl phenyl ether
Teric N5 [ICI Australia Ltd.]—POE (5) nonyl phenyl ether
Teric N8 [ICI Australia Ltd.]—POE (8) nonyl phenyl ether (8.5 EO)
Teric N9 [ICI Australia Ltd.]—POE (9) nonyl phenyl ether
Teric N10 [ICI Australia Ltd.]—POE (10) nonyl phenyl ether
Teric N11 [ICI Australia Ltd.]—POE (11) nonyl phenyl ether

Teric N12 [ICI Australia Ltd.]—POE (12) nonyl phenyl ether
Teric N13 [ICI Australia Ltd.]—POE (13) nonyl phenyl ether
Teric N15 [ICI Australia Ltd.]—POE (15) nonyl phenyl ether
Teric N20 [ICI Australia Ltd.]—POE (20) nonyl phenyl ether
Teric X5 [ICI Australia Ltd.]—POE (5) octyl phenyl ether
Teric X7 [ICI Australia Ltd.]—POE (7) octyl phenyl ether (7.5 moles EO)
Teric X8 [ICI Australia Ltd.]—POE (8) octyl phenyl ether (8.5 moles EO)
Teric X10 [ICI Australia Ltd.]—POE (10) octyl phenyl ether
Teric X11 [ICI Australia Ltd.]—POE (11) octyl phenyl ether
Teric X40, X40L [ICI Australia Ltd.]—POE (40) octyl phenyl ether
Tex-Wet 1001 [Intex]—Sodium dioctyl sulfosuccinate
Thorowet G-40, G-75 [Clough]—Sodium dioctyl sulfosuccinate
T-Maz 67 [Mazer]—POE (20) sorbitan monoisostearate
Tomah Tallow Amine [Tomah]—Tallow amine (tech.)
Tri-Ethane [PPG Industries]—Trichloroethane (inhibited)
Tris Amino (40% Concentrate) [Angus]—Tris (hydroxymethyl) amino methane
Tris Amino (Crystals) [Angus]—Tris (hydroxymethyl) amino methane
Triton GR-5M, GR-7M [Rohm & Haas]—Sodium dioctyl sulfosuccinate
Triton N-17 [Rohm & Haas]—POE (1) nonyl phenyl ether (1.5 EO)
Triton N-42 [Rohm & Haas]—POE (4) nonyl phenyl ether
Triton N-57 [Rohm & Haas]—POE (5) nonyl phenyl ether
Triton N-60 [Rohm & Haas]—POE (6) nonyl phenyl ether
Triton N-87 [Rohm & Haas]—POE (8) nonyl phenyl ether (8.5 EO)
Triton N-101 [Rohm & Haas]—POE (9) nonyl phenyl ether (9–10 EO)
Triton N-111 [Rohm & Haas]—POE (11) nonyl phenyl ether
Triton N-150 [Rohm & Haas]—POE (15) nonyl phenyl ether
Triton X-15 [Rohm & Haas]—POE (1) octyl phenyl ether
Triton X-35 [Rohm & Haas]—POE (3) octyl phenyl ether
Triton X-45 [Rohm & Haas]—POE (5) octyl phenyl ether
Triton X-100 [Rohm & Haas]—POE (9) octyl phenyl ether (9–10 EO)
Triton X-102 [Rohm & Haas]—POE (12) octyl phenyl ether (12–13 EO)
Triton X-114 [Rohm & Haas]—POE (8) octyl phenyl ether (7–8 moles EO)
Triton X-114SB [Rohm & Haas]—POE (8) octyl phenyl ether
Triton X-120 [Rohm & Haas]—POE (9) octyl phenyl ether (9–10 EO)
Triton X-305 [Rohm & Haas]—POE (30) octyl phenyl ether
Triton X-405 [Rohm & Haas]—POE (40) octyl phenyl ether
Triton X-705, X-705-50% [Rohm & Haas]—POE (70) octyl phenyl ether
Trona [Kerr-McGee]—Sodium borate (anhyd.)
Tronabor [Kerr-McGee]—Sodium borate (pentahydrate)
Troykyd Lecithin W.D. [Troy]—Lecithin
Trycol DNP-150, DNP-150/50 [Emery]—POE (150) dinonyl phenyl ether
Trycol NP-1 [Emery]—POE (1) nonyl phenyl ether

Trycol NP-4 [Emery]—POE (4) nonyl phenyl ether
Trycol NP-6 [Emery]—POE (6) nonyl phenyl ether
Trycol NP-9 [Emery]—POE (9) nonyl phenyl ether
Trycol NP-11 [Emery]—POE (11) nonyl phenyl ether
Trycol NP-20 [Emery]—POE (20) nonyl phenyl ether
Trycol OP-407 [Emery]—POE (40) octyl phenyl ether
Trycol SAL-20 [Emery]—POE (20) stearyl ether
Trycol TDA-9 [Emery]—POE (9) tridecyl ether
Trycol TDA-11 [Emery]— POE (11) tridecyl ether
Trylox CO-5 [Emery]—POE (5) castor oil
Trylox CO-10 [Emery]—POE (10) castor oil
Trylox CO-25 [Emery]—POE (25) castor oil
Trylox CO-30 [Emery]—POE (30) castor oil
Trylox CO-40 [Emery]—POE (40) castor oil
Trylox HCO-200/50 [Emery]—POE (200) hydrogenated castor oil
Ultra NCS Liquid [Witco/Organics]—Ammonium cumenesulfonate
Ultra NXS Liquid [Witco/Organics]—Ammonium xylenesulfonate
Ultra SCS Liquid [Witco/Organics]—Sodium cumenesulfonate
Ultra SXS Liquid, SXS Powder [Witco/Organics]—Sodium xylenesulfonate
Ultrawet 40SX [Arco]—Sodium xylenesulfonate
Unihib 305 [Lonza]—Aminotrimethylene phosphonic acid
Unitene D [Union Camp; Harrisons & Crosfield]—Dipentene
Varonic 400DL [Sherex]—POE (8) dilaurate
Varonic U-250 [Sherex]—POE (50) tallow amine
Varsulf OT [Sherex]—Sodium dioctyl sulfosuccinate
Volpo 5 [Croda]—POE (5) oleyl ether
Volpo CS3 [Croda Chem. Ltd.]—POE (3) cetyl/stearyl ether
Volpo CS10 [Croda Chem. Ltd.]—POE (10) cetyl/stearyl ether
Volpo CS15 [Croda Chem. Ltd.]—POE (15) cetyl/stearyl ether
Volpo CS20 [Croda Chem. Ltd.]—POE (20) cetyl/stearyl ether
Volpo O5 [Croda Chem. Ltd.]—POE (5) oleyl ether
Volpo S2 [Croda]—POE (2) stearyl ether
Volpo S10 [Croda]—POE (10) stearyl ether
Volpo S20 [Croda]—POE (20) stearyl ether
Volpo T10 [Croda]—POE (10) tridecyl ether
Volpo T15 [Croda]—POE (15) tridecyl ether
Witconate NCS [Witco/Organics]—Ammonium cumenesulfonate
Witconate NXS [Witco/Organics]—Ammonium xylenesulfonate
Witconate SCS [Witco/Organics]—Sodium cumenesulfonate
Witconate STS [Witco/Organics]—Sodium toluenesulfonate
Witconate SXS Liquid, SXS Powder [Witco/Organics]—Sodium xylenesulfonate

GENERIC CHEMICAL SYNONYMS AND CROSS REFERENCES

Acetate ester of ethylene glycol monoethyl ether. See Ethylene glycol monoethyl ether acetate
Acetic acid, butyl ester. See Butyl acetate
Acetic acid, ethyl ester. See Ethyl acetate
Acetic acid, 1-methylethyl ester. See Isopropyl acetate
Acetic acid, 2-methylpropyl ester. See Isobutyl acetate
Acetic ester. See Ethyl acetate
Acetic ether. See Ethyl acetate
Alcohol (CTFA). See Ethyl alcohol
Amides, coconut oil, with N-methyltaurine, sodium salts. See Sodium methyl cocoyl taurate
Amides, coconut oil, N-[2-[(sulfosuccinyl) oxy] ethyl], sodium salts. See Disodium cocoyl monoethanolamide sulfosuccinate
Amines, coco alkyl. See Coconut amine
Amines, hydrogenated tallow alkyl. See Tallow amine, hydrogenated
Amines, soya alkyl. See Soya amine
Amines, tallow alkyl. See Tallow amine
3-Aminobutanoic acid, N-coco alkyl derivs. See Cocaminobutyric acid
2-Amino-2-(hydroxymethyl)-1,3-propanediol. See Tris (hydroxymethyl) amino methane
2-Amino-2-methyl-1-propanol. See Aminomethyl propanol
Aminotri (methylenephosphonic acid), pentasodium salt. See Pentasodium aminotrimethylene phosphonate
Amino tris (methylene phosphonic acid). See Aminotrimethylene phosphonic acid
Ammonium lauryl benzene sulfonate. See Ammonium dodecylbenzenesulfonate
Ammonium lauryl ether sulfate. See Book I
Ammonium lauryl sulfate. See Book I
AMP. See Aminomethyl propanol
1,4-Anhydro-D-glucitol, 6-hexadecanoate. See Sorbitan monopalmitate
1,4-Anhydro-D-glucitol, 6-isooctadecanoate. See Sorbitan monoisostearate
Anhydrosorbitol monoisostearate. See Sorbitan monoisostearate
L-Aspartic acid, N-(3-carboxy-1-oxosulfopropyl)-N-octadecyl, tetrasodium salt. See Tetrasodium dicarboxyethyl stearyl sulfosuccinamate
Benzene, (1-methylethyl)-, monosulfo derivative, sodium salt. See Sodium cumenesulfonate
Benzenesulfonic acid, dimethyl-, ammonium salt. See Ammonium xylenesulfonate
Benzenesulfonic acid, dimethyl-, sodium salt. See Sodium xylenesulfonate
Benzenesulfonic acid, dodecyl-, ammonium salt. See Ammonium dodecylbenzenesulfonate

Benzenesulfonic acid, dodecyl-, compd. with 1-amino-2-propanol (1:1). See Monoisopropanolamine dodecylbenzenesulfonate
Benzenesulfonic acid, (1-methylethyl)-, ammonium salt. See Ammonium cumenesulfonate
Benzenesulfonic acid, methyl-, sodium salt. See Sodium toluenesulfonate
Benzoic acid, isostearyl ester. See Isostearyl benzoate
1,4-Bis (2-ethylhexyl) sulfobutanedioate, sodium salt. See Sodium dioctyl sulfosuccinate
Bis (1-methylamyl) sodium sulfosuccinate. See Sodium dihexyl sulfosuccinate
1,4 Bis (2-methylpropyl) sulfobutanedioate, sodium salt. See Sodium diisobutyl sulfosuccinate
Borax. See Sodium borate
Borax anhydrous. See Sodium borate
Borax decahydrate. See Sodium borate
Borax, dehydrated. See Sodium borate
Borax granular. See Sodium borate
Borax pentahydrate . See Sodium borate
Butanedioic acid, 4-(octadecylamino)-4-oxo-2-sulfo, disodium salt. See Disodium stearyl sulfosuccinamate
Butanedioic acid, sulfo-, 1,4-bis (2-ethylhexyl) ester, sodium salt. See Sodium dioctyl sulfosuccinate
Butanedioic acid, sulfo-, 1,4-bis (2-methylpropyl) ester, sodium salt. See Sodium diisobutyl sulfosuccinate
Butanedioic acid, sulfo-, C-(2-cocamidoethyl) esters, disodium salts. See Disodium cocoyl monoethanolamide sulfosuccinate
Butanedioic acid, sulfo-, 1,4-dihexyl ester, sodium salt. See Sodium dihexyl sulfosuccinate
Butanedioic acid, sulfo-, 1,4-dipentyl ester, sodium salt. See Sodium diamyl sulfosuccinate
Butanedioic acid, sulfo-, 1,4-ditridecyl ester, sodium salt. See Sodium ditridecyl sulfosuccinate
Butanedioic acid, sulfo-, 4-isodecyl ester, disodium salt. See Disodium isodecyl sulfosuccinate
Butanedioic acid, sulfo-, monooctadecyl ester, disodium salt. See Disodium stearyl sulfosuccinamate
Butanedioic acid, sulfo-, 1(or 4)- [2- [2- [(1-oxo-9-octadecenyl) amino] ethoxy] ethyl] ester, disodium salt. See Disodium oleamido PEG-2 sulfosuccinate
Butanedioic acid, sulfo-, C-[2[[2- [(1-oxo-9-octadecenyl) amino] ethoxy] ethyl] ester, disodium salt. See Disodium oleamido PEG-2 sulfosuccinate
Butanedioic acid, sulfo-, 2-[(1-oxotetradecyl) amino] ethyl ester, disodium salt. See Disodium myristamido MEA-sulfosuccinate
Butanoic acid, 3-amino-, N-coco alkyl derivs. See Cocaminobutyric acid
Butanoic acid, 4-(octadecylamino)-4-oxo-2-sulfo-, disodium salt. See Disodium stearyl

sulfosuccinamate
1-Butanol. See *n*-Butyl alcohol
n-Butanol. See *n*-Butyl alcohol
2-Butanone. See Methyl ethyl ketone
Butoxydiglycol (CTFA). See Diethylene glycol monobutyl ether
Butoxyethanol (CTFA). See Ethylene glycol monobutyl ether
2-Butoxyethanol. See Ethylene glycol monobutyl ether
2-(2-Butoxyethoxy) ethanol. See Diethylene glycol monobutyl ether
n-Butyl acetate. See Butyl acetate
Butyl acetate, normal. See Butyl acetate
Butyl alcohol, normal. See *n*-Butyl alcohol
Butyl stearate. See Book IV
Butyric alcohol. See *n*-Butyl alcohol
C_{9-11} linear primary alcohol 6 mole ethoxylate. See Pareth-91-6
Calcium dodecylbenzene sulfonate. See Book I
N-(2-Carboxyethyl)-N-(tallow acyl)-β-alanine. See Disodium tallowiminodipropionate
N-(3-Carboxy-1-oxosulfopropyl)-N-octadecyl-L-aspartic acid, tetrasodium salt. See Tetrasodium dicarboxyethyl stearyl sulfosuccinamate
Castor oil, hydrogenated. See Hydrogenated castor oil
Castor oil, sulfated. See Book III
Ceteareth-2 (CTFA). See POE (2) cetyl/stearyl ether
Ceteareth-3 (CTFA). See POE (3) cetyl/stearyl ether
Ceteareth-4 (CTFA). See POE (4) cetyl/stearyl ether
Ceteareth-6 (CTFA). See POE (6) cetyl/stearyl ether
Ceteareth-10 (CTFA). See POE (10) cetyl/stearyl ether
Ceteareth-12 (CTFA). See POE (12) cetyl/stearyl ether
Ceteareth-15 (CTFA). See POE (15) cetyl/stearyl ether
Ceteareth-20 (CTFA). See POE (20) cetyl/stearyl ether
Cetearyl alcohol (EO 20) polyethylene glycol ether. See POE (20) cetyl/stearyl ether
Cetyl alcohol. See Book IV
Cetyl-stearyl alcohol + 12 mol EO. See POE (12) cetyl/stearyl ether
Cetyl-stearyl alcohol + 20 mole EO. See POE (20) cetyl/stearyl ether
Cetyl trimethyl ammonium chloride. See Book IV
Cocamidopropylamine oxide. See Book III
Cocamine (CTFA). See Coconut amine
Coco amidopropyl betaine. See Book IV
Cocoamine. See Coconut amine
N-Coco-β-aminobutyric acid. See Cocaminobutyric acid
Cocoamphopropionate. See Book I
Coco betaine. See Book IV
Coco dimethyl amine. See Book I
Coco imidazoline. See Cocoyl imidazoline

Coconut diethanolamide. See Book III
Coconut imidazoline. See Cocoyl imidazoline
Coconut oil amine. See Coconut amine
Cocoyl hydroxyethyl imidazoline. See Cocoyl imidazoline
Cyclohexene, 1-methyl-4-(1-methylethenyl)-. See Dipentene
Decaglycerol tetraoleate. See Book III
Decaglyceryl decaoleate. See Decaglycerol decaoleate
Decaglyceryl decastearate. See Decaglycerol decastearate
Decaglyceryl octaoleate. See Decaglycerol octaoleate
Decanoic acid, monoester with 1,2,3-propanetriol. See Glyceryl monocaprate
Deceth-4 phosphate (CTFA). See POE (4) decyl ether phosphate
Decyl oleate. See Book IV
Diamyl sodium sulfosuccinate (CTFA). See Sodium diamyl sulfosuccinate
1,4:3,6-Dianhydro-2,5-di-O-methyl-D-glucitol. See Dimethyl isosorbide
N-(1,2-Dicarboxyethyl) N-alkyl (C_{18}) sulfosuccinamate. See Tetrasodium dicarboxyethyl stearyl sulfosuccinamate
Dichloroethane. See Ethylene dichloride
sym-Dichloroethane. See Ethylene dichloride
Dichloromethane. See Methylene chloride
Dicoco dimethyl ammonium chloride. See Book I
Dicyclohexyl sodium sulfosuccinate (CTFA). See Sodium dicyclohexyl sulfosuccinate
Diethylene glycol butyl ether. See Diethylene glycol monobutyl ether
Diethylene glycol ethyl ether. See Diethylene glycol monoethyl ether
Diethylene glycol methyl ether. See Diethylene glycol monomethyl ether
Diethylene glycol monolaurate. See Book I
Diethylene glycol monooleate. See Book I
Diethylene glycol monostearate. See Book I
Di-(2-ethylhexyl) sodium sulfosuccinate. See Sodium dioctyl sulfosuccinate
Dihexyl sodium sulfosuccinate (CTFA). See Sodium dihexyl sulfosuccinate
Dihexyl sulfosuccinate, sodium salt. See Sodium dihexyl sulfosuccinate
Diisobutyl sodium sulfosuccinate (CTFA). See Sodium diisobutyl sulfosuccinate
Diisopropyl adipate. See Book III
Dimethylbenzenesulfonic acid, sodium salt. See Sodium xylenesulfonate
Dimethylcarbinol. See Isopropyl alcohol
Dimethyl di(hydrogenated tallow) ammonium chloride. See Book IV
Dimethyl ketone. See Acetone
Dinonylphenol-150 mole ethylene oxide adduct. See POE (150) dinonyl phenyl ether
Dioctyl ester of sodium sulfosuccinic acid. See Sodium dioctyl sulfosuccinate
Dioctyl sodium sulfosuccinate (CTFA). See Sodium dioctyl sulfosuccinate
1,4-Dipentylsulfobutanedioic acid, sodium salt. See Sodium diamyl sulfosuccinate
Diphosphoric acid, tetrapotassium salt. See Tetrapotassium pyrophosphate
Diphosphoric acid, tetrasodium salt. See Tetrasodium pyrophosphate

Dipropylene glycol methyl ether. See Dipropylene glycol monomethyl ether
Dipropylmethane. See Heptane
Disodium cocamido MEA-sulfosuccinate (CTFA). See Disodium cocoyl monoethanolamide sulfosuccinate
Disodium deceth-6 sulfosuccinate. See Book I
Disodium isodecyl sulfosuccinate. See Book I
Disodium monomyristamido MEA-sulfosuccinate. See Disodium myristamido MEA-sulfosuccinate
Disodium monooleamido PEG-2 sulfosuccinate. See Disodium oleamido PEG-2 sulfosuccinate
Disodium N-octadecyl sulfosuccinamate. See Disodium stearyl sulfosuccinamate
Disodium N-oleyl sulfosuccinamate. See Book I
Disodium *n*-tallow-3,3′-iminodipropionate. See Disodium tallowiminodipropionate
Disodium N-tallow-β iminodipropionate. See Disodium tallowiminodipropionate
Disodium salt of a stearyl amide of sulfosuccinic acid. See Disodium stearyl sulfosuccinamate
Ditridecyl sodium sulfosuccinate (CTFA). See Sodium ditridecyl sulfosuccinate
DMI. See Dimethyl isosorbide
Docusate sodium. See Sodium dioctyl sulfosuccinate
Dodecylbenzene sulfonic acid. See Book I
Dodecylbenzenesulfonic acid, compd. with 1-amino-2-propanol (1:1). See Monoisopropanolamine dodecylbenzenesulfonate
Dutch oil. See Ethylene dichloride
Emulbesto [Lucas Meyer GmbH]. See Lecithin
Emulfluid [Lucas Meyer GmbH]. See Lecithin
Ethane, 1,2-dichloro-. See Ethylene dichloride
Ethanesulfonic acid, 2-[methyl (1-oxo-9-octadecenyl) amino]-, sodium salt. See Sodium methyl oleoyl taurate
Ethane, 1,1,1-trichloro-. See Trichloroethane
Ethanol, 2-butoxy-. See Ethylene glycol monobutyl ether
Ethanol, 2-(2-butoxyethoxy)-. See Diethylene glycol monobutyl ether
Ethanol, 2-ethoxy-. See Ethylene glycol monoethyl ether
Ethanol, 2-ethoxy-, acetate. See Ethylene glycol monoethyl ether acetate
Ethanol, 2-(2-ethoxyethoxy)-. See Diethylene glycol monoethyl ether
Ethanol, 2-(2-methoxyethoxy)-. See Diethylene glycol monomethyl ether
Ethanol, 2-(nonylphenoxy)-. See POE (1) nonyl phenyl ether
Ethanol, 2- [2- [2- [2- (4-nonylphenoxy) ethoxy] ethoxy] ethoxy]-. See POE (4) nonyl phenyl ether
Ethanol, 2-phenoxy-. See Ethylene glycol monophenyl ether
Ethanol, undenatured. See Ethyl alcohol
Ethoxydiglycol (CTFA). See Diethylene glycol monoethyl ether
Ethoxyethanol (CTFA). See Ethylene glycol monoethyl ether

Ethoxyethanol acetate (CTFA). See Ethylene glycol monoethyl ether acetate
2-Ethoxyethanol acetate. See Ethylene glycol monoethyl ether acetate
2-(2-Ethoxyethoxy) ethanol. See Diethylene glycol monoethyl ether
2-Ethoxyethyl acetate. See Ethylene glycol monoethyl ether acetate
Ethoxylated tridecyl alcohol (10 EO). See POE (10) tridecyl ether
Ethyl alcohol, undenatured. See Ethyl alcohol
Ethylene chloride. See Ethylene dichloride
Ethylene glycol butyl ether. See Ethylene glycol monobutyl ether
Ethylene glycol distearate. See Book I
Ethylene glycol ethyl ether. See Ethylene glycol monoethyl ether
Ethylene glycol monoethyl ether acetylated. See Ethylene glycol monoethyl ether acetate
Ethylene glycol monostearate. See Book III
Ethylene glycol nonyl phenyl ether. See POE (1) nonyl phenyl ether
Ethylene glycol octyl phenyl ether. See POE (1) octyl phenyl ether
Ethylene glycol phenyl ether. See Ethylene glycol monophenyl ether
4-Ethyl-2-(8-heptadecenyl)-4,5-dihydro-4-oxazolemethanol. See Ethyl hydroxymethyl oleyl oxazoline
Ethyl methyl ketone. See Methyl ethyl ketone
EtOH. See Ethyl alcohol
Fatty acids, coconut oil, sulfoethyl esters, sodium salts. See Sodium cocoyl isethionate
β–Δ–Fructofuranosyl-α-D-glucopyranoside, monooctadecanoate. See Sucrose monostearate
D-Glucitol, 1,4-anhydro-, 6-hexadecanoate. See Sorbitan monopalmitate
D-Glucitol, 1,4-anhydro-, 6-isooctadecanoate. See Sorbitan monoisostearate
D-Glucitol, 1,4:3,6-dianhydro-2,5-di-O-methyl. See Dimethyl isosorbide
α–Δ–Glucopyranoside, β-D-fructofuranosyl, dioctadecanoate. See Sucrose distearate
α–Δ–Glucopyranoside, β-D-fructofuranosyl, monooctadecanoate. See Sucrose monostearate
Glyceryl caprate (CTFA). See Glyceryl monocaprate
Glyceryl caprylate (CTFA). See Glyceryl monocaprylate
Glyceryl mono-di-caprylate/caprate. See Caprylic/capric glycerides
Glyceryl monolaurate. See Book I
Glyceryl monooleate. See Book I
Glyceryl monoricinoleate. See Book III
Glyceryl monostearate. See Book I
Glyceryl tri(12-hydroxystearate). See Hydrogenated castor oil
Graham's salt. See Sodium hexametaphosphate
Grain alcohol. See Ethyl alcohol
n-Heptane. See Heptane
Hexaglycerol distearate. See Hexaglyceryl distearate
Hexone. See Methyl isobutyl ketone
Hexyl laurate. See Book IV

Hydrogenated castor oil ethoxylate (200 EO). See POE (200) hydrogenated castor oil
Hydrogenated tallow amine (CTFA). See Tallow amine, hydrogenated
Hydroxyethane diphosphonic acid. See Etidronic acid
(1-Hydroxyethylidene) bisphosphonic acid. See Etidronic acid
1-(2-Hydroxyethyl)-2-norcoco-2-imidazoline. See Cocoyl imidazoline
12-Hydroxyoctadecanoic acid, methyl ester. See Methyl hydroxystearate
1H-Imidazole-1-ethanol, 4,5-dihydro-2-norcocoyl-. See Cocoyl imidazoline
IPA. See Isopropyl alcohol
Isobutanolamine. See Aminomethyl propanol
Isoceteth-20 (CTFA). See POE (20) isocetyl ether
Isocetyl stearate. See Book IV
Isopropanol. See Isopropyl alcohol
Isopropylacetone. See Methyl isobutyl ketone
Isopropylamine dodecylbenzenesulfonate. See Book I
Isopropyl lanolate. See Book IV
Isopropyl myristate. See Book IV
Isosorbide, dimethyl ether. See Dimethyl isosorbide
Isostearyl alcohol. See Book IV
Lanolin. See Book IV
Laureth-2 (CTFA). See POE (2) lauryl ether
Laureth-3 (CTFA). See POE (3) lauryl ether
Laureth-7 (CTFA). See POE (7) lauryl ether
Lauric diethanolamide. See Book III
Lauric monoethanolamide. See Book III
Lauric monoisopropanolamide. See Book III
Lauryl betaine. See Book I
Lauryl dimethyl amine. See Book I
Lauryl lactate. See Book IV
Lauryl trimethyl ammonium chloride. See Book I
Lignosulfonic acid, calcium salt. See Calcium lignosulfonate
Lignosulfonic acid, sodium salt. See Sodium lignosulfonate
DL-Limonene. See Dipentene
Linoleoyl diethanolamide. See Book III
Magnesium lauryl sulfate. See Book I
MEK (CTFA). See Methyl ethyl ketone
Metaphosphoric acid, hexasodium salt. See Sodium hexametaphosphate
Methane, dichloro-. See Methylene chloride
Methoxydiglycol (CTFA). See Diethylene glycol monomethyl ether
Methoxy dipropylene glycol. See Dipropylene glycol monomethyl ether
2-(2-Methoxyethoxy) ethanol. See Diethylene glycol monomethyl ether
Methoxypropanol (CTFA). See Propylene glycol monomethyl ether
1-Methoxy-2-propanol. See Propylene glycol monomethyl ether

Nonoxynol-20 (CTFA). See POE (20) nonyl phenyl ether
Nonyl nonoxynol-150 (CTFA). See POE (150) dinonyl phenyl ether
Nonyl phenol 4 polyglycol ether. See POE (4) nonyl phenyl ether
Nonyl phenol 5 polyglycol ether. See POE (5) nonyl phenyl ether
Nonyl phenol 6 polyglycol ether. See POE (6) nonyl phenyl ether
Nonyl phenol 8 polyglycol ether. See POE (8) nonyl phenyl ether
Nonyl phenol 9 polyglycol ether. See POE (9) nonyl phenyl ether
Nonyl phenol 10 polyglycol ether. See POE (10) nonyl phenyl ether
Nonyl phenol 15 polyglycol ether. See POE (15) nonyl phenyl ether
Nonyl phenol 20 polyglycol ether. See POE (20) nonyl phenyl ether
Nonyl phenol ethoxylate (2 moles EO). See POE (2) nonyl phenyl ether
Nonyl phenol ethoxylate (4 moles EO). See POE (4) nonyl phenyl ether
Nonyl phenol ethoxylate (5 moles EO). See POE (5) nonyl phenyl ether
Nonyl phenol ethoxylate (6 moles EO). See POE (6) nonyl phenyl ether
Nonyl phenol ethoxylate (8 moles EO). See POE (8) nonyl phenyl ether
Nonyl phenol ethoxylate (9 moles EO). See POE (9) nonyl phenyl ether
Nonyl phenol ethoxylate (10 moles EO). See POE (10) nonyl phenyl ether
Nonyl phenol ethoxylate (11 moles EO). See POE (11) nonyl phenyl ether
Nonyl phenol ethoxylate (12 moles EO). See POE (12) nonyl phenyl ether
Nonyl phenol ethoxylate (13 moles EO). See POE (13) nonyl phenyl ether
Nonyl phenol ethoxylate (15 moles EO). See POE (15) nonyl phenyl ether
Nonyl phenol ethoxylate (20 moles EO). See POE (20) nonyl phenyl ether
Nonylphenol polyethyleneglycol ether (14 moles EO). See POE (14) nonyl phenyl ether
Nonyl phenol polyglycol ether (11 moles EO). See POE (11) nonyl phenyl ether
Nonyl phenol polyglycol ether (12 moles EO). See POE (12) nonyl phenyl ether
Nonyl phenol polyglycol ether (13 moles EO). See POE (13) nonyl phenyl ether
2-(Nonylphenoxy) ethanol. See POE (1) nonyl phenyl ether
Nonylphenoxy polyethoxy ethanol (4 moles EO). See POE (4) nonyl phenyl ether
Nonylphenoxy polyethoxy ethanol (5 moles EO). See POE (5) nonyl phenyl ether
Nonylphenoxy polyethoxy ethanol (6 moles EO). See POE (6) nonyl phenyl ether
Nonylphenoxy polyethoxy ethanol (8 moles EO). See POE (8) nonyl phenyl ether
Nonylphenoxy polyethoxy ethanol (9 moles EO). See POE (9) nonyl phenyl ether
Nonylphenoxy polyethoxy ethanol (10 moles EO). See POE (10) nonyl phenyl ether
Nonylphenoxy polyethoxy ethanol (11 moles EO). See POE (11) nonyl phenyl ether
Nonylphenoxy polyethoxy ethanol (12 moles EO). See POE (12) nonyl phenyl ether
Nonylphenoxy polyethoxy ethanol (13 moles EO). See POE (13) nonyl phenyl ether
Nonylphenoxy polyethoxy ethanol (15 moles EO). See POE (15) nonyl phenyl ether
Nonylphenoxy polyethoxy ethanol (20 moles EO). See POE (20) nonyl phenyl ether
Nonylphenoxy poly(ethyleneoxy) ethanol (4 moles EO). See POE (4) nonyl phenyl ether
Nonylphenoxy poly(ethyleneoxy) ethanol (5 moles EO). See POE (5) nonyl phenyl ether
Nonylphenoxy poly(ethyleneoxy) ethanol (6 moles EO). See POE (6) nonyl phenyl ether
Nonylphenoxy poly(ethyleneoxy) ethanol (8 moles EO). See POE (8) nonyl phenyl ether

Nonylphenoxy poly(ethyleneoxy) ethanol (9 moles EO). See POE (9) nonyl phenyl ether
Nonylphenoxy poly(ethyleneoxy) ethanol (10 moles EO). See POE (10) nonyl phenyl ether
Nonylphenoxy poly(ethyleneoxy) ethanol (11 moles EO). See POE (11) nonyl phenyl ether
Nonylphenoxy poly(ethyleneoxy) ethanol (12 moles EO). See POE (12) nonyl phenyl ether
Nonylphenoxy poly(ethyleneoxy) ethanol (13 moles EO). See POE (13) nonyl phenyl ether
Nonylphenoxy poly(ethyleneoxy) ethanol (15 moles EO). See POE (15) nonyl phenyl ether
Nonylphenoxy poly(ethyleneoxy) ethanol (20 moles EO). See POE (20) nonyl phenyl ether
1-Octadecanamine. See Stearyl amine
Octadecanoic acid, 12-hydroxy-, methyl ester. See Methyl hydroxystearate
Octadecylamine. See Stearyl amine
n-Octadecylamine. See Stearyl amine
Octadecylamine, primary. See Stearyl amine
4-(Octadecylamino)-4-oxo-2-sulfobutanedioic acid, disodium salt. See Disodium stearyl sulfosuccinamate
Octanoic acid, monoester with 1,2,3-propanetriol. See Glyceryl monocaprylate
Octanoic/decanoic acid triglyceride. See Caprylic/capric triglyceride
Octoxynol-1 (CTFA). See POE (1) octyl phenyl ether
Octoxynol-3 (CTFA). See POE (3) octyl phenyl ether
Octoxynol-5 (CTFA). See POE (5) octyl phenyl ether
Octoxynol-7 (CTFA). See POE (7) octyl phenyl ether
Octoxynol-8 (CTFA). See POE (8) octyl phenyl ether
Octoxynol-9 (CTFA). See POE (9) octyl phenyl ether
Octoxynol-10 (CTFA). See POE (10) octyl phenyl ether
Octoxynol-11 (CTFA). See POE (11) octyl phenyl ether
Octoxynol-30 (CTFA). See POE (30) octyl phenyl ether
Octoxynol-40 (CTFA). See POE (40) octyl phenyl ether
Octoxynol-70 (CTFA). See POE (70) octyl phenyl ether
Octyl phenol ethoxylate (1 EO). See POE (1) octyl phenyl ether
Octyl phenol ethoxylate (3 EO). See POE (3) octyl phenyl ether
Octyl phenol ethoxylate (5 EO). See POE (5) octyl phenyl ether
Octyl phenol ethoxylate (6 EO). See POE (6) octyl phenyl ether
Octyl phenol ethoxylate (7 EO). See POE (7) octyl phenyl ether
Octyl phenol ethoxylate (8 EO). See POE (8) octyl phenyl ether
Octyl phenol ethoxylate (9 EO). See POE (9) octyl phenyl ether
Octyl phenol ethoxylate (10 EO). See POE (10) octyl phenyl ether
Octyl phenol ethoxylate (11 EO). See POE (11) octyl phenyl ether

Octyl phenol ethoxylate (12 EO). See POE (12) octyl phenyl ether
Octyl phenol ethoxylate (30 EO). See POE (30) octyl phenyl ether
Octyl phenol ethoxylate (40 EO). See POE (40) octyl phenyl ether
Octyl phenol ethoxylate (70 EO). See POE (70) octyl phenyl ether
Octyl phenol polyethoxy ethanol (1 EO). See POE (1) octyl phenyl ether
Octyl phenol polyethoxy ethanol (3 EO). See POE (3) octyl phenyl ether
Octyl phenol polyethoxy ethanol (5 EO). See POE (5) octyl phenyl ether
Octyl phenol polyethoxy ethanol (6 EO). See POE (6) octyl phenyl ether
Octyl phenol polyethoxy ethanol (7 EO). See POE (7) octyl phenyl ether
Octyl phenol polyethoxy ethanol (8 EO). See POE (8) octyl phenyl ether
Octyl phenol polyethoxy ethanol (9 EO). See POE (9) octyl phenyl ether
Octyl phenol polyethoxy ethanol (10 EO). See POE (10) octyl phenyl ether
Octyl phenol polyethoxy ethanol (11 EO). See POE (11) octyl phenyl ether
Octyl phenol polyethoxy ethanol (12 EO). See POE (12) octyl phenyl ether
Octyl phenol polyethoxy ethanol (30 EO). See POE (30) octyl phenyl ether
Octyl phenol polyethoxy ethanol (40 EO). See POE (40) octyl phenyl ether
Octyl phenol polyethoxy ethanol (70 EO). See POE (70) octyl phenyl ether
Octyl phenol polyglycol ether (1 EO). See POE (1) octyl phenyl ether
Octyl phenol polyglycol ether (3 EO). See POE (3) octyl phenyl ether
Octyl phenol polyglycol ether (5 EO). See POE (5) octyl phenyl ether
Octyl phenol polyglycol ether (6 EO). See POE (6) octyl phenyl ether
Octyl phenol polyglycol ether (7 EO). See POE (7) octyl phenyl ether
Octyl phenol polyglycol ether (8 EO). See POE (8) octyl phenyl ether
Octyl phenol polyglycol ether (9 EO). See POE (9) octyl phenyl ether
Octyl phenol polyglycol ether (10 EO). See POE (10) octyl phenyl ether
Octyl phenol polyglycol ether (11 EO). See POE (11) octyl phenyl ether
Octyl phenol polyglycol ether (12 EO). See POE (12) octyl phenyl ether
Octyl phenol polyglycol ether (30 EO). See POE (30) octyl phenyl ether
Octyl phenol polyglycol ether (40 EO). See POE (40) octyl phenyl ether
Octyl phenol polyglycol ether (70 EO). See POE (70) octyl phenyl ether
Octylphenoxy ethoxy ethanol (1 EO). See POE (1) octyl phenyl ether
Octylphenoxy polyethoxy ethanol (3 EO). See POE (3) octyl phenyl ether
Octylphenoxy polyethoxy ethanol (5 EO). See POE (5) octyl phenyl ether
Octylphenoxy polyethoxy ethanol (6 EO). See POE (6) octyl phenyl ether
Octylphenoxy polyethoxy ethanol (7 EO). See POE (7) octyl phenyl ether
Octylphenoxy polyethoxy ethanol (8 EO). See POE (8) octyl phenyl ether
Octylphenoxy polyethoxy ethanol (9 EO). See POE (9) octyl phenyl ether
Octylphenoxy polyethoxy ethanol (10 EO). See POE (10) octyl phenyl ether
Octylphenoxy polyethoxy ethanol (11 EO). See POE (11) octyl phenyl ether
Octylphenoxy polyethoxy ethanol (12 EO). See POE (12) octyl phenyl ether
Octylphenoxy polyethoxy ethanol (30 EO). See POE (30) octyl phenyl ether
Octylphenoxy polyethoxy ethanol (40 EO). See POE (40) octyl phenyl ether

Octylphenoxy polyethoxy ethanol (70 EO). See POE (70) octyl phenyl ether
Octylphenoxypoly (ethyleneoxy) ethanol (3 EO). See POE (3) octyl phenyl ether
Octylphenoxypoly (ethyleneoxy) ethanol (5 EO). See POE (5) octyl phenyl ether
Octylphenoxypoly (ethyleneoxy) ethanol (6 EO). See POE (6) octyl phenyl ether
Octylphenoxypoly (ethyleneoxy) ethanol (7 EO). See POE (7) octyl phenyl ether
Octylphenoxypoly (ethyleneoxy) ethanol (8 EO). See POE (8) octyl phenyl ether
Octylphenoxypoly (ethyleneoxy) ethanol (9 EO). See POE (9) octyl phenyl ether
Octylphenoxypoly (ethyleneoxy) ethanol (10 EO). See POE (10) octyl phenyl ether
Octylphenoxypoly (ethyleneoxy) ethanol (11 EO). See POE (11) octyl phenyl ether
Octylphenoxypoly (ethyleneoxy) ethanol (12 EO). See POE (12) octyl phenyl ether
Octylphenoxypoly (ethyleneoxy) ethanol (30 EO). See POE (30) octyl phenyl ether
Octylphenoxypoly (ethyleneoxy) ethanol (40 EO). See POE (40) octyl phenyl ether
Octylphenoxypoly (ethyleneoxy) ethanol (70 EO). See POE (70) octyl phenyl ether
Oleic diethanolamide. See Book III
Oleth-5 (CTFA). See POE (5) oleyl ether
Oleyl alcohol. See Book IV
Oleyl amide. See Book IV
Oleyl amine. See Book I
Oleyl dimethyl benzyl ammonium chloride. See Book IV
Oleyl imidazoline. See Book I
Oleyl monooleate. See Book IV
4-Oxazolemethanol, 4-ethyl-2-(8-heptadecenyl)-4,5-dihydro-. See Ethyl hydroxymethyl oleyl oxazoline
2-[(1-Oxotetradecyl) amino] ethyl ester of sulfobutanedioic acid, disodium salt. See Disodium myristamido MEA-sulfosuccinate
Palmityl dimethyl amine. See Book I
PEG-5 castor oil (CTFA). See POE (5) castor oil
PEG (5) castor oil. See POE (5) castor oil
PEG-10 castor oil (CTFA). See POE (10) castor oil
PEG-25 castor oil (CTFA). See POE (25) castor oil
PEG (25) castor oil. See POE (25) castor oil
PEG-30 castor oil (CTFA). See POE (30) castor oil
PEG (30) castor oil. See POE (30) castor oil
PEG-40 castor oil (CTFA). See POE (40) castor oil
PEG (100) castor oil. See POE (100) castor oil
PEG 500 castor oil. See POE (10) castor oil
PEG 2000 castor oil. See POE (40) castor oil
PEG-100 castor oil (CTFA). See POE (100) castor oil
PEG-40 castor oil, hydrogenated. See POE (40) hydrogenated castor oil
PEG-2 cetyl/stearyl ether. See POE (2) cetyl/stearyl ether
PEG-3 cetyl/stearyl ether. See POE (3) cetyl/stearyl ether
PEG (3) cetyl/stearyl ether. See POE (3) cetyl/stearyl ether

PEG-4 cetyl/stearyl ether. See POE (4) cetyl/stearyl ether
PEG-6 cetyl/stearyl ether. See POE (6) cetyl/stearyl ether
PEG-10 cetyl/stearyl ether. See POE (10) cetyl/stearyl ether
PEG-12 cetyl/stearyl ether. See POE (12) cetyl/stearyl ether
PEG-15 cetyl/stearyl ether. See POE (15) cetyl/stearyl ether
PEG (15) cetyl/stearyl ether. See POE (15) cetyl/stearyl ether
PEG-20 cetyl/stearyl ether. See POE (20) cetyl/stearyl ether
PEG 100 cetyl/stearyl ether. See POE (2) cetyl/stearyl ether
PEG 200 cetyl/stearyl ether. See POE (4) cetyl/stearyl ether
PEG 300 cetyl/stearyl ether. See POE (6) cetyl/stearyl ether
PEG 500 cetyl/stearyl ether. See POE (10) cetyl/stearyl ether
PEG 600 cetyl/stearyl ether. See POE (12) cetyl/stearyl ether
PEG 1000 cetyl/stearyl ether. See POE (20) cetyl/stearyl ether
PEG-5 cocoate (CTFA). See POE (5) monococoate
PEG-15 cocoate (CTFA). See POE (15) monococoate
PEG-4 decyl ether phosphate. See POE (4) decyl ether phosphate
PEG 200 decyl ether phosphate. See POE (4) decyl ether phosphate
PEG-4 dilaurate (CTFA). See POE (4) dilaurate
PEG-6 dilaurate (CTFA). See POE (6) dilaurate
PEG-8 dilaurate (CTFA). See POE (8) dilaurate
PEG-12 dilaurate (CTFA). See POE (12) dilaurate
PEG 200 dilaurate. See POE (4) dilaurate
PEG 300 dilaurate. See POE (6) dilaurate
PEG 400 dilaurate. See POE (8) dilaurate
PEG 600 dilaurate. See POE (12) dilaurate
PEG-150 dinonyl phenyl ether. See POE (150) dinonyl phenyl ether
PEG (150) dinonyl phenyl ether. See POE (150) dinonyl phenyl ether
PEG-6-32 dioleate (CTFA). See POE 1500 dioleate
PEG 1500 dioleate. See POE 1500 dioleate
PEG-32 distearate (CTFA). See POE (32) distearate
PEG 1540 distearate. See POE (32) distearate
PEG-12 ditallate (CTFA). See POE (12) ditallate
PEG 600 ditallate. See POE (12) ditallate
PEG-8 ditriricinoleate (CTFA). See POE (8) ditriricinoleate
PEG 400 ditriricinoleate. See POE (8) ditriricinoleate
PEG-40 hydrogenated castor oil (CTFA). See POE (40) hydrogenated castor oil
PEG-50 hydrogenated castor oil (CTFA). See POE (50) hydrogenated castor oil
PEG (50) hydrogenated castor oil. See POE (50) hydrogenated castor oil
PEG-60 hydrogenated castor oil (CTFA). See POE (60) hydrogenated castor oil
PEG (60) hydrogenated castor oil. See POE (60) hydrogenated castor oil
PEG-100 hydrogenated castor oil (CTFA). See POE (100) hydrogenated castor oil
PEG (100) hydrogenated castor oil. See POE (100) hydrogenated castor oil

PEG-200 hydrogenated castor oil (CTFA). See POE (200) hydrogenated castor oil
PEG (200) hydrogenated castor oil. See POE (200) hydrogenated castor oil
PEG 2000 hydrogenated castor oil. See POE (40) hydrogenated castor oil
PEG (50) hydrogenated tallow amide. See POE (50) hydrogenated tallow amide
PEG-20 isocetyl ether. See POE (20) isocetyl ether
PEG 1000 isocetyl ether. See POE (20) isocetyl ether
PEG-1 lauryl ether. See Book I
PEG-2 lauryl ether. See POE (2) lauryl ether
PEG-3 lauryl ether. See POE (3) lauryl ether
PEG (3) lauryl ether. See POE (3) lauryl ether
PEG-7 lauryl ether. See POE (7) lauryl ether
PEG (7) lauryl ether. See POE (7) lauryl ether
PEG 100 lauryl ether. See POE (2) lauryl ether
PEG (5) monococoate. See POE (5) monococoate
PEG (15) monococoate. See POE (15) monococoate
PEG (5) monooleate. See POE (5) monooleate
PEG (10) monooleate. See POE (10) monooleate
PEG 100 monoricinoleate. See Diethylene glycol monoricinoleate
PEG (5) monostearate. See POE (5) monostearate
PEG (10) monostearate. See POE (10) monostearate
PEG (100) monostearate. See POE (100) monostearate
PEG 6000 monostearate. See POE (150) monostearate
PEG-1 nonyl phenyl ether. See POE (1) nonyl phenyl ether
PEG-2 nonyl phenyl ether. See POE (2) nonyl phenyl ether
PEG-4 nonyl phenyl ether. See POE (4) nonyl phenyl ether
PEG-5 nonyl phenyl ether. See POE (5) nonyl phenyl ether
PEG (5) nonyl phenyl ether. See POE (5) nonyl phenyl ether
PEG-6 nonyl phenyl ether. See POE (6) nonyl phenyl ether
PEG-8 nonyl phenyl ether. See POE (8) nonyl phenyl ether
PEG-9 nonyl phenyl ether. See POE (9) nonyl phenyl ether
PEG-10 nonyl phenyl ether. See POE (10) nonyl phenyl ether
PEG-11 nonyl phenyl ether. See POE (11) nonyl phenyl ether
PEG (11) nonyl phenyl ether. See POE (11) nonyl phenyl ether
PEG-12 nonyl phenyl ether. See POE (12) nonyl phenyl ether
PEG-13 nonyl phenyl ether. See POE (13) nonyl phenyl ether
PEG (13) nonyl phenyl ether. See POE (13) nonyl phenyl ether
PEG-14 nonyl phenyl ether. See POE (14) nonyl phenyl ether
PEG (14) nonyl phenyl ether. See POE (14) nonyl phenyl ether
PEG-15 nonyl phenyl ether. See POE (15) nonyl phenyl ether
PEG (15) nonyl phenyl ether. See POE (15) nonyl phenyl ether
PEG-20 nonyl phenyl ether. See POE (20) nonyl phenyl ether
PEG 100 nonyl phenyl ether. See POE (2) nonyl phenyl ether

PEG 200 nonyl phenyl ether. See POE (4) nonyl phenyl ether
PEG 300 nonyl phenyl ether. See POE (6) nonyl phenyl ether
PEG 400 nonyl phenyl ether. See POE (8) nonyl phenyl ether
PEG 450 nonyl phenyl ether. See POE (9) nonyl phenyl ether
PEG 500 nonyl phenyl ether. See POE (10) nonyl phenyl ether
PEG 600 nonyl phenyl ether. See POE (12) nonyl phenyl ether
PEG 1000 nonyl phenyl ether. See POE (20) nonyl phenyl ether
PEG-1 octyl phenyl ether. See POE (1) octyl phenyl ether
PEG-3 octyl phenyl ether. See POE (3) octyl phenyl ether
PEG (3) octyl phenyl ether. See POE (3) octyl phenyl ether
PEG-5 octyl phenyl ether. See POE (5) octyl phenyl ether
PEG (5) octyl phenyl ether. See POE (5) octyl phenyl ether
PEG-6 octyl phenyl ether. See POE (6) octyl phenyl ether
PEG (6) octyl phenyl ether. See POE (6) octyl phenyl ether
PEG-7 octyl phenyl ether. See POE (7) octyl phenyl ether
PEG (7) octyl phenyl ether. See POE (7) octyl phenyl ether
PEG-8 octyl phenyl ether. See POE (8) octyl phenyl ether
PEG-9 octyl phenyl ether. See POE (9) octyl phenyl ether
PEG (9) octyl phenyl ether. See POE (9) octyl phenyl ether
PEG-10 octyl phenyl ether. See POE (10) octyl phenyl ether
PEG-11 octyl phenyl ether. See POE (11) octyl phenyl ether
PEG (11) octyl phenyl ether. See POE (11) octyl phenyl ether
PEG-12 octyl phenyl ether. See POE (12) octyl phenyl ether
PEG (12) octyl phenyl ether. See POE (12) octyl phenyl ether
PEG-30 octyl phenyl ether. See POE (30) octyl phenyl ether
PEG (30) octyl phenyl ether. See POE (30) octyl phenyl ether
PEG-40 octyl phenyl ether. See POE (40) octyl phenyl ether
PEG-70 octyl phenyl ether. See POE (70) octyl phenyl ether
PEG (70) octyl phenyl ether. See POE (70) octyl phenyl ether
PEG 400 octyl phenyl ether. See POE (8) octyl phenyl ether
PEG 500 octyl phenyl ether. See POE (10) octyl phenyl ether
PEG 2000 octyl phenyl ether. See POE (40) octyl phenyl ether
PEG-5 oleamide (CTFA). See POE (5) oleyl amide
PEG-2 oleamine (CTFA). See POE (2) oleyl amine
PEG-5 oleamine (CTFA). See POE (5) oleyl amine
PEG-15 oleamine (CTFA). See POE (15) oleyl amine
PEG-5 oleate (CTFA). See POE (5) monooleate
PEG-10 oleate (CTFA). See POE (10) monooleate
PEG (5) oleyl amide. See POE (5) oleyl amide
PEG (5) oleyl amine. See POE (5) oleyl amine
PEG (15) oleyl amine. See POE (15) oleyl amine
PEG 100 oleyl amine. See POE (2) oleyl amine

PEG-5 oleyl ether. See POE (5) oleyl ether
PEG (5) oleyl ether. See POE (5) oleyl ether
PEG-2 ricinoleate (CTFA). See Diethylene glycol monoricinoleate
PEG-20 sorbitan isostearate (CTFA). See POE (20) sorbitan monoisostearate
PEG-20 sorbitan monoisostearate. See POE (20) sorbitan monoisostearate
PEG 1000 sorbitan monoisostearate. See POE (20) sorbitan monoisostearate
PEG-40 sorbitan peroleate (CTFA). See POE (40) sorbitan peroleate
PEG 2000 sorbitan peroleate. See POE (40) sorbitan peroleate
PEG-10 stearamine (CTFA). See POE (10) stearyl amine
PEG-15 stearamine (CTFA). See POE (15) stearyl amine
PEG-5 stearate (CTFA). See POE (5) monostearate
PEG-10 stearate (CTFA). See POE (10) monostearate
PEG-100 stearate (CTFA). See POE (100) monostearate
PEG-150 stearate (CTFA). See POE (150) monostearate
PEG (15) stearyl amine. See POE (15) stearyl amine
PEG 500 stearyl amine. See POE (10) stearyl amine
PEG-2 stearyl ether. See POE (2) stearyl ether
PEG-10 stearyl ether. See POE (10) stearyl ether
PEG-15 stearyl ether. See POE (15) stearyl ether
PEG (15) stearyl ether. See POE (15) stearyl ether
PEG-20 stearyl ether. See POE (20) stearyl ether
PEG-21 stearyl ether. See POE (21) stearyl ether
PEG (21) stearyl ether. See POE (21) stearyl ether
PEG-40 stearyl ether. See POE (40) stearyl ether
PEG 500 stearyl ether. See POE (10) stearyl ether
PEG 100 stearyl ether. See POE (2) stearyl ether
PEG 1000 stearyl ether. See POE (20) stearyl ether
PEG 2000 stearyl ether. See POE (40) stearyl ether
PEG-50 tallow amide (CTFA). See POE (50) hydrogenated tallow amide
PEG-50 tallow amine (CTFA). See POE (50) tallow amine
PEG (50) tallow amine. See POE (50) tallow amine
PEG (50) tallow amine, hydrogenated. See POE (50) tallow amine
PEG-3 tallow aminopropylamine (CTFA). See POE (3) tallow aminopropylamine
PEG (3) tallow aminopropylamine. See POE (3) tallow aminopropylamine
PEG-9 tridecyl ether. See POE (9) tridecyl ether
PEG-10 tridecyl ether. See POE (10) tridecyl ether
PEG-11 tridecyl ether. See POE (11) tridecyl ether
PEG (11) tridecyl ether. See POE (11) tridecyl ether
PEG-15 tridecyl ether. See POE (15) tridecyl ether
PEG (15) tridecyl ether. See POE (15) tridecyl ether
PEG 450 tridecyl ether. See POE (9) tridecyl ether
PEG 500 tridecyl ether. See POE (10) tridecyl ether

PEG-6 tridecyl ether phosphate. See POE (6) tridecyl ether phosphate
PEG 300 tridecyl ether phosphate. See POE (6) tridecyl ether phosphate
2-Pentanone, 4-methyl-. See Methyl isobutyl ketone
Pentasodium amino tris (methylene phosphonate). See Pentasodium aminotrimethylene phosphonate
Pentasodium [nitrilotris (methylene)] tris phosphonate. See Pentasodium aminotrimethylene phosphonate
Phenoxyethanol (CTFA). See Ethylene glycol monophenyl ether
2-Phenoxyethanol. See Ethylene glycol monophenyl ether
Phenoxytol. See Ethylene glycol monophenyl ether
Phosphonic acid, (1-hydroxyethylidene) bis-. See Etidronic acid
Phosphonic acid, [nitrilotris (methylene)] tris-. See Aminotrimethylene phosphonic acid
Phosphonic acid, [nitrilotris (methylene)] tris-, pentasodium salt. See Pentasodium aminotrimethylene phosphonate
POE (4) cetearyl ether. See POE (4) cetyl/stearyl ether
POE (3) ceto stearyl ether. See POE (3) cetyl/stearyl ether
POE 10 ceto stearyl ether. See POE (10) cetyl/stearyl ether
POE 15 ceto stearyl ether. See POE (15) cetyl/stearyl ether
POE 20 ceto stearyl ether. See POE (20) cetyl/stearyl ether
POE (2) cetyl ether. See Book I
POE (20) cetyl ether. See Book I
POE (4) cetyl/stearyl alcohol. See POE (4) cetyl/stearyl ether
POE (6) cetyl/stearyl alcohol. See POE (6) cetyl/stearyl ether
POE (10) cetyl/stearyl alcohol. See POE (10) cetyl/stearyl ether
POE (15) cetyl/stearyl alcohol. See POE (15) cetyl/stearyl ether
POE (30) cetyl/stearyl ether. See Book I
POE (50) cetyl/stearyl ether. See Book I
POE (6) coconut amide. See Book III
POE (2) coconut amine. See Book I
POE (5) coconut amine. See Book I
POE (15) coconut amine. See Book I
POE (20) dilaurate. See Book III
POE (32) dilaurate. See Book III
POE (75) dilaurate. See Book III
POE (150) dilaurate. See Book III
POE (150) dinonyl phenol. See POE (150) dinonyl phenyl ether
POE (4) dioleate. See Book I
POE (8) dioleate. See Book I
POE (12) dioleate. See Book I
POE (4) distearate. See Book I
POE (8) distearate. See Book I
POE (12) distearate. See Book I

POE (150) distearate. See Book III
POE (7) glyceryl monococoate. See Book IV
POE (20) glyceryl monostearate. See Book I
POE (5) hydrogenated castor oil. See Book I
POE (20) isohexadecyl alcohol. See POE (20) isocetyl ether
POE (20) isohexadecyl ether. See POE (20) isocetyl ether
POE (2) isostearyl ether. See Book I
POE (20) isostearyl ether. See Book I
POE (75) lanolin. See Book IV
POE (7) lauryl alcohol. See POE (7) lauryl ether
POE (4) lauryl ether. See Book I
POE (12) lauryl ether. See Book I
POE (23) lauryl ether. See Book I
POE (120) methyl glucose dioleate. See Book III
POE (8) monococoate. See Book I
POE (4) monolaurate. See Book I
POE (6) monolaurate. See Book I
POE (8) monolaurate. See Book I
POE (12) monolaurate. See Book I
POE (4) monooleate. See Book I
POE (8) monooleate. See Book I
POE (12) monooleate. See Book I
POE (20) monooleate. See Book I
POE (2) monoricinoleate. See Diethylene glycol monoricinoleate
POE (4) monostearate. See Book I
POE (8) monostearate. See Book I
POE (12) monostearate. See Book I
POE (20) monostearate. See Book I
POE (40) monostearate. See Book I
POE (75) monostearate. See Book I
POE (1500) monostearate. See Book I
POE (8) monotallate. See Book I
POE (1) nonyl phenol. See POE (1) nonyl phenyl ether
POE (4) nonyl phenol. See POE (4) nonyl phenyl ether
POE (5) nonyl phenol. See POE (5) nonyl phenyl ether
POE (6) nonyl phenol. See POE (6) nonyl phenyl ether
POE (8) nonyl phenol. See POE (8) nonyl phenyl ether
POE (9) nonyl phenol. See POE (9) nonyl phenyl ether
POE (10) nonyl phenol. See POE (10) nonyl phenyl ether
POE (11) nonyl phenol. See POE (11) nonyl phenyl ether
POE (12) nonyl phenol. See POE (12) nonyl phenyl ether
POE (13) nonyl phenol. See POE (13) nonyl phenyl ether

POE (15) nonyl phenol. See POE (15) nonyl phenyl ether
POE (20) nonyl phenol. See POE (20) nonyl phenyl ether
POE (30) nonyl phenyl ether. See Book III
POE (40) nonyl phenyl ether. See Book III
POE (50) nonyl phenyl ether. See Book III
POE (100) nonyl phenyl ether. See Book III
POE (10) octadecylamine. See POE (10) stearyl amine
POE (15) octadecylamine. See POE (15) stearyl amine
POE (3) octyl phenol. See POE (3) octyl phenyl ether
POE (5) octyl phenol. See POE (5) octyl phenyl ether
POE (6) octyl phenol. See POE (6) octyl phenyl ether
POE (7) octyl phenol. See POE (7) octyl phenyl ether
POE (8) octyl phenol. See POE (8) octyl phenyl ether
POE (9) octyl phenol. See POE (9) octyl phenyl ether
POE (10) octyl phenol. See POE (10) octyl phenyl ether
POE (11) octyl phenol. See POE (11) octyl phenyl ether
POE (12) octyl phenol. See POE (12) octyl phenyl ether
POE (30) octyl phenol. See POE (30) octyl phenyl ether
POE (40) octyl phenol. See POE (40) octyl phenyl ether
POE (70) octyl phenol. See POE (70) octyl phenyl ether
POE (30) octyl phenyl alcohol. See POE (30) octyl phenyl ether
POE (40) octyl phenyl alcohol. See POE (40) octyl phenyl ether
POE (5) oleamide. See POE (5) oleyl amide
POE (2) oleyl ether. See Book I
POE (3) oleyl ether. See Book IV
POE (4) oleyl ether. See Book I
POE (10) oleyl ether. See Book I
POE (20) oleyl ether. See Book I
POE (12) POP (6) decyltetradecyl ether. See POP (6) POE (12) tetradecyl ether
POE (20) POP (6) decyltetradecyl ether. See POP (6) POE (20) tetradecyl ether
POE (30) POP (6) decyltetradecyl ether. See POP (6) POE (30) tetradecyl ether
POE (12) POP (6) tetradecyl ether. See POP (6) POE (12) tetradecyl ether
POE (20) POP (6) tetradecyl ether. See POP (6) POE (20) tetradecyl ether
POE (30) POP (6) tetradecyl ether. See POP (6) POE (30) tetradecyl ether
POE (4) sorbitan monolaurate. See Book I
POE (20) sorbitan monolaurate. See Book I
POE (80) sorbitan monolaurate. See Book III
POE (5) sorbitan monooleate. See Book I
POE (20) sorbitan monooleate. See Book I
POE (20) sorbitan monopalmitate. See Book I
POE (4) sorbitan monostearate. See Book I
POE (20) sorbitan monostearate. See Book I

POE (20) sorbitan trioleate. See Book I
POE (20) sorbitan tristearate. See Book I
POE (40) sorbitol septaoleate. See POE (40) sorbitan peroleate
POE (5) soya amine. See Book I
POE (10) soya amine. See Book I
POE (15) soya amine. See Book I
POE (5) soya sterol. See Book IV
POE (10) soya sterol. See Book IV
POE (25) soya sterol. See Book IV
POE (2) stearyl alcohol. See POE (2) stearyl ether
POE (10) stearyl alcohol. See POE (10) stearyl ether
POE (15) stearyl alcohol. See POE (15) stearyl ether
POE (20) stearyl alcohol. See POE (20) stearyl ether
POE (21) stearyl alcohol. See POE (21) stearyl ether
POE (40) stearyl alcohol. See POE (40) stearyl ether
POE (2) stearyl amine. See Book I
POE (5) stearyl amine. See Book I
POE (50) stearyl amine. See Book I
POE (2) tallow amine. See Book I
POE (5) tallow amine. See Book I
POE (15) tallow amine. See Book I
POE (9) tridecyl alcohol. See POE (9) tridecyl ether
POE (3) tridecyl ether. See Book I
POE (6) tridecyl ether. See Book I
POE (12) tridecyl ether. See Book I
Polyethoxylated nonylphenol (11 EO). See POE (11) nonyl phenyl ether
Polyethoxylated nonylphenol (12 EO). See POE (12) nonyl phenyl ether
Polyethoxylated octyl phenol (40 EO). See POE (40) octyl phenyl ether
Polyethoxylated octyl phenol (70 EO). See POE (70) octyl phenyl ether
Polyethoxylated octylphenol (9 EO). See POE (9) octyl phenyl ether
Polyglyceryl-10 decaoleate (CTFA). See Decaglycerol decaoleate
Polyglyceryl-10 decastearate (CTFA). See Decaglycerol decastearate
Polyglyceryl-6 distearate (CTFA). See Hexaglyceryl distearate
Polyoxyethylene. See POE
Polyoxypropylene. See POP
Polypropylene glycol. See PPG
POP (9) methyl diethyl ammonium chloride. See POP (9) methyl diethyl ammonium chloride
POP (25) methyl diethyl ammonium chloride. See POP (25) methyl diethyl ammonium chloride
POP (2) methyl ether. See Dipropylene glycol monomethyl ether
POP (3) methyl ether. See Tripropylene glycol monomethyl ether

POP (10) methyl glucose ether. See Book IV
POP (20) methyl glucose ether. See Book IV
POP (6) POE (12) tetradecyl ether. See POP (6) POE (12) tetradecyl ether
POP (6) POE (20) tetradecyl ether. See POP (6) POE (20) tetradecyl ether
POP (6) POE (30) tetradecyl ether. See POP (6) POE (30) tetradecyl ether
Potassium pyrophosphate. See Tetrapotassium pyrophosphate
PPG-6-decyltetradeceth-12 (CTFA). See POP (6) POE (12) tetradecyl ether
PPG-6-decyltetradeceth-20 (CTFA). See POP (6) POE (20) tetradecyl ether
PPG-6-decyltetradeceth-30 (CTFA). See POP (6) POE (30) tetradecyl ether
PPG-9 diethylmonium chloride (CTFA). See POP (9) methyl diethyl ammonium chloride
PPG-25 diethylmonium chloride (CTFA). See POP (25) methyl diethyl ammonium chloride
PPG-3-isosteareth-9. See Book I
PPG-2 methyl ether (CTFA). See Dipropylene glycol monomethyl ether
PPG (2) methyl ether. See Dipropylene glycol monomethyl ether
PPG-3 methyl ether (CTFA). See Tripropylene glycol monomethyl ether
PPG (3) methyl ether. See Tripropylene glycol monomethyl ether
PPG (15) stearyl ether. See Book IV
Primary hydrogenated tallow amine. See Tallow amine, hydrogenated
1,3-Propanediol, 2-amino-2-(hydroxymethyl)-. See Tris (hydroxymethyl) amino methane
2-Propanol. See Isopropyl alcohol
1-Propanol, 2-amino-2-methyl-. See Aminomethyl propanol
2-Propanol, 1-methoxy-. See Propylene glycol monomethyl ether
2-Propanol, 1-(2-methoxypropoxy)-. See Dipropylene glycol monomethyl ether
2-Propanone. See Acetone
sec-Propyl alcohol. See Isopropyl alcohol
Propyl carbinol. See *n*-Butyl alcohol
Propylene glycol monolaurate. See Book I
Propylene glycol monooleate. See Book I
Propylene glycol monoricinoleate. See Book III
Propylene glycol monostearate. See Book I
Puffed borax. See Sodium borate
Quaternary ammonium compounds, tallow alkyl trimethyl, chlorides. See Tallow trimethyl ammonium chloride
Quaternary ammonium compounds, (3-tallowamidopropyl) (2-hydroxy-3-sulfopropyl) dimethyl, hydroxide, inner salt. See Tallowamidopropyl hydroxysultaine
Quaternium-6. See POP (9) methyl diethyl ammonium chloride
Quaternium-20. See POP (25) methyl diethyl ammonium chloride
Silicic acid, disodium salt. See Sodium metasilicate
Sodium bistridecyl sulfosuccinate. See Sodium ditridecyl sulfosuccinate
Sodium borax, anhydrous. See Sodium borate

Stearic acid amide. See Book IV
Stearyl primary amine. See Stearyl amine
Stearyl-cetyl alcohol ethoxylate (10 moles EO). See POE (10) cetyl/stearyl ether
Succinic acid, sulfo-, 1,4, dicyclohexyl ester, sodium salt. See Sodium dicyclohexyl sulfosuccinate
Sucrose cocoate (CTFA). See Sucrose monococoate
Sucrose distearate, acetates. See Acetylated sucrose distearate
Sucrose stearate (CTFA). See Sucrose monostearate
Sulfated nonoxynol-4, ammonium salt. See Ammonium nonoxynol-4 sulfate
Sulfobutanedioic acid, C-(2-cocamidoethyl) esters, disodium salts. See Disodium cocoyl monoethanolamide sulfosuccinate
Sulfobutanedioic acid, 4-isodecyl ester, disodium salt. See Disodium isodecyl sulfosuccinate
Sulfobutanedioic acid, monooctadecyl ester, disodium salt. See Disodium stearyl sulfosuccinamate
Sulfobutanedioic acid, 1(or 4)- [2- [2- [(1-oxo-9-octadecenyl) amino] ethoxy] ethyl] ester, disodium salt. See Disodium oleamido PEG-2 sulfosuccinate
Sulfobutanedioic acid, C- [2- [2- [(1-oxo-9-octadecenyl) amino] ethoxy] ethyl] ester, disodium salt. See Disodium oleamido PEG-2 sulfosuccinate
Sulfobutanedioic acid, 2 [1-oxotetradecyl) amino] ethyl ester, disodium salt. See Disodium myristamido MEA-sulfosuccinate
Sulfosuccinic acid, 1,4-dicyclohexyl ester, sodium salt. See Sodium dicyclohexyl sulfosuccinate
N-[2-[(Sulfosuccinyl) oxy] ethyl] coconut oil amides, sodium salts. See Disodium cocoyl monoethanolamide sulfosuccinate
(3-Tallowamidopropyl) (2-hydroxy-3-sulfopropyl) dimethyl quaternary ammonium compounds, hydroxide, inner salt. See Tallowamidopropyl hydroxysultaine
Tallow dimethyl amine oxide. See Book IV
Tallow primary amine. See Tallow amine
Tallowtrimonium chloride (CTFA). See Tallow trimethyl ammonium chloride
Tetrachloroethylene. See Perchloroethylene
Tetrasodium N-(1,2-dicarboxyethyl)-N-octadecyl sulfosuccinate. See Tetrasodium dicarboxyethyl stearyl sulfosuccinamate
THAM. See Tris (hydroxymethyl) amino methane
Tincal. See Sodium borate
TKPP. See Tetrapotassium pyrophosphate
Toluene sulfonic acid, sodium salt. See Sodium toluenesulfonate
1,1,1-Trichloroethane. See Trichloroethane
Trideceth-6 phosphate (CTFA). See POE (6) tridecyl ether phosphate
Trideceth-9 (CTFA). See POE (9) tridecyl ether
Trideceth-10 (CTFA). See POE (10) tridecyl ether
Trideceth-11 (CTFA). See POE (11) tridecyl ether

Trideceth-15 (CTFA). See POE (15) tridecyl ether
Tridecyl alcohol ethylene oxide adduct (9 EO). See POE (9) tridecyl ether
Tridecyl alcohol, ethoxylated (9 EO). See POE (9) tridecyl ether
Tridecyloxypoly (ethyleneoxy) ethanol (10 EO). See POE (10) tridecyl ether
Tridecyloxypoly (ethyleneoxy) ethanol (15 EO). See POE (15) tridecyl ether
Tri (hydroxymethyl) amino methane. See Tris (hydroxymethyl) amino methane
Trimethyl tallow ammonium chloride. See Tallow trimethyl ammonium chloride
Triphosphoric acid, pentasodium salt. See Pentasodium triphosphate
Tris amine buffer. See Tris (hydroxymethyl) amino methane
Tromethamine (CTFA). See Tris (hydroxymethyl) amino methane
TSPP. See Tetrasodium pyrophosphate
Vinegar naphtha. See Ethyl acetate
Xylene sulfonate, ammonium salt. See Ammonium xylenesulfonate

TRADENAME PRODUCT MANUFACTURERS

Akzo Chemie America
300 S. Wacker Dr.
Chicago, IL 60606

Akzo Chemie B.V.
Stationsstraat 48, PO Box 247
3800 AE-Amersfoort, Netherlands

Albright & Wilson (Australia) Ltd.
610 St. Kilda Rd., PO Box 4544
Melbourne 3001, Australia

Albright & Wilson/
Detergents Div., Marchon Works
PO Box 15, Whitehaven
Cumbria CA28 9QQ, UK

Albright & Wilson/Phosphates Div.
PO Box 3, Hagley Rd. W.
Oldbury, Warley
W. Midlands B68 0NN, UK

Alco Chemical Corp.
909 Mueller Dr.
Chattanooga, TN 37406

Alcolac Inc.
3440 Fairfield Rd.
Baltimore, MD 21226

Alkaril Chemicals Inc.
Industrial Pkwy., PO Box 1010
Winder, GA 30680

Allied Colloids Ltd.
PO Box 38, Low Moor, Bradford
Yorkshire BD12 OJZ, UK

American Can Co.
American Lane
Greenwich, CT 06830

American Color & Chemical Corp.
11400 Westinghous Blvd.
Charlotte, NC 28232

American Cyanamid Co.
Berdan Ave.
Wayne, NJ 07470

American Hoechst
Rt. 202–206 North
Somerville, NJ 08876

American Lecithin Co., Inc.
32–34 61st St.
Woodside, NY 11377

Angus Chemical Co.
2211 Sanders Rd.
Northbrook, IL 60062

Arco Chemical Co./
Div. Atlantic Richfield Co.
1500 Market St.
Philadelphia, PA 19101

Armak Co.
8401 West 47th St.
McCook, IL 60525

Armak Ind. Chem. /
Div. Akzo Chemie America
300 S. Wacker Dr.
Chicago, IL 60606

Ashland Chemical Co.
Box 2219
Columbus, OH 43216

Atlas Chem. Ind. N.V.
Imperial Chemical House
Milbank, London, SW1P 3JF, UK

BASF AG
Carl-Bosch-Strasse 38
D-6700 Ludwigshafen, F.R.G.

BASF Wyandotte Corp.
100 Cherry Hill Rd.
Parsippany, NJ 07054

Bofors Lakeway
5025 Evanston Ave.
PO Box 328
Muskegon, MI 49443

Borg-Warner Chemicals Inc.
International Center
Parkersburg, WV 26102

Bostik South/Div. of USM Corp.
PO Box 5695
Greenville, SC 29606

Canada Packers Ltd.
5100 Timberlea Blvd.
Mississauga, Ontario L4W 2S5, Canada

Capital City Products/
Div. of Stokely-Van Camp
PO Box 569
Columbus, OH 43216

Carson Chemicals, Inc.
2779 East El Presidio
Long Beach, CA 90810

CasChem Inc.
40 Avenue A
Bayonne, NJ 07002

Central Soya Inc./Chemurgy Div.
1300 Ft. Wayne Bank Bldg., POB 1400
Fort Wayne, IN 46801

Chemax, Inc.
PO Box 6067, Highway 25 South
Greenville, SC 29606

Chemform Corp.
141 S.W. 8th St.
Pompano Beach, FL 33061

Ciba-Geigy Corp.
PO Box 18300
Greensboro, NC 27419

Cleary Chemical Corp., W.A.
1049 Somerset St.
Somerset, NJ 08873

Clough Chemical Co., Ltd.
178 St. Pierre
St.-Jean, Quebec J3B 7B5, Canada

Colloids Inc.
394 Frelinghuysen Ave.
Newark, NJ 07114

Continental Oil Co.
5 Greenway Plaza East, PO Box 2197
Houston, TX 77001

Croda Chem. Ltd.
Cowick Hall, Snaith Goole
North Humberside DN14 9AA, UK

Croda Inc.
51 Madison Ave.
New York, NY 10010

TRADENAME PRODUCT MANUFACTURERS

Croda Surfactants Inc.
183 Madison Ave.
New York, NY 10016

Crown Zellerbach Corp.
1 Bush St.
San Francisco, CA 94104

Cyanamid B.V.
Postbus 1523, BM 3000
Rotterdam, The Netherlands

Cyclo Chemical Corp.
7500 N.W. 66th St.
Miami, FL 33166

Diamond Shamrock/Process Chem. Div.
350 Mt. Kemble Ave.
Morristown, NJ 07960

Domtar Inc./CDC Div.
1136 Matheson Blvd.
Mississauga, Ontario L4W 2V4, Canada

Dow Chemical Co.
1703 S. Saginaw Rd.
Midland, MI 48640

Dow Chemical Europe SA
Leland I. Doan Str. 3 CH-8810
Horgen, Switzerland

Drew Produtos Quimicos Ltds.
Rua Sampaio Viana, 425
04004, Sao Paulo, Brazil-CP4885

Durkee Industrial Foods/SCM Corp.
900 Union Commerce Bldg.
Cleveland, OH 44115

Dutton & Reinisch Ltd.
Crown House, London Rd., Morden
Surrey SM45DU, UK

Dynamit Nobel
10 Link Dr.
Rockleigh, NJ 07647

Eastman Chemical Products, Inc.
PO Box 431
Kingsport, TN 37662

Emery Industries Inc.
1501 W. Elizabeth Ave.
Linden, NJ 07036

Emulsion Systems Inc.
215 Kent Ave.
Brooklyn, NY 11211

Finetex Inc.
418 Falmouth Ave.
Elmwood Park, NJ 07407

FMC Corp.
2000 Market St.
Philadelphia, PA 19103

GAF Corp.
1361 Alps Rd.
Wayne, NJ 07470

Georgia-Pacific Corp.
300 W. Laurel St.
Bellingham, WA 98225

Geronazzo Ind. Chim. SpA, Mario
78, Ospiate Di Bollate
Milano, Italy 20021

Glyco Chemicals Inc.
PO Box 700, 51 Weaver St.
Greenwich, CT 06830

Goldschmidt AG, Th.
Goldschmidtstr. 100
4300 Essen 1, Postfach 101461
West Germany

Graden Chem. Co., Inc.
426 Bryan St.
Havertown, PA 19083

Chemische Fabrik Grunau GmbH
Robert-Hansen Str. 1, Postfach 1063
D-7918 Jllertissen
Bavaria, West Germany

Hall, C.P. Co.
7300 South Central Ave.
Chicago, IL 60638

Harrisons & Crosfield (Canada) Ltd.
2 Banigan Drive
Toronto, Ontario M4H 1E9 Canada

Hart Chem. Ltd.
256 Victoria Rd. South
Guelph, Ontario N1H 6K8, Canada

Hart Products Corp.
173 Sussex St.
Jersey City, NJ 07302

Hefti Ltd.
PO Box 1623, CH-8048
Zurich, Switzerland

Henkel Chem. (Canada) Ltd.
9550 Ray Lawson Blvd.
Ville d'Anjou, Quebec H1J 1L3, Canada

Henkel Inc.
480 Alfred Ave.
Teaneck, NJ 07666

Henkel KGaA
Postfach 1100, D-4000
Dusseldorf 1, West Germany

Heterene Chemical Co., Inc.
POB 247, 792 21 Ave.
Paterson, NJ 07513

High Point Chem. Corp.
601 Taylor St., PO Box 2316
High Point, NC 27261

Hodag Chem. Corp.
7247 N. Central Park Ave.
Skokie, IL 60076

Hoechst AG
Verhaufkanststoffe, D-6230
Frankfurt (M) 80, West Germany

Hüls AG, Chemische Werke
Postfach 1320 D-4370
Marl 1, West Germany

Humko Sheffield Chem./Div. Kraft Inc.
PO Box 398
Memphis, TN 38101

ICI Australia Ltd.
ICI House, 1 Nicholson St.
Melbourne 3000, Australia

ICI Ltd. Organics Div.
Smith's Rd., Bolton
Lancs, BL3 2QJ, UK

ICI Specialty Chemicals
Everslann 45 B-3078
Kortenberg, Belgium

ICI United States Inc.
New Murphy Rd. & Concord Pike
Wilmington, DE 19897

IMC Chemicals
666 Garland Pl.
Des Plaines, IL 60016

TRADENAME PRODUCT MANUFACTURERS

Inolex Chem.Co.
4221 S. Western Blvd.
Chicago, IL 60609

Intex Products Inc.
PO Box 6648
Greenville, SC 29606

ITT Rayonier Inc.
1177 Summer St.
Stamford, CT 06905

Jefferson Chemical Co. Inc.
PO Box 4128
Austin, TX 78765

Jetco Chem.
PO Box 1898
Corsicana, TX 75110

Kao Corp.
14-10 Nihonbashi, Kayabacho 1-chome, Chuo-ku
Tokyo 103, Japan

Elektrochemische Fabrik Kempen GmbH
Postfach 100 260
D-4152 Kempen 1, West Germany

Kenobel S.A.
Rue Gachard 88, Bte. 9
B-1050 Bruxelles, Belgium

Kerr-McGee Chemical Corp.
PO Box 25861
Oklahoma City, OK 73125

Lankro Chem. Ltd.
PO Box 1, Eccles
Manchester M30 0BH, UK

Lilachim S.A.
37-33 Rue de la Loi
1040 Brussels, Belgium

Lipo Chemicals Inc.
207 19th Ave.
Paterson, NJ 07504

Lonza Inc.
22-10 Route 208
Fair Lawn, NJ 07410

Manchem Ltd.
Aston New Rd.
Manchester M11 4AT, UK

Manro Products Ltd.
Bridge St., Stalybridge
Cheshire SK15 1PH, UK

Mazer Chem. Inc.
3938 Porett Dr.
Gurnee, IL 60031

McIntyre Chem. Co., Ltd.
4851 S. St. Louis Ave.
Chicago, IL 60632

Merck Chemical Div./Merck & Co., Inc.
PO Box 2000
Rahway, NJ 07065

Miranol Chem. Corp.
68 Culver Rd., PO Box 411
Dayton, NJ 08810

Mona Industries, Inc.
PO Box 425, 76 E. 24th St
Paterson, NJ 07544

Monsanto Co.
800 N. Lindbergh Blvd.
St. Louis, MO 63167

Nease Chemical Co.
PO Box 221
State College, PA 16801

Nikko Chem. Co., Ltd.
1-4-8 Nihonbashi-Bakurocho, Chuo-ku
Tokyo 103, Japan

Nippon Nyukazai Co., Ltd.
19-9, 3-chome, Ginza, Chuo-ku
Tokyo 104, Japan

Nippon Oil & Fats Co., Ltd.
5-1 chome, Yurakucho, Chiyoda-ku
Tokyo, Japan

NL Industries
PO Box 700
Hightstown, NJ 08520

Norman, Fox & Co.
5511 S. Boyle Ave.
PO Box 58727
Vernon, CA 90058

Oleofina S.A.
Rue de Science 37 Wetenschapsstraat
1040 Brussels, Belgium

Olin Chemicals
120 Long Ridge Rd.
Stamford, CT 06904

Onyx Chem. Co./
Millmaster Onyx Group
190 Warren St.
Jersey City, NJ 07302

Phillips Chemical Co.
PO Box 792
Pasadena, TX 77501

Pilot Chem. Co.
11756 Burke St.
Santa Fe Springs, CA 90670

PPG Industries
One Gateway Center
Pittsburgh, PA 15222

PQ Corp.
PO Box 840
Valley Forge, PA 19482

Procter & Gamble Co.
301 E. 6th St., PO Box 599
Cincinnati, OH 45201

Procter Chemical Co.
PO Box 399
Salisbury, NC 28144

PVO International Inc.
416 Division St.
Boonton, NJ 07005

Quimigal-Quimica de Portugal E.P.
Av. Infante Santo No. 2
1300 Lisboa, Portugal

Reed Chemical Corp.
413 Wanaque Ave.
Pompton Lakes, NJ 07442

Reed Lignin Inc.
81 Holly Hill Lane
Greenwich, CT 06830

Reed Ltd.
PO Box 2025
Quebec, P.Q., Canada

Rewo Chemicals Inc.
107B Allen Blvd.
E. Farmingdale, NY 11735

TRADENAME PRODUCT MANUFACTURERS

Rewo Chemische Werke GmbH
Postfach 1160, Industriegebiet West
D-6497 Steinau an der Strasse
West Germany

Richardson Co.
2400 Devon Ave.
Des Plaines, IL 60018

Rohm & Haas Co.
Independence Mall West
Philadelphia, PA 19105

Ruetgers Nease Chemical Co., Inc.
Box 221, 201 Struble Rd.
State College, PA 16801

Ryoto Co. Ltd.
Sun Bldg.
13-12 Ginza 5-chome, chuo-ku
Tokyo 105, Japan

Sanyo Chem. Industries Ltd.
11-1 Ikkyo Nomoto-cho Higashiyama-ku
Kyoto 605, Japan

Scher Chem. Inc.
Industrial West & Styertowne Rd.
Clifton, NJ 07012

Seppic
70 Champs Elysees
75008 Paris, France

Servo Chemische Fabriek B.V.
PO Box 1, 7490 AA
Delden, Holland

Shell Chem. Co.
One Shell Plaza
Houston, TX 77001

Sherex Chem. Co.
5777 Frantz Rd., PO Box 646
Dublin, OH 43017

Stauffer Chem. Co.
Nyala Farm Rd.
Westport, CT 06880

Stepan Co.
Edens & Winnetka Rds.
Northfield, IL 60093

Stepan Europe
BP127
38340 Voreppe, France

Stokely-Van Camp Inc.
PO Box 569
Columbus, OH 43216

Taiwan Surfactant Corp.
8-1 Floor, No. 106, Sec. 2
Changan East Road
Taipei, Taiwan, R.O.C.

Temfibre Inc.
C.P. 3000 Temiscaming
Quebec J0Z 3R0 Canada

Texaco Chemical Co.
PO Box 15730
Austin, TX 78761

Thompson-Hayward Chemical Co.
PO Box 2383
5200 Speaker Rd.
Kansas City, KS 66110

Toho Chem. Industry Co., Ltd.
14-9, 1-chome, Kakigara-cho
Nihonbashi, Chuo-ku
Tokyo 103, Japan

Tomah Products Inc.
1012 Terra Dr., PO Box 388
Milton, WI 53563

Troy Chemical Corp., Inc.
One Avenue L
Newark, NJ 07105

Union Camp Corp.
1600 Valley Rd.
Wayne, NJ 07470

Union Carbide Corp.
39 Old Ridgebury Rd.
Danbury, CT 06817

Union Oil of California
1 California St.
San Francisco, CA 94111

United States Borax & Chemical Corp.
3075 Wilshire Blvd.
Los Angeles, CA 90010

United States Industrial Chemicals Co.
99 Park Ave.
New York, NY 10016

Vanderbilt & Co., Inc., R.T.
30 Winfield St.
Norwalk, CT 06855

Westvaco Chem. Div.
PO Box 70848
Charleston Hts, SC 29415

Witco/Organics Div. & Sonneborn Div.
277 Park Ave.
New York, NY 10017